Quantum Biocomputing in Quantum Biology
Volume II

Hafiz Md. Hasan Babu

Quantum Biocomputing in Quantum Biology Volume II

Memory Devices, Programmable Logic Devices, Nanoprocessors, Heat, Speed, and Data Related Issues

Springer

Hafiz Md. Hasan Babu
Department of Computer Science
and Engineering
University of Dhaka
Dhaka, Bangladesh

ISBN 978-981-97-5348-2 ISBN 978-981-97-5349-9 (eBook)
https://doi.org/10.1007/978-981-97-5349-9

This Springer imprint is published by the registered company Springer Nature Singapore Pte Ltd.
The registered company address is: 152 Beach Road, #21-01/04 Gateway East, Singapore 189721, Singapore

To my respected great parents and also to my lovely wife, daughter, and son who made me possible to write this book

Preface

Quantum biology is the field of study that investigates processes in living organisms that cannot be accurately described by the classical laws of physics. This means that quantum theory has to be applied to understand those processes. All matter, including living matter, is subject to the laws of physics. Biology and biological processes often deal with electrons and protons that are continuously being transferred between different parts of a cell or a macromolecular system. These transfer processes can only take place when the system exchanges energy with its environment in the form of molecular vibrations and phonons. Such a system is called an 'open quantum system', and special physical laws apply to it. Good examples of biological processes in which quantum effects are visible are the transport of electrons and protons in photosynthesis, respiration, vision, catalysis, olfaction, and in basically every other biological transport process. Further examples include the transfer of electronic and/or vibrational energy, and magnetic field effects in electron transfer and bird migration. Without the laws of quantum mechanics, we cannot understand life and life processes. The challenge is to understand how in a wet and noisy environment (such as a protein, a membrane, a cell, and an entire organism) the 'perfect' laws of quantum physics survive. In the near future we will see new experiments that will study, for example, the effects of strong magnetic fields, single molecule/system analysis, and femtosecond coherent microscopy. One challenge is to understand how quantum effects, clearly present at some level of functional description, translate into observations at a higher level of complexity. We will see new systems being investigated, such as neurons, neural networks, and maybe the entire brain. We will see a closer connection between our further understanding of life, and our understanding of quantum informatics, quantum computing, artificial intelligence, and various other technologies. Computing in Quantum Biology is the combination of quantum computing and DNA computing which is introduced here for the first time. This is a new platform for the computing system, which consists of a combination of quantum computing and DNA computing—it can be called the cross-platform system of quantum and DNA computing which can be done in two ways, namely, "quantum-DNA computing or quantum biocomputing or quantum biological computing" and "DNA-quantum computing or bioquantum computing or biological quantum computing" The book "Quantum Biocomputing: Computing in Quantum Biology" starts with the basics of Quantum Computing, Biocomputing, Quantum-DNA Computing, and DNA-Quantum Computing in Volume I. Volume I discussed the fundamental operations in quantum computing, different types of quantum arithmetic circuits, fundamental operations in biocomputing, DNA arithmetic circuits, quantum-DNA arithmetic circuits, and DNA-quantum arithmetic circuits such as basic and universal gate operations, half-adder, full-adder, half subtractor, full subtractor, N-qubit adders, multipliers, and dividers. Different quantum and DNA combinational circuits, such as quantum, quantum-DNA and DNA-quantum encoder, decoder, multiplexer, and demultiplexer, have also been discussed in Volume I of this book. Different types of sequential circuits, such as SR-latch, SR flip-flop, D flip-flop, T flip-flop, JK flip-flop, shift register, ripple counter, synchronous counter in quantum, quantum-DNA, and DNA-quantum computing. Computing in quantum biology means all logical computations in quantum-DNA and DNA-quantum computing modes. This volume (Volume II) will discuss the architecture and applications of different types of memory devices, such as Random-Access Memory (RAM), Read-Only Memory (ROM), and Programmable Read-Only Memory (PROM), cache memory in quantum, quantum-DNA, and DNA-quantum computing. Different types of programmable logic devices such as Programmable Logic Array (PLA), Field Programmable Gate Array (FPGA), Complex Programmable Logic Device (CPLD) in quantum, quantum-DNA, and DNA-quantum computing are also constructed. The readers will get knowledge about the designs of quantum, quantum-DNA, and DNA-quantum nanoprocessors and their components such as RAM, Instruction Register (IR), Program Counter (PC), incrementor circuit, decoder, multiplexer, Arithmetic Logic Unit (ALU), and accumulator in quantum computing and DNA computing. The readers will get the idea of quantum computing, quantum-DNA computing, and DNA-quantum computing from basic to advanced levels which will help them to design new quantum, quantum-DNA, and DNA-quantum circuits. The last part of the book will discuss the data, heat, and speed-related issues of quantum biocomputing.

Dhaka, Bangladesh Hafiz Md. Hasan Babu

Acknowledgments

I would like to express my sincerest gratitude and special appreciation to the various researchers in the field of quantum-DNA computing. The contents of the quantum-DNA computing book have been compiled from a wide variety of research works, where the researchers are pioneer in their respective fields. All the research articles related to the contents are listed at the end of each chapter.

I am grateful to my great parents and dear family members for their endless support. Most of all, I want to thank my lovely wife Mrs. Sitara Roshan, sweet daughter Ms. Fariha Tasnim, and sweet son Md. Tahsin Hasan for their invaluable cooperation to complete this book.

Finally, I am also thankful to all of them, especially to my beloved students Nitish Biswas, Md. Tareq Hasan, and Rownak Borhan Himel, who have provided their immense support and important time to finish this book.

Contents

About the Author

Dr. Hafiz Md. Hasan Babu is currently working as a Professor in the Department of Computer Science and Engineering, University of Dhaka as well as the Dean of the Faculty of Engineering and Technology of the University of Dhaka, Bangladesh. In addition, at present, he is a member (part-time) of Bangladesh Accreditation Council, Ministry of Education of the Government of the People's Republic of Bangladesh. He is also the Director of the Board of Directors of Bangladesh Submarine Cable Company Limited. He was the Chairman of the Department of Computer Science and Engineering of the University of Dhaka from 19-02-2003 to 18-02-2006 and Pro-Vice-Chancellor of the National University of Bangladesh from 12-07-2016 to 11-07-2020. He was also a Professor and the founding Chairman of the Department of Robotics and Mechatronics Engineering, University of Dhaka, Bangladesh. He obtained his Ph.D. degree in Electronics and Computer Science from Japan under the Japanese Government Scholarship and received his M.Sc. degree in Computer Science and Engineering from Czech Republic under the Czech Government Scholarship. He also received the DAAD Research Fellowship from Germany.

Dr. Hafiz Md. Hasan Babu was awarded Dr. M. O. Ghani Memorial Gold Medal by the Bangladesh Academy of Sciences in 2015 for his excellent research work in the progress of Physical Sciences in Bangladesh. In addition, he was awarded the UGC Gold Medal Award-2017 in Mathematics, Statistics and Computer Science category for his research work on quantum multiplier-accumulator device. He is currently an Associate Editor of the famous research journal titled "IET Computers and Digital Techniques" published by the Institution of Engineering and Technology of the UK. He was a member of Prime Minister's ICT Task Force in Bangladesh. He was also the President of Bangladesh Computer Society for the session 2017–2020. At present, he is the President of International Internet Society, Bangladesh.

Professor Dr. Hafiz Md. Hasan Babu published more than a hundred research papers. Three of his research papers have received the best research awards in the International Conferences.

In addition, he has published the following four textbooks by four famous publishers of the United Kingdom (UK), Singapore, and the United States of America (USA) for the graduate and post-graduate students:

1. Hafiz Md. Hasan Babu, "Quantum Computing: A Pathway to Quantum Logic Design," IOP (Institute of Physics) Publishing, 2020, Bristol, UK.
2. Hafiz Md. Hasan Babu, "Reversible and DNA Computing," Wiley Publishers, 2021, UK.
3. Hafiz Md. Hasan Babu, "VLSI Circuits and Embedded Systems," CRC Press (A Publication of Taylor & Francis Group), July 2022, USA.
4. Md. Jahangir Alam, Guoqing Hu, Hafiz Md. Hasan Babu and Huazhong Xu, "Control Engineering Theory and Applications," CRC Press (A Publication of Taylor & Francis Group), September 2022, USA.
5. Hafiz Md. Hasan Babu, "Multiple-Valued Computing in Quantum Molecular Biology", Volume I, CRC Press, 2023, USA
6. Hafiz Md. Hasan Babu, "Multiple-Valued Computing in Quantum Molecular Biology", Volume II, CRC Press, 2023, USA
7. Hafiz Md. Hasan Babu, "DNA Logic Design: Computing with DNA", World Scientific Publishing Company, May 2024, Singapore.

Acronyms

ALU	Arithmetic Logic Unit
BCD	Binary Coded Decimal
CLB	Configurable Logic Block
CPLD	Complex Programmable Logic Device
CPU	Central Processing Unit
DNA	Deoxyribonucleic Acid
EMR	Electron Magnetic Resonance
FPGA	Field-Programmable Gate Array
IR	Instruction Register
LUT	Look-Up Table
MUX	Multiplexer
NMR	Nuclear Magnetic Resonance
NTI	Negative Ternary Inverter
PAL	Programmable Array Logic
PCR	Polymerase Chain Reaction
PLA	Programmable Logic Array
PROM	Programmable Read-Only Memory
PTI	Positive Ternary Inverter
QB	Quantum Biology
RAM	Random-Access Memory
RF	Radio Frequency
SIPO	Serial-In Parallel-Out
SISO	Serial-In Serial-Out
SPLD	Simple Programmable Logic Devices
STI	Standard Ternary Inverter
XNOR	Exclusive NOR
XOR	Exclusive OR

List of Figures

List of Tables

Chapter 1
Basic Operations in Quantum Computing and Biocomputing

Quantum computing is a multidisciplinary field comprising aspects of computer science, physics, and mathematics that utilizes quantum mechanics to solve complex problems faster than on classical computers. The field of quantum computing includes hardware research and application development. Quantum computers are able to solve certain types of problems faster than classical computers by taking advantage of quantum mechanical effects, such as superposition and quantum interference. Some applications where quantum computers can provide such a speed boost include machine learning (ML), optimization, and simulation of physical systems. Eventual use cases could be portfolio optimization in finance or the simulation of chemical systems, solving problems that are currently impossible for even the most powerful supercomputers on the market. Quantum computing is a computational method based on quantum mechanics and quantum physics. It is a beautiful combination of physics, mathematics, computer science, and information theory. By controlling the behavior of small physical particles such as atoms, electrons, photons, and other microscopic particles, they achieved exponentially greater energy efficiency, lower power consumption, and extremely faster than traditional computers. Quantum computers can measure multiple phenomena simultaneously through quantum entanglement. DNA computing is an emerging branch of computing that utilizes DNA and molecular biology hardware instead of traditional electronic computing. It involves executing molecular reaction techniques on DNA molecules, which provide high computation power and storage capacity. The main challenge in implementing DNA computing is conducting wet lab experiments in a controlled manner. DNA computing has the potential to integrate with quantum computing and nanotechnology, expanding the scope of research in this field. DNA computing or biocomputing or biological computing is performing calculations with biological molecules instead of traditional silicon chips. The idea that single molecules or even atoms can be used for calculations dates back to 1959, when the American physicist Richard Feynman presented his ideas on nanotechnology. The characteristics of the DNA molecule aid in the induction of quantum features such as superposition, tunneling, coherence,

H. Md. Hasan Babu, *Quantum Biocomputing in Quantum Biology Volume II*,
https://doi.org/10.1007/978-981-97-5349-9_1

and entanglement. Superposition is the set of quantum particle states in which a particle can be in either a single or mixed state. Quantum computing is based on quantum bits, sometimes that is known as "qubits," which may represent either |0> or |1> and the fact that qubits may acquire a mixed state known as superposition, in which they can be both |1> and |0> at the same time is fascinating. Consider the following scenario: it's a coin. When a coin is tossed over the head, it begins to revolve at random. Because it spins randomly, there is a chance of being head, tail, or both at the same moment throughout the rotation. Tunneling occurs when a particle can pass through a potential energy barrier that is typically stronger than the particle's kinetic energy. In the quantum universe, a particle can pass through a barrier if it does not have any kinetic energy. In the actual world, for example, a ball with 100 J energy may readily overcome a barrier (hill) with just 70 J energy. However, if the ball's energy is 100 J and the barrier's energy is 200 J, the ball will never pass through the barrier. The ball will go up to a distance of 100 J before returning. However, in the quantum realm, a particle with less energy than the barrier can also permeate it, which is a fascinating concept. Quantum coherence refers to the concept of superpositioning, which is central to quantum physics and quantum computing. Quantum coherence considers a scenario in which an object's wave property is divided into the two waves that coherently interfere with one another. The theory behind quantum coherence is that all things exhibit wave-like qualities. It's related to quantum entanglement in that it includes the sharing states of two quantum particles rather than two quantum waves of a single particle. Entanglement is a term used to describe a relationship between two or more particles that interact in such a manner that makes it difficult to characterize each particle separately. Measurements of the particles, on the other hand, show correlations, therefore the particles must always be characterized as a quantum state of the entire system. Nuclear Magnetic Resonance (NMR) is a physical phenomenon in which electromagnetic radiation is absorbed and emitted by the nucleus in a magnetic field. NMR uses Radio Frequency (RF) to produce a strong magnetic field that excites the nuclei of molecules, causing them to exist in superposition. The superposition principle underpins both quantum coherence and quantum entanglement. In NMR, coherence is a physical condition in which numerous spins line up and revolve at the same speed around the magnetic field direction. NMR relaxation is the reversal of the NMR process. To get the original molecule, NMR relaxation qubits are set in the ground state. In this situation, generating EMR must be halted, and the molecules must then lose energy in order to reach their ground state configuration. This procedure is carried out at two temperatures: room temperature and zero kelvin and is named Reverse Nuclear Magnetic Resonance (RNMR). NMR and RNMR may help to convert a DNA sequence to qubits and qubits to a DNA sequence, respectively. Part I contains the block diagram, architecture, and applications of memory devices such as Random-Access Memory (RAM), Read-Only Memory (ROM), and Programmable Read-Only Memory (PROM), cache memory in quantum, DNA, quantum-DNA and DNA-quantum computing. In Part II, programmable logic devices such as Programmable Logic Array (PLA), Field Programmable Gate Array (FPGA), and Complex Programmable Logic Device (CPLD) in quantum, DNA, quantum-DNA and DNA-quantum computing are described with their architectures and applications. Quantum, DNA, quantum-DNA, and DNA-quantum nanoprocessors are designed in the last part which is Part III. Part IV discusses the heat calculation, heat transfer, speed calculation, and data management issues in quantum biocomputing. The computing in quantum biology means the combination of quantum computing and DNA computing. Computing in quantum biology is completely new thing that is going to be introduced here in this book.

1.1 Introduction

Compared with today's classical CMOS-based systems, quantum computers guarantee an exponential increase in power. As a result, it is difficult for the human mind to perceive this increasing magnitude. So there is real excitement that quantum computers will deliver benefits that are not possible with today's systems. The design of quantum computing-based systems begin with reversible logic circuit synthesis, low-power CMOS circuits, and nanotechnology-based systems. Quantum mechanical processes have been proved to be a good choice for constructing reversible gates, and these gates are known as quantum gates because quantum mechanics is basically reversible. Reversible computing is a form of unconventional computing. Because of the unity of quantum mechanics, quantum circuits are rather reversible unless they simply "collapse" the quantum states in which they operate. Quantum computing is an emergent field of cutting-edge computer science harnessing the unique qualities of quantum mechanics to solve problems beyond the ability of even the most powerful classical computers. The field of quantum computing contains a range of disciplines, including quantum hardware and quantum algorithms. While still in development, quantum technology will soon be able to solve complex problems that supercomputers can't solve, or can't solve fast enough. By taking advantage of quantum physics, fully realized quantum computers would be able to process massively complicated problems at orders of magnitude faster than modern machines. For a quantum computer, challenges that might take a classical computer thousands of years to complete might be reduced to a matter of minutes. The study of subatomic particles, also known as quantum mechanics, reveals unique and fundamental natural principles. Quantum computers harness these fundamental phenomena to compute probabilistically and quantum mechanically.

Biocomputing or DNA computing or biological computing is a field that uses DNA, biochemistry, and molecular biology rather than traditional silicon chips for computations. Recently researchers have emphasized exploring this field because of several advantages of DNA computing. For example, Boolean logic uses two input or output states, and binary logic faces challenges from non-Boolean logic when processing uncertain or imprecise information. Biocomputing is an emerging branch of computing that replaces traditional electronic computing with deoxyribonucleic acid (DNA), biochemistry, and molecular biology hardware. It may seem strange that computation may be done in a test tube using biological molecules rather than semiconductor chips. Key takeaways of biocomputing:

1. Biological computers (or mini-brains) are 3D cultures of brain tissue and neurones that mimic the structure and main functions of our brains.
2. This technology will make it possible to combine the computational performance of the best computers with the energy efficiency of the human brain.
3. In the future, "bio-computers" could become invaluable tools for research, particularly the study of certain diseases.
4. The development of organoid intelligence has been made possible by three technological breakthroughs: electrophysiology, AI and cerebral organoids.

While biocomputing is in an early phase, biocomputers have the potential to enable far more powerful computing than today's best computers - while using less energy and generating less heat. Furthermore, biocomputers will be able to use parallel computing, which will represent a significant improvement upon regular computing, and will be able to better self-organize and self-repair. While authoritative estimates of the eventual environmental impact of biocomputing do not yet exist, biocomputing could potentially reduce our reliance on the silicon and rare earth minerals that power today's computers.

Millions of natural supercomputers exist inside living organisms. DNA (deoxyribonucleic acid) molecules, which make up human genes, have the ability to execute calculations hundreds of times quicker than even the most powerful human-built computers. DNA could one day be incorporated into a computer chip to produce a "biochip" that will allow computers to run even quicker. Complex mathematical problems have already been solved using DNA molecules.

Quantum computing and DNA computing can be combined to achieve the advantages of both. This combination is called quantum biology. Computing in quantum biology means quantum-DNA computing and DNA-quantum computing. Quantum-DNA computing or quantum biological computing or quantum biocomputing explores the potential of using DNA's structure and properties, including its quantum mechanical behavior, to perform computations. While DNA computing focuses on using biological molecules for computation, quantum computing utilizes quantum mechanical principles, such as superposition and entanglement, to revolutionize information processing. The idea is to leverage DNA's unique characteristics to build quantum computers or to enhance the performance of existing ones. It also investigate the potential for harnessing biological systems and processes for quantum computation, or for using biological materials to build quantum computers. This field investigates if living organisms naturally utilize quantum effects for computation and if we can engineer biological systems to perform quantum tasks. The biology-Inspired quantum computing focuses on using biological materials (like proteins, DNA, or RNA) to build quantum computers or to create quantum computing hardware. It also includes hybrid systems where biological systems interact with quantum devices. For instance, researchers are exploring using biological scaffolds to hold qubits or using biomolecules as qubits themselves. On the other hand, in the world of computing, scientists and researchers constantly seek new ways to enhance computational power and solve complex problems more efficiently. Traditional computing, based on classical bits, has made remarkable progress, but it is reaching its limits. However, a fascinating field of research is emerging at the intersection of biology and quantum physics, known as bioquantum computing or biological quantum computing or DNA-quantum computing. This revolutionary approach harnesses the principles of quantum mechanics within biological systems, promising unprecedented computational capabilities. This computing refers to speculative ideas that biological systems might perform quantum computations or that we could harness biological processes to implement quantum computing. This paradigm is highly exploratory and not yet realized in any form, lying at the intersection of quantum physics, biology, and computer science. There are two main interpretations:

1. Biology as the computer: Certain processes in living organisms might naturally exploit quantum effects to compute or process information. For example, it has been hypothesized that the brain could be a quantum computer, or that plants perform quantum optimizations in photosynthesis. These ideas suggest that evolution might have stumbled upon quantum mechanisms to enhance functionality (like efficiency of energy transfer or perhaps even consciousness via quantum processes in neurons).

2. Biology-inspired hardware: Using biological materials or biologically derived structures to build quantum computers. For instance, using proteins, DNA, or other biomolecules as qubits or as scaffolds to hold and manipulate qubits. This also covers hybrid approaches where biological systems interface with quantum systems (like a living organism that interacts with a quantum device). The computer that will perform all these operations is quantum biocomputer. This chapter will present how basic operations can be performed in quantum computing and DNA computing.

1.2 Basic Gates in Quantum Computing

If the input and output assignments of a function or circuit are one-to-one, the function or circuit is said to be reversible. That is, the outputs of a reversible circuit can be calculated uniquely from the inputs, and the inputs may be recovered from the outputs.

A quantum logic gate, also known as a quantum gate, is a fundamental quantum circuit that works with a minimal number of qubits. They are equivalent to classical logic gates in conventional digital computers for quantum computers. Unlike many classical logic gates, quantum logic gates are reversible. Unitary matrices are used to represent them. The most common quantum gates work with one- or two-qubit spaces. This means that quantum gates can be characterized by orthonormal 2×2 or 4×4 matrices.

There are only three fundamental gates in quantum computing based on the synthetization of reversible logic circuits are given as follows:

1. Controlled NOT (CNOT) gate.
2. Controlled-V gate.
3. Controlled-V+ gate.

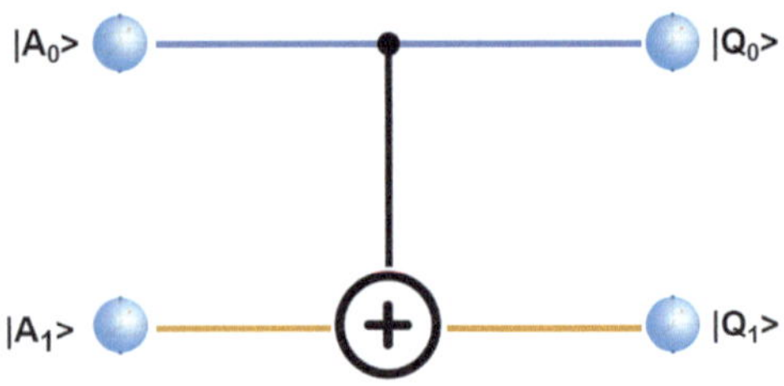

Fig. 1.1 Quantum CNOT gate

1.2.1 Quantum Controlled NOT Gate

The quantum controlled-NOT gate, or simply CNOT gate is a two-qubit operation in which the first qubit is known as the control qubit and the second qubit is known as the target qubit. Using a combination of CNOT gates and single-qubit rotations, any quantum circuit can be simulated to an arbitrary degree of accuracy.

The CNOT gate utilizes a quantum register with two qubits. The CNOT gate flips the second qubit (the target qubit) if and only if the first qubit (the control qubit) is $|\mathbf{1}>$. That means it performs a classical NOT on the target whenever the control is in the state of $|\mathbf{1}>$. It is also known as the controlled-x or CX gate. Table 1.1 shows the functionality of a quantum controlled-NOT gate.

Here, the input bits $|\mathbf{A}_0>$ and $|\mathbf{A}_1>$ are the control qubit and the target qubit. Quantum CNOT gate produces two outputs where the controlled output is $|Q_1>$ which gets flipped only if the controlled input $|\mathbf{A}_0>$ is $|\mathbf{1}>$. During this time the output $|\mathbf{Q}_0>$ always remains the same as per the input $|\mathbf{A}_0>$.

The next question is, how to draw a circuit diagram of a quantum CNOT gate? This is pretty simple. In Table 1.1, CNOT is the quantum NOT operation which is controlled by a qubit. Therefore, Pauli X-gate with a controlled qubit is used to design the quantum CNOT gate. Figure 1.1 shows the logic circuit diagram of the quantum CNOT gate.

Table 1.1 Operations in quantum CNOT gate

Input		Output	
$\|\mathbf{A}_0>$	$\|\mathbf{A}_1>$	$\|\mathbf{Q}_0>$	$\|\mathbf{Q}_1>$
$\|\mathbf{0}>$	$\|0>$	$\|0>$	$\|0>$
$\|\mathbf{0}>$	$\|1>$	$\|0>$	$\|1>$
$\|\mathbf{1}>$	$\|0>$	$\|1>$	$\|1>$
$\|\mathbf{1}>$	$\|1>$	$\|1>$	$\|0>$

1.2.1.1 Design Procedure of Quantum CNOT Gate

In quantum logic circuits, this gate has the simplest architectural design. Easy to construct this only in three steps.

1. First, two superposition-mode qubits are drawn as $|\mathbf{A_0}>$ and $|\mathbf{A_1}>$, where $|\mathbf{A_0}>$ is the control qubit, and $|\mathbf{A_1}>$ is the target qubit.
2. Then two lines are drawn from these inputs to the target outputs $|\mathbf{Q_0}>$ and $|\mathbf{Q_1}>$.
3. Finally, on the $|\mathbf{A_1}>$ line, a NOT gate is drawn and connects it to the $|\mathbf{A_0}>$ line, because $|\mathbf{A_0}>$ controls the NOT gate of the target output $|\mathbf{Q_1}>$.

1.2.1.2 Working Principle of Quantum CNOT Gate

The only input to the NOT gate is $|\mathbf{A_1}>$, and the gate control input is $|\mathbf{A_0}>$, which are termed as CNOT gate. The block diagram of the quantum CNOT gate is shown in Fig. 1.2.

The working procedure of the quantum CNOT gate is given below:

1. The output value $|\mathbf{Q_0}>$ is always the same as the value of input $|\mathbf{A_0}>$ because there is no operation through this line.
2. In the case of the output value of $|\mathbf{Q_1}>$,
   (a) When the control qubit $|\mathbf{A_0}>$ is $|\mathbf{0}>$, the gate will close and will work as a buffer. That means the output value of $|\mathbf{Q_1}>$ is the same as the input value $|\mathbf{A_1}>$. According to CNOT table, when $|\mathbf{A_1}> = |\mathbf{0}>$ **or** $|\mathbf{1}>$ with control qubit $|\mathbf{A_0}> = |\mathbf{0}>$, the output $|\mathbf{Q_1}>$ is $|\mathbf{0}>$ and $|\mathbf{1}>$, respectively.
   (b) When the control qubit $|\mathbf{A_0}>$ is $|\mathbf{1}>$, the gate will open and works as an inverter. That means the output value of $|\mathbf{Q_1}>$ is opposite to the input value $|\mathbf{A_1}>$. According to CNOT table, when the $|\mathbf{A_1}> = |\mathbf{0}>$ **or** $|\mathbf{1}>$ with control qubit $|\mathbf{A_0}> = |\mathbf{1}>$, it flips the $|\mathbf{A_1}>$ qubit and produces $|\mathbf{1}>$ and $|\mathbf{0}>$ **as** outputs in $|\mathbf{Q_1}>$, respectively.

From the input and output relation, it is observed that this gate works exactly like a classical XOR operation. Therefore, CNOT gate can be used in the quantum XOR operation.

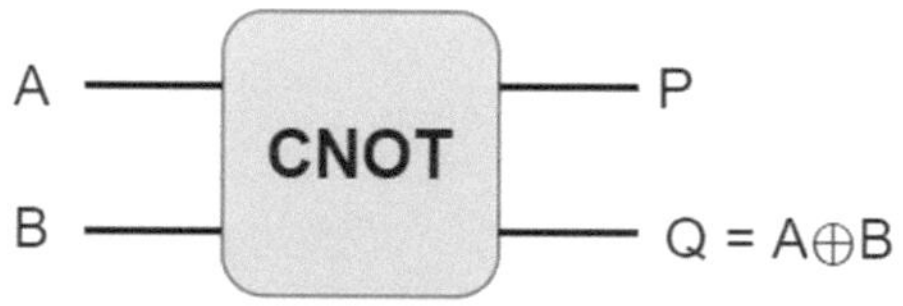

Fig. 1.2 The block diagram of quantum CNOT gate

And therefore, the CNOT gate can be described as the gate that maps the basis states of input ($|\mathbf{A}_0>$, $|\mathbf{A}_1>$) to output ($|\mathbf{Q}_0> = |\mathbf{A}_0>$, & $|\mathbf{Q}_1> = (|\mathbf{A}_0>$ **XOR** $|\mathbf{Q}_1>$)). Figure 1.2 shows the block diagram of the quantum CNOT gate.

The first experimental realization of a CNOT gate was accomplished, in 1995. A single Beryllium ion in a trap was employed in this experiment. The two qubits were encoded into the optical state and the vibrational state of the trapped ion. The CNOT-operation reliability was estimated to be on the order of 90% at the time of the trial.

1.2.2 *Quantum Controlled-V Gate*

The second fundamental gate in quantum computing is the controlled-V gate (or V gate) which is shown in Fig. 1.3.

When the control signal $|\mathbf{A}_0> = |0>$, the qubit $|\mathbf{A}_1>$ will pass through the controlled part and remain unchanged, i.e., $|\mathbf{Q}_1> = |\mathbf{A}_1>$. When $|\mathbf{A}_0> = |1>$, then the unitary operation $V = \frac{i+1}{2}\begin{pmatrix} 1 & -i \\ -i & 1 \end{pmatrix}$ is applied to the input $|\mathbf{A}_1>$, that is, $|\mathbf{Q}_1> = V(|\mathbf{A}_1>)$.

1.2.2.1 Design Procedure of Quantum Controlled-V Gate

Figure 1.3 shows the circuit diagram of the quantum controlled-V gate. This gate is also a two-input gate and the number of output is the same here as well.

Three steps to construct a quantum controlled-V gate are given below.

1. First, two superposition-mode qubits are drawn, $|\mathbf{A}_0>$ and $|\mathbf{A}_1>$. And in this gate also, the control qubit is $|\mathbf{A}_0>$, and the target qubit is $|\mathbf{A}_1>$.
2. From these inputs, two lines are drawn to output $|\mathbf{Q}_0>$ and $|\mathbf{Q}_1>$.
3. Then **V** gate is drawn on the $|\mathbf{A}_1>$ line and connect it to the $|\mathbf{A}_0>$ line because $|\mathbf{A}_0>$ is the control qubit that controls the V gate of output $|\mathbf{Q}_1>$.

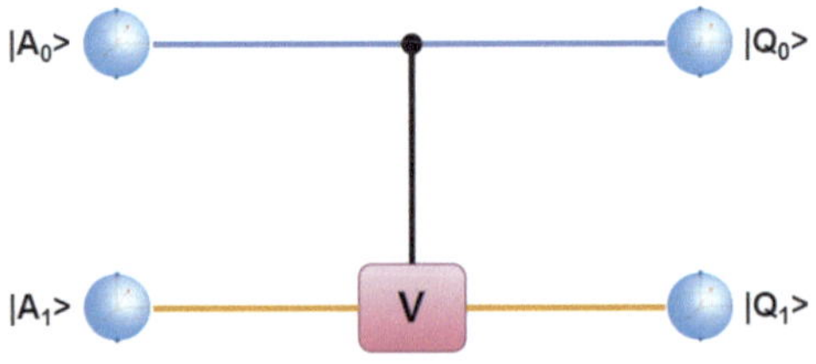

Fig. 1.3 Quantum controlled-V gate

1.2.2.2 Working Principle of Quantum Controlled-V Gate

The quantum controlled-V gate's working principle is not as straightforward as the quantum CNOT gate. Consider the operational table of the controlled-V gate as shown in Table 1.2.

It is called quantum controlled-V gate because, as Table 1.2 shows here the outputs of the gate depend on the control qubit $|\mathbf{A}_0>$.

The working procedure of the quantum controlled-V gate is listed below:

1. Considering the controlled-V gate, input $|\mathbf{A}_1>$ values as $|0>$, $|1>$, $|\mathbf{v}> = |0>$, $|\mathbf{V}> = |1>$, $|\mathbf{w}> = |0>$ and $|\mathbf{W}> = |1>$ and for control input $|\mathbf{A}_0>$ values are $|0>$ and $|1>$.
2. The output value $|\mathbf{Q}_0>$ is always the same as the input value $|\mathbf{A}_0>$ due to the fact that this line does not have any operations.
3. For the output value of $|\mathbf{Q}_1>$,
   (a) The controlled-V gate will not open if the control qubit $|\mathbf{A}_0>$ is $|0>$, eventually will work as a buffer. That means the output value of $|\mathbf{Q}_1>$ will be the same as the input value of $|\mathbf{A}_1>$ when the input value of $|\mathbf{A}_0>$ is $|0>$.
   (b) When the control qubit $|\mathbf{A}_0>$ is $|1>$, the quantum controlled-V gate will open and will work with the input values of $|\mathbf{A}_1>$ to produce output $|\mathbf{Q}_1>$ according to the operational table (Table 1.2) of the controlled-V gate. Two intermediate inputs $|\mathbf{v}>$ and $|\mathbf{V}>$ are used for controlled-V gate output values. Here the $|\mathbf{v}>$ value is assigned to the V gate output if its input $|\mathbf{A}_1>$ is $|0>$ and $|\mathbf{V}>$ value if its input $|\mathbf{A}_1>$ is $|1>$. Now, if the input of V gate is $|\mathbf{v}>$ then the output will $|1>$ (flipped) because the $|\mathbf{v}>$ value is the result of applying input $|0>$ to the V gate. In the same way, applying $|\mathbf{V}>$ input to a V gate results in an output of $|0>$. If the input of a V gate is $|\mathbf{w}>$ then its output will $|0>$, and if the input of a V gate is $|\mathbf{W}>$ then the target output will be $|1>$. The inputs $|\mathbf{w}>$ and $|\mathbf{W}>$ are the properties of quantum controlled-V^+ gate.

Table 1.2 Operations in the quantum controlled-V gate

Input		Output	
$\|\mathbf{A}_0>$	$\|\mathbf{A}_1>$	$\|\mathbf{Q}_0>$	$\|\mathbf{Q}_1>$
$\|\mathbf{0}>$	$\|X>$	$\|0>$	$\|X>$
$\|\mathbf{1}>$	$\|0>$	$\|1>$	$\|v>$
$\|\mathbf{1}>$	$\|1>$	$\|1>$	$\|V>$
$\|\mathbf{1}>$	$\|v>$	$\|1>$	$\|1>$
$\|\mathbf{1}>$	$\|V>$	$\|1>$	$\|0>$
$\|\mathbf{1}>$	$\|w>$	$\|1>$	$\|0>$
$\|\mathbf{1}>$	$\|W>$	$\|1>$	$\|1>$

1.2.3 Quantum Controlled-V+ Gate

The last of three fundamental quantum gates is the quantum controlled-V^+ gate (or simply, V+ gate). In the controlled-V+ gate when the control signal $|\mathbf{A}_0>$ = |0>, the qubit $|\mathbf{A}_1>$ will pass through the controlled part unchanged, that is, $|\mathbf{Q}_1> = |\mathbf{A}_1>$. When $|\mathbf{A}_0>$ = |1>, the unitary operation V+ = V^{-1} is applied to the input $|\mathbf{A}_1>$, that is, $|\mathbf{Q}_1> = V^+ (|\mathbf{A}_1>)$. The controlled-$V^+$ gate is shown in Fig. 1.4.

1.2.3.1 Design Procedure of Quantum Controlled-V+ Gate

The design architecture of the quantum controlled-V+ gate is shown in Fig. 1.4. The construction is similar to the quantum controlled-V gate, but the difference is in the operational part.

The construction of this fundamental gate is easy which has three steps like the other two.

1. At first, two superposition-mode qubits are drawn $|\mathbf{A}_0>$ and $|\mathbf{A}_1>$. And again, the control qubit is $|\mathbf{A}_0>$, and the target qubit is $|\mathbf{A}_1>$.
2. Then two lines from these inputs to produce outputs $|\mathbf{Q}_0>$ and $|\mathbf{Q}_1>$.
3. Subsequently, **V+** gate is drawn on the $|\mathbf{A}_1>$ line and connect it to the $|\mathbf{A}_0>$ line because $|\mathbf{A}_0>$ controls the **V+** gate of output $|\mathbf{Q}_1>$ depending on the input $|\mathbf{A}_1>$.

1.2.3.2 Working Principle of Quantum Controlled-V+ Gate

The controlled-V and the controlled-V+ gates are the two types of square-root-of-not gates. If two controlled-V or two controlled-V+ gates are triggered in series, they act as an inverter.

As similar to the controlled-V gate, when the control input is |1>, the corresponding unitary operator is propagated to the second output, where the unitary operation for the controlled-V+ is, V+ $=\frac{1}{i+1}\begin{pmatrix} 1 & \frac{-1}{i} \\ i & 1 \end{pmatrix}$. The operation on this gate is shown in Table 1.3.

The quantum controlled-V gate's operations are described below.

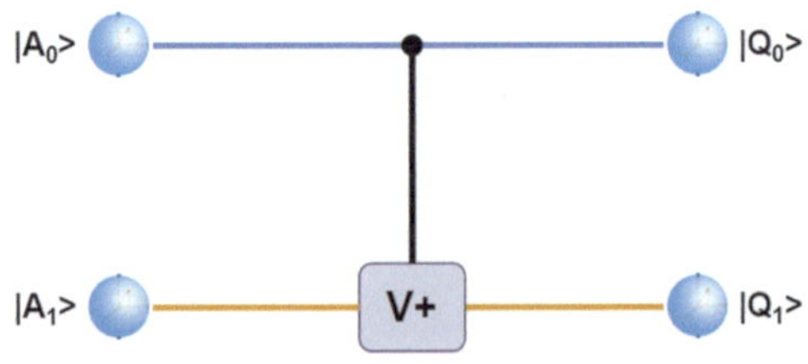

Fig. 1.4 Quantum controlled-V+ gate

Table 1.3 Operations in the quantum controlled-V+ gate

Input		Output	
$\|\mathbf{A}_0 >$	$\|\mathbf{A}_1 >$	$\|\mathbf{Q}_0 >$	$\|\mathbf{Q}_1 >$
|**0** >	|X>	|0>	|X>
|**1** >	|0>	|1>	|w>
|**1** >	|1>	|1>	|W>
|**1** >	|v>	|1>	|0>
|**1** >	|V>	|1>	|1>
|**1** >	|w>	|1>	|1>
|**1** >	|W>	|1>	|0>

1. The input qubits $|\mathbf{A}_0 >$ and $|\mathbf{A}_1 >$ are the target qubits.
2. The input qubit can be one of from $|\mathbf{A}_1 >$ values as |0>, |1>, |**w** > = |0>, |**W** > = |1>, |**v** > = |0>, and |**V** > = |1> and for control input $|\mathbf{A}_0 >$ values are |0> and |1>.
3. Again, the output value $|\mathbf{Q}_0 >$ is always the same as the input value $|\mathbf{A}_0 >$ because there is no operation across this line.
4. For the value of $|Q_1>$'s output,

   (a) When the control qubit $|\mathbf{A}_0 >$ is |0> the controlled-V+ gate will not open, consequently, the input $|\mathbf{A}_1 >$ will work as a buffer. That means the output value of $|\mathbf{Q}_1 >$ is the same as the input value of $|\mathbf{A}_1 >$.

   (b) When the control qubit $|\mathbf{A}_0 >$ is |1>, the controlled-V+ gate will work as per Table 1.3. So, with the input qubits of $|\mathbf{A}_1 >$, the output $|\mathbf{Q}_1 >$ will be produced according to Table 1.3. The intermediate inputs variables |**w** > and |**W** > as output from the controlled-V+ gate. Here the |**w** > value is assigned to the controlled-V+ gate output if the input $|\mathbf{A}_1 >$ is |0> and |**W** > value if the input $|\mathbf{A}_1 >$ is |1>. Now, if the input of the controlled-V+ gate is |**w** > then the output will be |1> because the |**w** > value is the result of applying input |0> to a controlled-V+ gate. Correspondingly, applying |**W** > input to controlled-V+ gate results in an output of |0>. And finally, if the input of a controlled-V+ gate is |**v** > (might come from a controlled-V gate as output) then |0> is obtained as output, and if the input is |**V** > then the target output will be generated as |1>.

1.3 Basic Operations in Quantum Computing

To perform logical operations, classical computing has three fundamental gates, namely OR gate, AND gate, and NOT gate. Also, four additional logic gates are

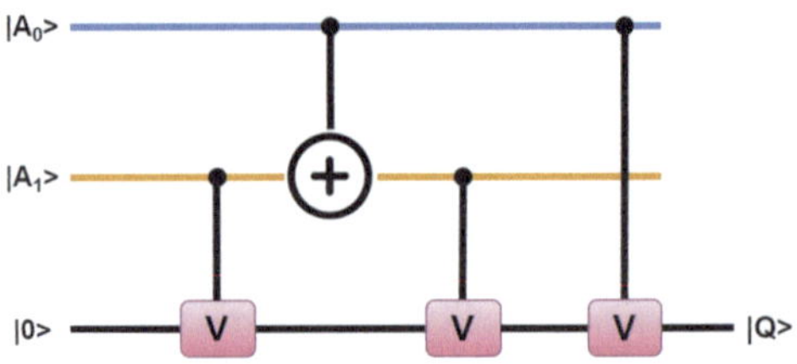

Fig. 1.5 Circuit diagram of quantum OR operation

Table 1.4 Truth table of a quantum OR operation

$\|A_0>$	$\|A_1>$	$\|Q>$
$\|0>$	$\|0>$	$\|0>$
$\|0>$	$\|1>$	$\|1>$
$\|1>$	$\|0>$	$\|1>$
$\|1>$	$\|1>$	$\|1>$

made up of the fundamental gates such as NOR gate, NAND gate, XOR gate, and XNOR gate. In a circuit, logic gates produce decisions based on a combination of digital signals coming from its inputs. All classical computations are performed at the hardware level using those seven quantum operations (see Fig. 1.5 to Fig. 1.10 and their corresponding Table 1.4 to Table 1.9 for seven basic quantum operations).

Specifically, it is possible to construct the classical fundamental gates from quantum fundamental gates so that they behave exactly as quantum computing requires, and with them, perform all necessary logical operations in the quantum realm. This section will present how to construct the classical fundamental gates for quantum computing to perform quantum operations.

1.3.1 Quantum OR Operation

A circuit that performs quantum OR operation in quantum logic can be described as $|Y> = |A_0>$ QOR $|A_1>$, or $|Y> = |A_0> + |A_1>$. And if OR is outputted value for more than two inputs, then merge them as chain-OR which will give the output as follows: $|Y> = |A_0> + |A_1> + \ldots + |A_n>$.

Figure 1.5 shows the circuit diagram of quantum OR operations which includes three quantum controlled-V gates that are connected linearly, one quantum CNOT gate, two input qubits $|A_0>$ and $|A_1>$, and a constant qubit (also known as Ancilla qubit) $|0>$ which is the actual input of the 1st controlled-V gate. Ancilla bits are extra bits that are used in computation to achieve certain goals. All these are configured in the way depicted in Fig. 1.5. And $|Q>$ is the resulting qubit which contains the quantum value of the quantum OR operations of $|A_0>$ and $|A_1>$. The truth table of quantum OR operation is shown in Table 1.4.

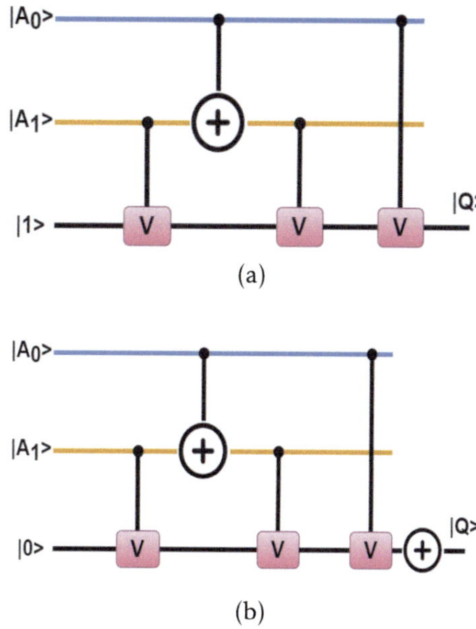

Fig. 1.6 Circuit diagrams of quantum NOR operation

Table 1.5 Truth table of a quantum NOR operation

$\|A_0>$	$\|A_1>$	$\|Q>$
|0 >	|0>	|1>
|0 >	|1>	|0>
|1 >	|0>	|0>
|1 >	|1>	|0>

1.3.2 Quantum NOR Operation

The quantum NOR operation is nothing but an inverted operation of the quantum OR operation. This implies that it will give the output |1> only when both of the input qubits are |0>. Otherwise, the output is always |0>. Therefore, an extra quantum NOT operation is needed along with the quantum OR circuit to construct the quantum NOR operation circuit diagram. The circuit diagram of quantum NOR operation with the above idea is shown in Fig. 1.6a. But have a close look at Fig. 1.6b. Why these two circuits? Well, any of them can be used to conduct the quantum NOR operation. In Fig. 1.6b, |1> is used as the constant qubit instead, which makes the differences in the two circuits, but the result is always the same.

As mentioned earlier, the operation in Fig. 1.6a is exactly the same as the quantum OR operation including a quantum NOT operation that inverts the outputs of the quantum OR operation and produces the output of the quantum NOR operation as shown in Table 1.5.

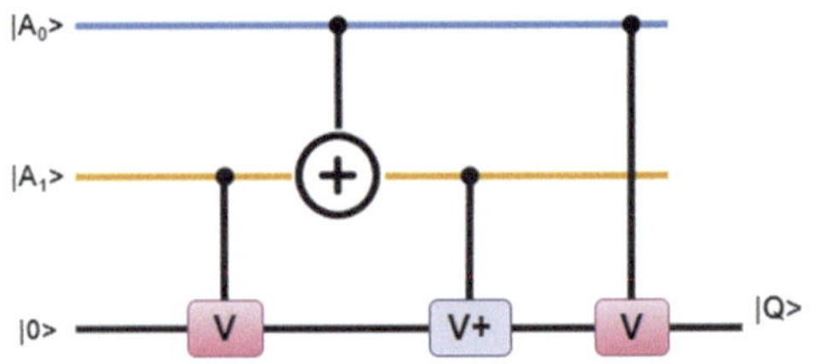

Fig. 1.7 The circuit diagram of quantum AND operation

This section will describe the working procedure of Fig. 1.6b and will show how the quantum NOR results are formed with that circuit for each pattern of input qubits of $|A_0>$ and $|A_1>$.

1.3.3 Quantum AND Operation

A circuit that performs quantum AND or QAND operation in Boolean quantum logic can be described as $|Y> = |A_0>$ QAND $|A_1>$, or $|Y> = |A_0>.|A_1>$ and to do AND operation for more than two input values, it is needed to combine them as chain-AND which will give the output as $|Y> = |A_0>.|A_1>. \ldots .|A_n>$.

Figure 1.7 shows the circuit diagram of the quantum AND operation, which includes two controlled-V gates, one CNOT gate, and one controlled-V+ gate. And it requires three input qubits $|A_o>$, $|A_1>$, and a constant qubit of $|0>$, respectively, where the 1st V gate is controlled by the value of $|A_1>$, CNOT gate and the 2nd controlled-V (which produces the final output) is controlled by the value of $|A_0>$, and the controlled-V+ gate is controlled by the output value of CNOT gate.

The quantum AND operation's functionality is shown in the truth table as shown in Table 1.6. It will always produce output $|0>$ if either one of two input qubits or both of the input qubits is $|0>$. The output is $|1>$ only when both input qubits are $|1>$.

1.3.4 Quantum NAND Operation

The quantum NAND operation is simply the opposite of the quantum AND operation. This implies that it will give the output $|0>$ only when both of the input qubits are $|1>$. Otherwise, the output is always $|1>$. Therefore, to construct the quantum NAND operation circuit diagram, it is needed to apply an extra quantum NOT operation in addition to the quantum AND circuit. The circuit diagram of quantum NAND operation using the above concept is shown in Fig. 1.8a.

Let's have a close look at how to construct quantum NAND in a different way.

Both (a) and (b) of Fig. 1.8 show operational circuit diagrams of quantum NAND operation. The quantum NAND operation can be done by using any of them. So far,

Table 1.6 Truth table of a quantum AND operation

|A$_0$>	|A$_1$>	|Q>
|0 >	|0>	|0>
|0 >	|1>	|0>
|1 >	|0>	|0>
|1 >	|1>	|1>

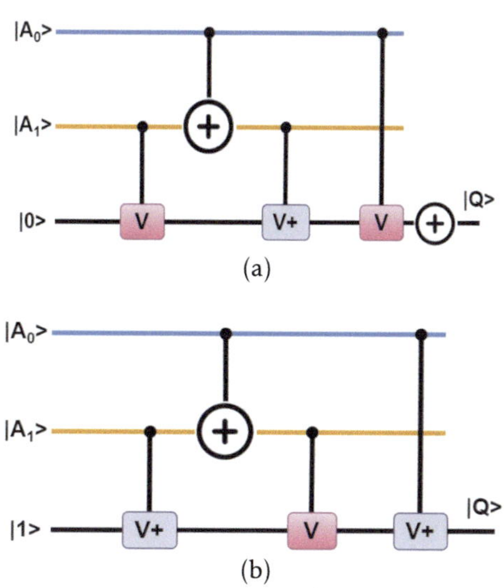

Fig. 1.8 Circuit diagrams of quantum NAND operation

the most familiar one is shown in Fig. 1.8 (a). But take a perceptive look at Fig. 1.8 (b). In Fig. 1.8 (b), the circuit got changed completely. Here two V+ gates, one CNOT gate, and one V gate are required. And |1> is used as the ancilla bit instead of |0>.

As mentioned earlier, the operation in Fig. 1.8 (a) is exactly the same as the quantum AND operation including a quantum NOT operation that inverts the outputs of the quantum AND operation and produces the output of the quantum NAND operation as shown in Table 1.7.

1.3.5 Quantum XOR Operation

The logic behind quantum XOR or simply QXOR for a two-valued system (binary system) is very simple; if the qubits are the same, the result is |0> and if the qubits are different, the result is |1>. This operational circuit has already been designed because the quantum XOR is exactly the same as the quantum CNOT gate. Figure 1.9 shows

Table 1.7 Truth table of a quantum NAND operation

$\|A_0>$	$\|A_1>$	$\|Q>$
$\|0>$	$\|0>$	$\|1>$
$\|0>$	$\|1>$	$\|1>$
$\|1>$	$\|0>$	$\|1>$
$\|1>$	$\|1>$	$\|0>$

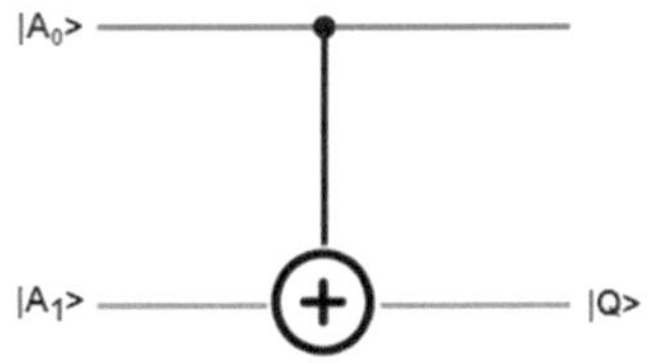

Fig. 1.9 Circuit diagram of quantum XOR operation

Table 1.8 Truth table for quantum XOR operation

$\|A_0>$	$\|A_1>$	$\|Q>$
$\|0>$	$\|0>$	$\|0>$
$\|0>$	$\|1>$	$\|1>$
$\|1>$	$\|0>$	$\|1>$
$\|1>$	$\|1>$	$\|0>$

the circuit diagram of XOR operation and Table 1.8 shows the truth table for quantum XOR operation.

As it is equivalent to the CNOT gate, its working procedure is already shown.

In short, $|A_0>$ is the control input and the operation will occur only if the value of $|A_0>$ is $|1>$. Therefore, for four input patterns, in two cases the operation will be held and for the other two the output will be the same as the input value of $|A_1>$. As a result, for input value $|A_0> = |1>$, $|A_1> = |0>$ it will produce $|1>$ as output, and for input value $|A_0> = |1>$, $|A_1> = |1>$ it will produce $|0>$ as output accordingly.

1.3.6 *Quantum XNOR Operation*

Quantum exclusive NOR or quantum XNOR or simply QXNOR is the inverted operation of quantum XOR operation. It inverts the output of the QXOR operation. Therefore, adding a quantum NOT gate to the output of the QXOR can construct the QXNOR operational circuit. Figure 1.10 shows the circuit diagram of quantum XNOR operation and Table 1.9 shows the **t**ruth table for the quantum XNOR operation.

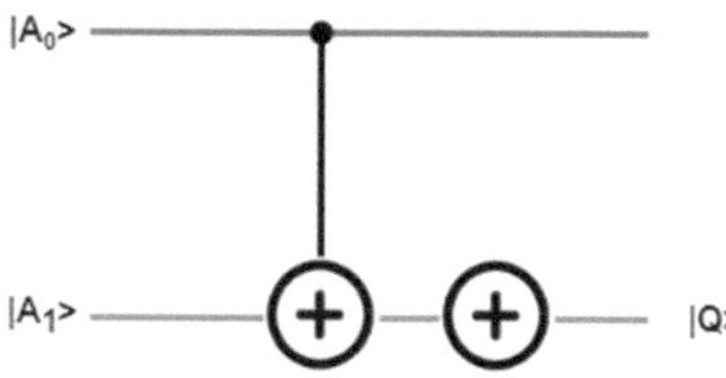

Fig. 1.10 Circuit diagram of a quantum XNOR operation

Table 1.9 Truth table for quantum XNOR operation

$\|A_0>$	$\|A_1>$	$\|Q>$
|0 >	|0>	|1>
|0 >	|1>	|0>
|1 >	|0>	|0>
|1 >	|1>	|1>

From Table 1.9, it is observed that the output will be |1> only if both inputs are the same. Otherwise the output will be |0>.

1.4 Basic Operations in Biocomputing

In classical computing, data storage device stores data by converting them into binary digits. But in DNA computing, instead of binary digits, information or data will now be kept in the form of nitrogen bases **A**, **T**, **G**, and **C**. These bases will make sequences to store data, and encoding and decoding are required to operate with those DNA sequences so that the outcomes become meaningful.

The capacity to generate short DNA sequences artificially allows these sequences to be used as inputs for algorithms. DNA has properties that allow it to be used to simulate classical logic processes. Single-stranded DNA naturally migrates toward complementary sequences to form double-stranded complexes, whereas double-stranded DNA wants to be in double-stranded form.

A program on a DNA computer is executed as a series of synthesizing, extracting, modifying, and cloning the DNA strands. Instead of using electrical impulses to represent bits of information, the DNA computer uses the chemical properties of DNA molecules by examining the patterns of combination or growth of the molecules or strings. DNA can do this through the manufacture of enzymes, which are biological catalysts that could be called the "software," used to execute the desired calculation. Enzymes do not function sequentially, working on one DNA at a time. Rather, numerous copies of the enzyme can act massively parallel on many DNA molecules concurrently. DNA computers work by encoding the problem to be solved in DNA's language: the base-four number system, which includes the base-four values **A**, **T**,

C, and **G**, which are more than enough when compared to an electronic computer, which only requires two numbers, 0 and 1.

DNA has copying, pasting, repairing, and many other operations, just like a CPU has addition, bit-shifting, logical operators, and so on, that allow it to accomplish even the most complex computations. The right sequences are sorted out using genetic algorithm methods in a DNA computer, which computes in test tubes or on a glass slide coated in 24K gold.

As mentioned earlier, the sequence of base patterns will store pieces of information. In two-valued (binary) DNA operations, consider

ACCTAG = true, which is equivalent to binary “1”; and

TGGATC = false, which is equivalent to binary “0.”

These base sequences represent data. With these base sequences all fundamental gates and operations will be done in the following sections.

As mentioned earlier, the sequence of base patterns will store pieces of information. In two-valued (binary) DNA operations, consider

ACCTAG = true, which is equivalent to binary “1” and

TGGATC = false, which is equivalent to binary “0.”

These base sequences represent data. With these base sequences all fundamental gates and operations will be done in the following sections.

1.4.1 DNA NOT Operation

The basic binary logical operation is fully conversant to us. As a result, the logical function of binary logic gates does not need to be explained again. Only how to construct them in the DNA computer to perform DNA computing will be discussed. Figure 1.11 shows the operational diagram of the DNA NOT operation. To design this, a test tube is needed with the DNA mixture, the annealing temperature is less than 60 °C. To perform the operation, the DNase enzyme is needed. And the base sequence ACCTAG will be used.

The DNA NOT operation inverts the input base sequence. The operational table shows the input-output mapping in Table 1.10. Remember, the base sequences TGGATC, and ACCTAG represents the Boolean false and true respectively.

The operational logic is pretty straightforward. The DNA bases make pairs only if the sequences meet the conditions A-T or C-G. Here, the sequence ACCTAG is treated as the base sequence. Now if the input sequence makes a pair with the base sequence in the test tube as shown in Fig. 1.12, then they will return the output as true. And if they do not make a pair then the output will be false.

So, how to detect whether the input sequence and the base sequence make pairs or not? The answer will be detected by the DNase enzyme.

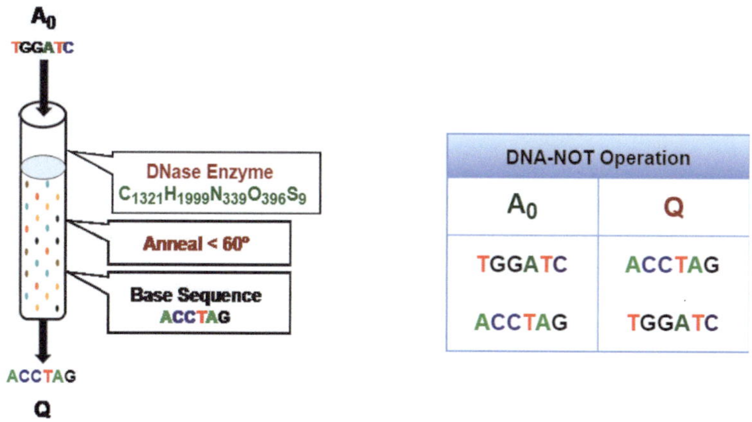

Fig. 1.11 The circuit architecture of a DNA NOT operation

Table 1.10 The truth table of a DNA NOT operation

A0	Q
TGGATC	ACCTAG
ACCTAG	TGGATC

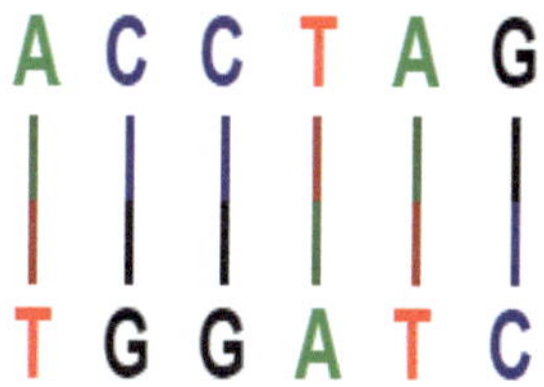

Fig. 1.12 The pair matching between the DNA base sequences

1.4.2 DNA OR Operation

Figure 1.13 shows the operational diagram of DNA OR operation. The logic is as same as the binary OR operation in classical computing. The input-output combinations of DNA OR operation are shown in Table 1.11.

To design DNA OR operation, a test tube with the DNA mixture is needed, the annealing temperature is 60 °C approximately. To perform the DNA OR operations DNase enzyme is needed. And here the base sequence TGGATC will be used.

Here the base sequence in the test tube is TCCATC. And the input to the DNA OR operation is two DNA sequences. Let the two input sequences be A_0 and A_1 and the output be Q.

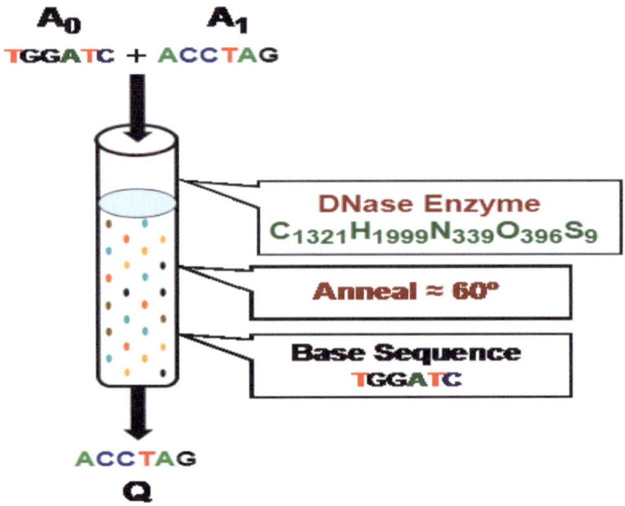

Fig. 1.13 The circuit architecture of a DNA OR operation

Table 1.11 The truth table of a DNA OR operation

A1	A0	Q
TGGATC	TGGATC	TGGATC
TGGATC	ACCTAG	ACCTAG
ACCTAG	TGGATC	ACCTAG
ACCTAG	ACCTAG	ACCTAG

1.4.3 DNA NOR Operation

The NOR operation is nothing but the inverted output of the OR operation, both the DNA OR and DNA NOT operational systems are designed already; therefore, it's easy to design the operational system for the DNA NOR operation, which is shown in Fig. 1.14.

From the above circuit, it is easy to understand the architecture of the DNA NOR operation. To perform DNA NOR operation, first DNA OR operation is needed to perform. Then the output of the DNA NOR operation will be inverted by the DNA NOT operation.

The input-output mapping for the DNA NOR operation is shown in Table 1.12. And one input-output mapping in Table 1.12, where the inputs are TGGATC and ACCTAG, and they produce TGGATC as output which is the inverted output from the DNA OR gate.

As the working procedure of DNA OR is explained before and DNA NOT operations too, therefore it's a doodle for us to understand the working procedure of DNA NOR operation.

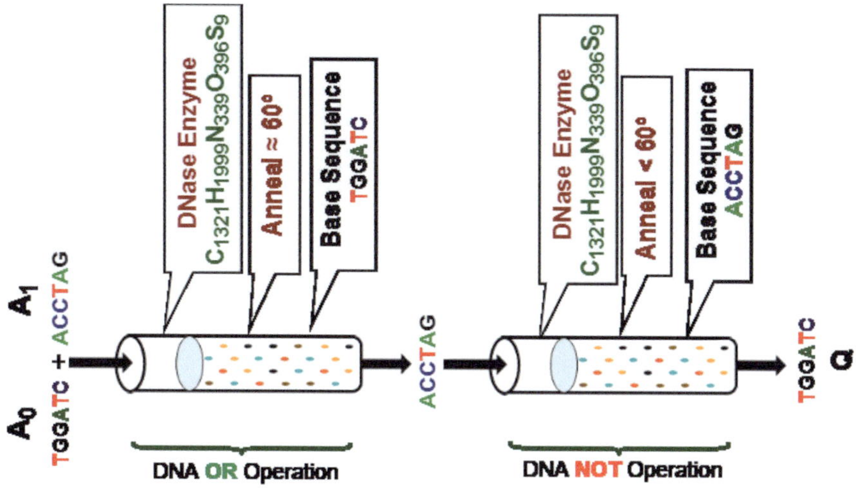

Fig. 1.14 The circuit architecture of a DNA NOR operation

Table 1.12 The truth table of a DNA NOR operation

A1	A0	Q
TGGATC	TGGATC	ACCTAG
TGGATC	ACCTAG	TGGATC
ACCTAG	TGGATC	TGGATC
ACCTAG	ACCTAG	TGGATC

Table 1.13 The truth table of a DNA NAND operation

A1	A0	Q
TGGATC	TGGATC	ACCTAG
TGGATC	ACCTAG	ACCTAG
ACCTAG	TGGATC	ACCTAG
ACCTAG	ACCTAG	TGGATC

1.4.4 DNA NAND Operation

In the structure of DNA OR, TGGATC is used as the base sequence in the test tube. The DNA computing system which will perform the operations of DNA NAND operation is containing ACCTAG as the base sequence instead of TGGATC. And the annealing temperature should be more than 60 °C to perform this operation.

The output for the given input sequences that will be produced by the DNA NAND operation is shown in Table 1.13.

From the above table, it is clear that the DNA NAND will produce an output sequence TGGATC only if the given input sequences both are ACCTAG. Otherwise, it will always generate ACCTAG as an output. Figure 1.15 shows the architecture of the DNA NAND operation.

As always, consider the set of input sequence patterns, and observe the behavior of the system of the DNA NAND operation. Let the two input sequences be A_0 and A_1. And the output be Q.

1.4.5 DNA AND Operation

Normally, the AND operation is implemented first, then invert the output of the AND operation, and get the result of the NAND operation. But, in DNA computing, it is totally opposite.

This is because it is easy to implement the NAND operation first in DNA computing. Then by inverting the output of NAND operations, easy to get the output of the AND operations. Figure 1.16 is to understand how to get the output of the DNA AND operations from the DNA NAND operation. Table 1.14 shows the truth table of DNA AND operation.

From the above table, it is clear that the DNA AND operation will generate the output sequence ACCTAG only if the input base sequences both are ACCTAG; otherwise, the output sequence will always be TGGATC.

DNA AND operation is nothing but the inverted value of the DNA NAND operation. So, at first DNA NAND operation will perform, and then DNA NOT operation will be performed to invert the output value of the DNA NAND operation.

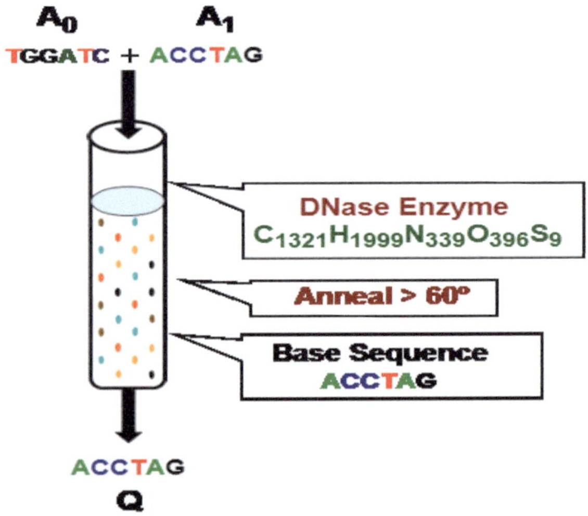

Fig. 1.15 The circuit architecture of a DNA NAND operation

1.4.6 DNA XOR Operation

Table 1.15 shows the truth table of the DNA XOR operation. It will produce ACCTAG when the given input sequences are not the same sequence pattern. And when both the inputs are the same, the DNA XOR will generate TGGATC as output. To design the architecture of the DNA XOR operation, there is no need for any base mixture as none of the sequences produces the DNA XOR output which is shown in Table 1.15. The input sequences must be complementary in order to have opposite values, and they will bind together to form a double-stranded sequence.

The logic is very simple. If the input sequences are the same, then they will not be able to make bonds together. If bonds are not created, the DNase enzyme will destroy them. And the output will be false (TGGATC). And when the input sequences are not the same, they will create the DNA double strands, and therefore the DNase enzyme

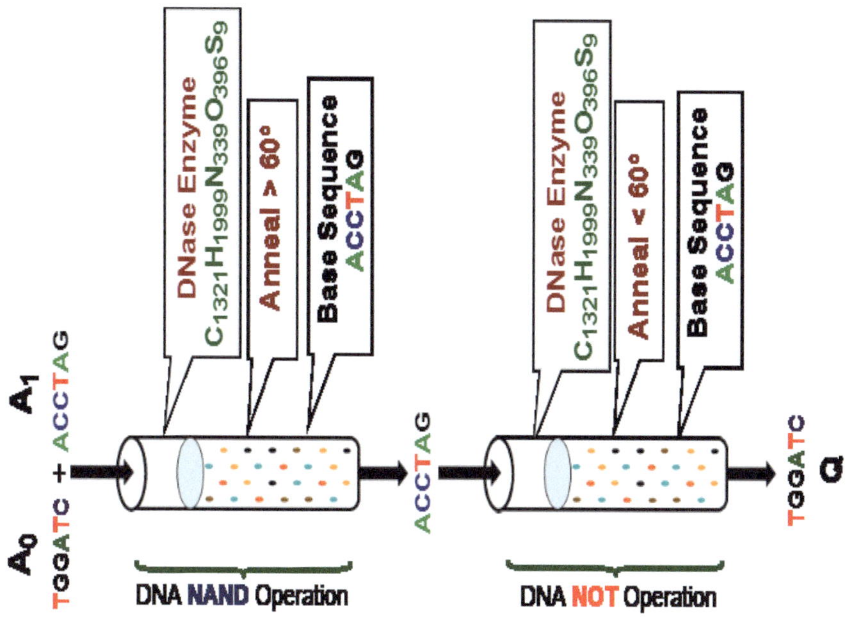

Fig. 1.16 The circuit architecture of a DNA AND operation

Table 1.14 The truth table of a DNA AND operation

A1	A0	Q
TGGATC	TGGATC	TGGATC
TGGATC	ACCTAG	TGGATC
ACCTAG	TGGATC	TGGATC
ACCTAG	ACCTAG	ACCTAG

Table 1.15 The truth table of a DNA XOR operation

A1	A0	Q
TGGATC	TGGATC	TGGATC
TGGATC	ACCTAG	ACCTAG
ACCTAG	TGGATC	ACCTAG
ACCTAG	ACCTAG	TGGATC

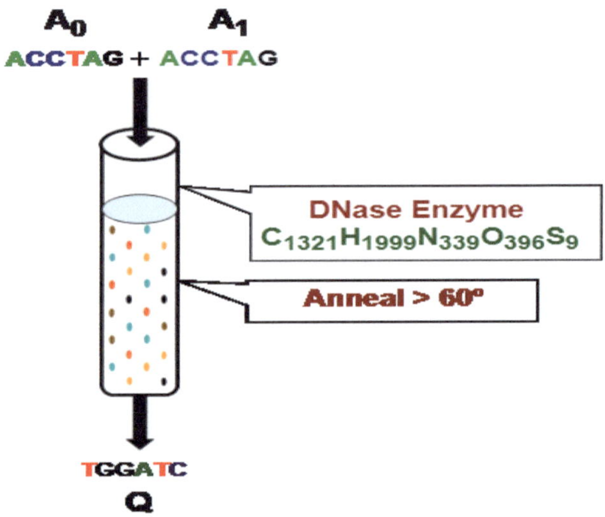

Fig. 1.17 The circuit architecture of a DNA XOR operation

Table 1.16 The truth table of a DNA XNOR operation

A1	A0	Q
TGGATC	TGGATC	ACCTAG
TGGATC	ACCTAG	TGGATC
ACCTAG	TGGATC	TGGATC
ACCTAG	ACCTAG	ACCTAG

will not affect them. As a result, the output will be true (ACCTAG). The architecture of the DNA XOR operation is shown in Fig. 1.17.

There is no need for a base sequence to perform DNA XOR operations. And the required ideal annealing temperature is more than 60 °C.

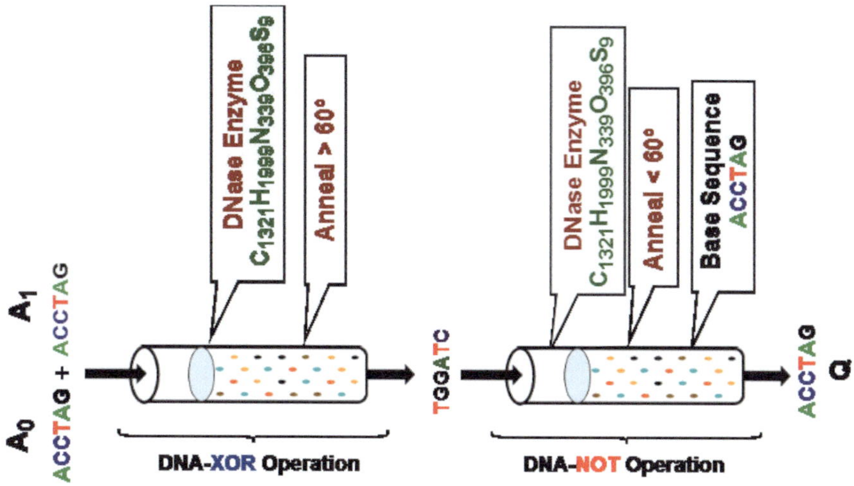

Fig. 1.18 The circuit architecture of a DNA XNOR operation

1.4.7 DNA XNOR Operation

As the name suggests DNA XNOR is the inverted output of the DNA XOR operation. Therefore, a DNA NOT operational system should be added to the output of the DNA XOR operational system to get the output result of the DNA XNOR operation. Table 1.16 shows the operations in the DNA XNOR operations. Figure 1.18 shows the circuit architecture of the DNA XNOR operations.

1.5 Summary

Quantum reversible logic has several fundamental gates such as Pauli gate, Pauli X-gate, Hadamard gate, Toffoli gate, Fredkin gate, Deutsch gate, and Swap gate. But the synthesized fundamental quantum gates are the controlled-NOT gate, the controlled-V gate, and the controlled-V+ gate. The quantum logical expression and the classical logical expression are the same but the working procedure is different as bit operations and qubit operations are not the same. The classical logic gates and operations in quantum circuits are presented in this chapter with their truth tables and working principles.

The architecture of the DNA computer is not like the classical or quantum computer, it uses chemical reactions performed in a test tube to produce the output. The processes to perform DNA computation are preparing, mixing and annealing, melting, amplifying, separating, extracting, cutting, ligating, substituting, marking and destroying sequences, and detecting and reading sequences. The DNA operation's truth tables are performed in the same way as the classical computation. But the design of the architecture is totally different. The total execution time is the required time to perform the largest pipeline in the operation. The heat required to perform a regular operation in DNA computing is certain. The required heat is 284–490 °C. All basic logic operations are presented in this chapter with their architecture, truth table, and working principles.

Part I
Memory Devices in Quantum Biocomputing

Overview

Both DNA computers and quantum computers have the potential to exceed the power of conventional digital computers, though substantial technical difficulties first must be overcome. Through coherent superposition of states, quantum computers are more powerful than classical Turing machines. DNA computers are evolvable through biotechnology techniques. By combining DNA and quantum computers, both of these characteristics might be captured. DNA computers could be used to self-assemble quantum logic circuits from gates attached to DNA strands. Moreover, quantum computers might be implemented directly using the physical characteristics of the DNA molecule. Because of quantum mechanics, quantum computers operate in an entirely different way from ordinary computers. For quantum computers, the most important principle is quantum superposition. In the classical world, when the card falls, one of two outcomes is observed, either the card lands face up or face down. In the quantum world, both events happen simultaneously, and therefore, the outcome is a superposition of the classical states. But more than that, the classical states become entangled so that they are correlated with and affect each other. In analogy with conventional "bits" of information, quantum mechanical bits of information are termed qubits, and carry two possible values, $|0>$ and $|1>$. Macromolecules of nucleic acids are the prime conveyers of genetic information. They are composed of nucleotide building blocks. In DNA, the nucleotides are the purines adenine (A) and guanine (G), and the pyrimidines thymine (T) and cytosine (C). Single-stranded (ss) DNA molecules, or oligonucleotides, are formed when the nucleotides are connected together with phosphodiester bonds. The single strands of DNA form a double-stranded (ds) molecule when the nucleotides hydrogen bonds to their Watson-Crick complements are, A = T and G = C, and vice versa. In the DNA helix, the intertwined strands are complementary, and one strand serves as the template for the replication of the other. The base pairing of one oligonucleotide to another is called hybridization. Because of its importance in biology and medicine, DNA is a well-characterized molecule, and many standard laboratory techniques exist for its manipulation. Therefore, it is a good choice for nanotechnology and unconventional computing systems. In this section, several ideas for combining DNA and quantum computing in memory devices have been given. The benefit to such a union might be a computationally efficient computer that can be evolved and adapted using DNA manipulation techniques from biotechnology. Of course, much work, both theoretical and applied, would be involved in approaching such a solution. This section of the book will cover memory devices used in quantum computing, Quantum-DNA computing, and DNA-Quantum computing. In separate chapters, memory devices such as RAM, ROM, PROM, and cache memory will be described in these quantum biocomputing modes.

Chapter 2
Memory Devices in Quantum Computing

2.1 Introduction

Theoretical physics, functional analysis, and algorithmic computer science are all bridge fields in quantum computation. Too far, the primary goal of quantum computation research has been to demonstrate that the time required to solve particular tasks is less for a quantum computer than for a conventional computer. Quantum memory is the capability of a quantum system to store and retrieve quantum information encoded in the quantum states of particles such as electrons or photons. It is the quantum-mechanical counterpart to conventional computer memory. While ordinary memory operates with binary states, quantum memory preserves quantum states for subsequent access. These states contain valuable computational data called qubits.

Unlike the classical memory found in everyday computers, the states stored in quantum memory can exist in a quantum superposition. Quantum superposition is a fundamental principle of quantum mechanics that describes the ability of quantum systems to exist in multiple states simultaneously. Mathematically, it means that a quantum state can be represented as a linear combination of two or more basis states. This characteristic offers significantly greater practical flexibility in quantum algorithms compared to classical information storage methods.

Quantum memory research can be traced back to the early days of quantum information science, which emerged in the 1980s and 1990s with foundational work by physicists like Richard Feynman, David Deutsch, and Peter Shor, among others. Quantum memory is required for the formation of a synchronization tool that can match the multiple procedures in a quantum computer, a quantum gate that retains the identity of any state, and a method for turning preset photons into on-demand photons, among other quantum information processing devices. Quantum memory may be utilized in a variety of applications, including quantum computing and quantum communication. Continuous research and experimentation have enabled the storing

H. Md. Hasan Babu, *Quantum Biocomputing in Quantum Biology Volume II*,
https://doi.org/10.1007/978-981-97-5349-9_2

of qubits in quantum memory. Quantum memory is the quantum-mechanical equivalent of conventional computer memory in quantum computing. Unlike conventional memory, which stores information as binary states (represented by "1's" and "0's"), quantum memory saves a quantum state for subsequent retrieval. Qubits (represented by "$|1>$" and "$|0>$"), which provide important computing information, are stored in these states. Unlike traditional computer memory, the states saved in quantum memory can be in a quantum superposition, providing far more practical flexibility in quantum algorithms than traditional information storage. This chapter will present four common memory devices in quantum computing, those are Quantum Random-Access Memory (QRAM), Quantum Read-Only Memory (QROM), Quantum Programmable Read-Only Memory (QPROM), and Quantum Cache Memory. Quantum memory devices are a mandatory part of quantum computation. So the details of those four memory devices will be shown in this chapter.

2.2 Quantum Random-Access Memory

A fundamental ability of any computing device is the capacity to store information in an array of memory cells. The most flexible architecture for memory arrays is random-access memory, or RAM, in which any memory cell can be addressed. Semiconductor memories are classified into Random-Access Memories (RAMs) and Sequential Access Memories (SAMs) based on access time. A RAM is composed of a memory array, an input register ("address register"), and an output register. Each cell of the array is associated with a unique numerical address. When the address register is initialized with the address of a memory cell, the content of the cell is returned to the output register ("decoding"). Just as RAM forms an essential component of classical computers, quantum random-access memory (QRAM) will make up an essential component of quantum computers, should large quantum computers eventually be built. Quantum Random Access Memory (QRAM) is a type of memory that uses quantum mechanical principles to store and access data. Unlike classical RAM, which uses bits (0 or 1), QRAM uses qubits, which can exist in a superposition of states, allowing for the simultaneous access of multiple memory locations. This enables potentially faster and more efficient data processing in quantum computers. It has the same three basic components as the conventional RAM, but the address and output registers are composed of qubits instead of bits. [The memory array can be either quantum or classical, depending on the RAM's usage.] The QRAM can then perform memory accesses in coherent quantum superposition: if the quantum computer needs to access a superposition of memory cells, the address register must contain a superposition of addresses and the QRAM will return a superposition of data in a data register, correlated with the address register. The possibility of efficiently implementing these devices would yield an exponential speedup for pattern recognition algorithms, period finding, discrete logarithm, and quantum Fourier transform algorithms over classical data. Moreover, QRAMs are required for the implementation of various algorithms, such as quantum searching on a classical database, collision finding, element distinctness in the classical and quantum settings, and the quantum algorithm for the evaluation of general NAND trees. Finally, QRAMs permit the introduction of new quantum computation primitives, such as quantum cryptographic database searches or the coherent routing of signals through a quantum network of quantum computers.

2.2.1 History

For major memory tasks, early computers employed relays, mechanical counters, or delay lines. Serial devices, such as ultrasonic delay lines, could only replicate data in the sequence in which it was written. Drum memory could be increased at a minimal cost, but retrieving memory objects will need the understanding of the drum's physical arrangement to maximize performance efficiently. For smaller and quicker memory like registers, latches made of vacuum tube triodes, and subsequently discrete transistors, were utilized. Such registers were big and expensive to employ for significant quantities of data; in most cases, only a few dozen or a few hundred bits of memory could be given. Starting in 1947, the Williams tube was the first viable kind of random-access memory. It used electrically charged dots on the front of a cathode ray tube to store data. Memory had random access because the CRT's electron beam could read and write the spots on the tube in any sequence. The Williams tube had a capacity of a few hundred to a thousand bits, but it was far smaller, quicker, and less power-hungry than individual vacuum tube latches. The Williams tube, developed at the University of Manchester in England, supplied the medium for the first electronically stored program to be implemented in the Manchester Baby computer, which successfully executed a program for the first time on June 21, 1948. Rather than being created for the Baby, the Williams tube memory was used to show the memory's dependability. Magnetic-core memory was developed until the mid-1970s after it was conceived in 1947. Using an array of magnetic rings, it became a popular kind of random-access memory. Data might be stored with one bit per ring by altering the sense of each ring's magnetism. Access to every memory place in any sequence was feasible since each ring contained a combination of address wires to select and read or write it. Until the early 1970s, when solid-state MOS (metal-oxide-silicon) semiconductor memory superseded magnetic core memory in integrated circuits (ICs), magnetic core memory was the conventional kind of computer memory system. Permanent (or read-only) random-access memory was often built before the introduction of integrated read-only memory circuits, employing diode matrices controlled by address decoders or specifically wound core rope memory planes. Bipolar memory, which utilized bipolar transistors, was the first form of semiconductor memory in the 1960s.

2.2.2 Basic Definition

Quantum Random-Access (QRA) memory is the hardware of a computing device that stores the operating system (OS), application programs, and data in use so that the CPU can access them fast. The main memory in a computer is RAM, or volatile memory. This indicates that data is stored in RAM while the computer is turned on, but it is lost when the machine is turned off. The OS and other data are reloaded into RAM when the machine is restarted, generally from an HDD or SSD. When the RAM is full, the computer CPU must go back and forth between the RAM and the hard drive to unload the data. Therefore, the smaller the amount of RAM, the slower the computer's operations.

The internal structure of a memory unit is specified by the number of words it contains and the number of bits in each word. Special input lines called address lines select one particular word. Each word in memory is assigned an identification number, called an address, starting from 0 and continuing with 1, 2, 3, up to $2^{\mathrm{k}} - 1$ where k is the number of address lines. The selection of a specific word inside the memory is done by applying the k-bit binary address to the address lines. A decoder inside the memory accepts this address and opens the paths needed to select the bits of the specified word. Computer memories may range from 1024 words, requiring an address of 10 bits, to 2^{32} words, requiring 32 address bits. It is customary to refer to the number of words (or bytes) in memory with one of the letters: K (Kilo) is equal to 2^{10}, M (Mega) is equal to 2^{20} and G (Giga) is equal to 2^{30}. 2^{k}-to-n RAM architecture is shown in Fig. 2.1.

Communication between memory and its environment is achieved through data input and output lines, address selection lines, and control lines that specify the direction of transfer. The n data input lines provide the information to be stored in memory, and the n data output lines supply the information coming out of a particular word chosen among the 2^k available inside the memory. The two control inputs specify the direction of transfer desired. When the value of address lines $k = 2$ and the input line $n = 1$, it is called 2^2-to-1, 4-to-1 RAM.

2.2.3 Advantages

There are many advantages of RAM. Some benefits of quantum RAM include:

1. RAM is a must to store the data for processing on the CPU. RAM is a component that is mandatory to have in a system to allow for the storage of data, which will be processed by the CPU.
2. More RAM is a factor to increase the speed of a computer.
3. If the CPU wants to read the data from RAM, then it is fast as compared to data access from hard disk, CD, DVD, floppy, and USB.
4. Increases the computer system speed, essentially, the more RAM a system has the faster it will operate.

5. It's efficient. It's extremely faster compared to hard drive storage for a CPU to read data from.
6. It can write and read operations.
7. RAM is power efficient.
8. The cost is less than SSDs and it operates faster than them.

2.2.4 *Disadvantages*

There are many disadvantages of RAM, which are given below:

1. Less RAM is a factor in decreasing the speed of a computer.
2. If the CPU wants to read the data from RAM, then it is slow as compared to data access from registers and cache.
3. RAM is volatile, which means it is difficult to store data for a long time. Unplanned circumstances like a power outage can result in data loss.
4. It is expensive.

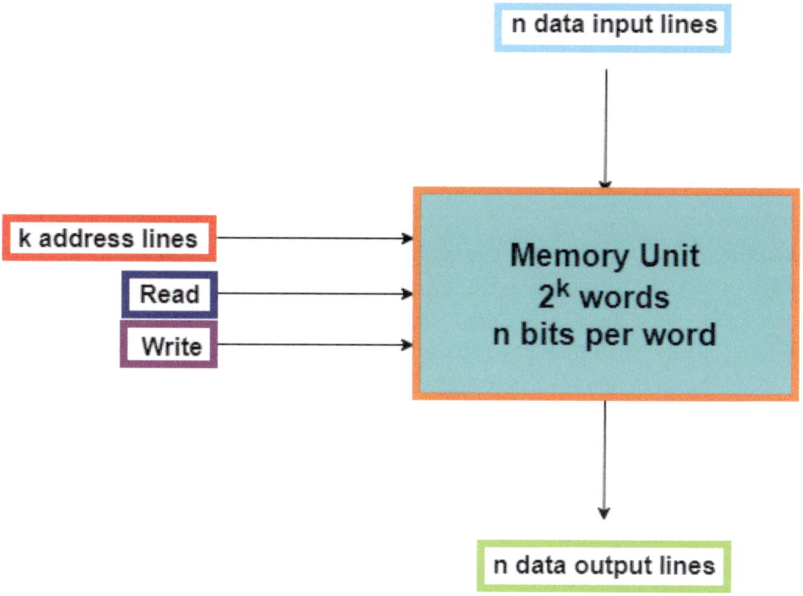

Fig. 2.1 2^k-to-n RAM

2.2.5 Basic Functions

The two operations that a random-access memory can perform are the write and read operations. The write signal specifies a transfer-in operation and the read signal specifies a transfer-out operation. On accepting one of these control signals. The internal circuits inside the memory provide the desired function. The steps that must be taken to transfer a new word to be stored into memory are as follows:

1. Apply the binary address of the desired word to the address lines
2. Apply the data bits that must be stored in memory to the data input lines.
3. Activate the write input.

The memory unit will then take the bits presently available in the input data lines and store them in the specified by the address lines. The steps that must be taken to transfer a stored word out of memory are as follows:

1. Apply the binary address of the desired word to the address lines; and
2. Activate the read input.

The memory unit will then take the bits from the word that has been selected by the address and apply them to the output data lines. The content of the selected word does not change after reading.

RAM's primary function in a computer is to read and write any data. RAM interacts with the hard disk of the computer. An example can be given. When open a Word file, the Word file is saved on the computer's hard disk before it is opened, and the Word file is stored in the computer's RAM as soon as it is opened.

There are many basic and main functions of computer memory RAM, which are given below.

1. **Reading Files:** Hard drives can store a vast number of files, but compared to other computer components, drives run very slowly. Accessing hard drive files especially when those files are scattered across the drive due to fragmentation requires the drive to move its mechanical read/write head back and forth and to wait for the spinning platters to spin into the correct position. Even though drives spin at thousands of rotations per minute, this process causes a noticeable delay when reading files. To lessen the slowdown, computers store files in RAM after the files are first to read from the drive. RAM has no moving parts (and runs at a higher speed than even a solid-state drive) so the files can load very quickly during subsequent uses.
2. **Temporary Storage:** In addition to storing files read from the hard drive, RAM also stores data that programs are actively using but that doesn't need to be saved permanently. By keeping this data in RAM, programs can work with it quickly, improving speed and responsiveness.
3. **Loading Applications:** Loading a software application is also the main function of RAM. Any software or application opens in the computer using RAM itself.
4. **Speed:** RAM speed is measured in Megahertz (MHz), millions of cycles per second so that it can be compared to any processor's clock speed.

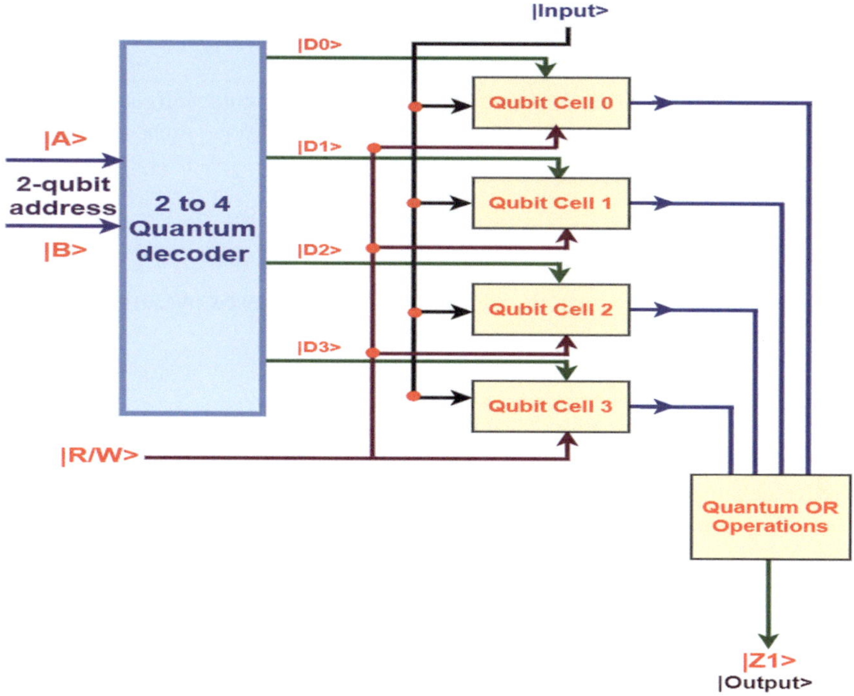

Fig. 2.2 Block diagram of a quantum 4-to-1 RAM

2.2.6 Block Diagram

Quantum 4-to-1 RAM chip has a memory capacity of four words of one qubit per word. This requires a 2-qubit address and a 1-bit bidirectional data bus. The 1-bit bidirectional data bus allows the transfer of data either from memory to CPU during a read operation or from CPU to memory during a write operation. The read-and-write inputs specify the memory operation, and the two chip select (CS) control inputs are for enabling the chip only when the microprocessor selects it.

Figure 2.2 represents the Quantum 4-to-1 RAM general organization of the block diagram. This quantum RAM consists of four separate "Words" of memory and each is single qubits wide. Quantum 4-to-1 RAM block diagram is shown in Fig. 2.2.

2.2.7 Design Architecture of a 4-to-1 RAM

RAM consists of three basic components such as decoder, qubit cells, and quantum OR circuits. To execute quantum 4-to-1 RAM, the following operations are required:

1. A quantum 2-to-4 decoder.
2. Qubit cells and
3. Quantum OR operations for corresponding minterms.

Quantum decoder and Quantum OR operation are discussed in earlier chapters. Quantum qubit cell needs to be discussed here.

2.2.7.1 Circuit Design of a Qubit Cell

The fundamental design of this qubit cell is based on the R-S flip-flop (Fig. 2.3). Each cell has three inputs and a single output. The inputs are labeled "|Select >," "|R/W >," and "|Input >." The output line is labeled "|Q >." To perform the qubit cell output, two quantum NOT, six quantum AND, and two quantum NOR operations are needed to perform. The circuit design of quantum single-qubit cell is shown in Fig. 2.3.

Step 1: First draw three input qubits |Input >, |R/W >, and |Select >. Two possible states for a qubit are the states |**0** > false, and |**1** > true.

Step 2: Then, draw quantum NOT operation with the |input > and |R/W > qubits.

Step 3: Two-input qubits (NOT operation of |input > and |select >) will go through quantum AND operations and the output of these operations will go to another quantum AND operation with input qubit |R/W >.

Step 4: Again, two inputs (|input > and |select >) will go through quantum AND operations and the output of these operations will go through another quantum AND operation with input qubit |R/W >.

Step 5: The outputs of Steps 3 and 4 will go to the R-S flip-flop as input.

Step 6: Finally, the output of R-S flip-flop and |select > will go to quantum AND operation, then this output with NOT of |R/W > will go through another quantum AND operations to produce desired RAM cell qubit output.

2.2.7.2 Working Procedure of Qubit Cell

In a sequential device as simple as an R-S flip-flop could be used to remember one qubit of data. To develop a complete memory cell, called a qubit cell, we need to perform the flip-flop for storage. The number of total quantum cells per word will be $m \times n$, where m represents words with n qubits. The cell has three inputs and a single output. The inputs are labeled "|Select >," "|R/W >," and "|Input >." The output line is labeled "|Q >."

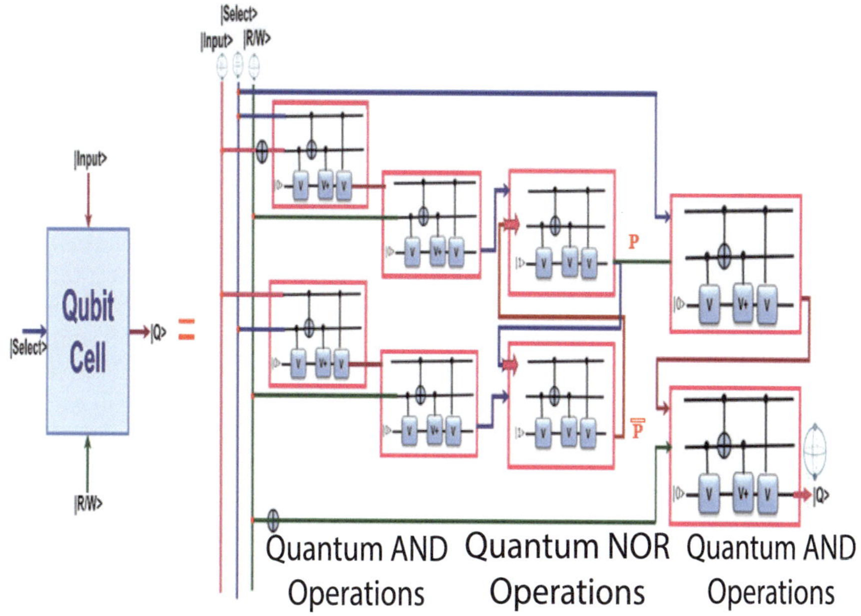

Fig. 2.3 Quantum single-qubit cell

The "|Select >" input is used to access the cell, either for reading or writing. When the select line is high, "|1 >," a memory operation can be performed on this cell. When the select line of the binary cell is low, "|**0** >," the contents of the cell are not currently of interest, i.e., at present, the cell is not being read from or written to. It can be seen that how "|Select >" is given this power by noting that both the inputs and the output of the underlying R-S flip-flop are routed through and gates and that |Select > is one of the inputs to each of these gates. Thus, if "|Select >" is low |**0** >, the inputs to the R-S flip-flop will stay low |**0** > (meaning that its stored value will not change) and the output produced by the cell will be low (regardless of whether the actual qubit held in the flip-flop is "|**0** >" or "|**1** >").

The next input to examine is "|R/W >." A system clock will drive this input. As the case is with the clocked R-S flip-flop, a low, "|**0** >," will signify "|R >" while a high, "|**1** >," will signify "|W >." During the read phase, it will not be possible to write to the cell. Likewise, during the writing phase, it will not be possible to read the contents of the cell.

Assume the cell has been selected (i.e., "|Select >" is high signifying that a memory access operation is to be performed on this cell). Furthermore, assume that the clock value on the "|R/W >" line is low (forcing the "negated |R/W >" to high) indicating the cell contents are to be read. In this case, the value output by the cell will depend solely on the P-valued of the flip-flop. If P is low, the cell outputs a "|**0** >," if P is high, the cell outputs a "|**1** >." This is because the AND gate attached to the cell's output line has three inputs: "|Select >," "negated |R/W >," and P; and both "|Select >" and "negated |R/W >" are currently high.

As mentioned earlier, when the cell is being read its contents cannot be modified. The reason for this is that the same low value on the "|R/W >" line that allows the cell to be read is fed into the AND gates guarding the inputs to the flip-flop. Thus during reads, the inputs to R and S are guaranteed to be low preventing the value of the flip-flop from being modified. When the cell is selected and the "|R/W >" line is set to high, signifying a "write" operation, the value placed into the cell will depend solely on the state of the "|Input >" line. The reason for this is that the and gates that guard the R and S inputs of the flip-flop will both have two of their inputs set high: the "|Select >" and "|R/W >" inputs. Thus, if "Input" is high, S (set) will receive a high and the flip-flop will store a "|**1** >." If, on the other hand, "|Input >" is low, then R (reset) which receives a negated version of "|Input >" will go high and the flip-flop will reset to "|**0** >." Note that having a negated version of the input line run into R is a clever idea since it prevents the R-S flip-flop from ever entering into its invalid state. (Recall that if R and S are ever set to "|**1** >" at the same time the flip-flop enters a stable, but invalid state.) It is worth mentioning that during write operations, reading is prohibited. This is easy to see since the AND operation guarding the "|Q >" line receives one of its inputs from "negated |R/W >" which is held low during write operations. Hence, output from the cell will always be low, "|**0** >," during writes, regardless of the actual value on the P line of R-S flip-flop. The circuit of quantum qubit cells is shown in Fig. 2.4.

To perform the quantum 4-to-1 RAM, four selection lines are needed to select four quantum qubit cells. The output of quantum 2-to-4 decoder with four output qubits will perform as selection input qubit of qubit cells as |Select >. Figure 2.4 shows the 4-qubit cells to perform operations on qubit cell outputs from |Q0 > to |Q3 > for further minterms OR operation.

2.2.8 Working Principle of a Quantum RAM Memory

Figure 2.5 presents an implementation of a quantum 4-to-1 RAM. Quantum RAM consists of four separate "Words" of memory, where each one is a single-qubit wide. The quantum RAM cell has three inputs and one output. The complete circuit of a quantum RAM cell is described in Fig. 2.4 with proper explanation. A word consists of two quantum RAM cells arranged in such a way so that both qubits can be accessed simultaneously. Four words of memory need two address lines. |A > and |B > are the two-qubit address lines input that goes through a 2-to-4 decoder that selects one of the four words. The memory-enabled input enables the decoder. If the memory enables is |**0** >, all outputs of the decoder will be |**0** > and in that case, none of the memory addresses will be selected. But when the memory enables is |**1** > one of the four words is selected. The word is selected by the value in the two address lines. When a word has been selected, the read/write input determines the operation. During the read operation, the four qubits of the selected word pass to the quantum

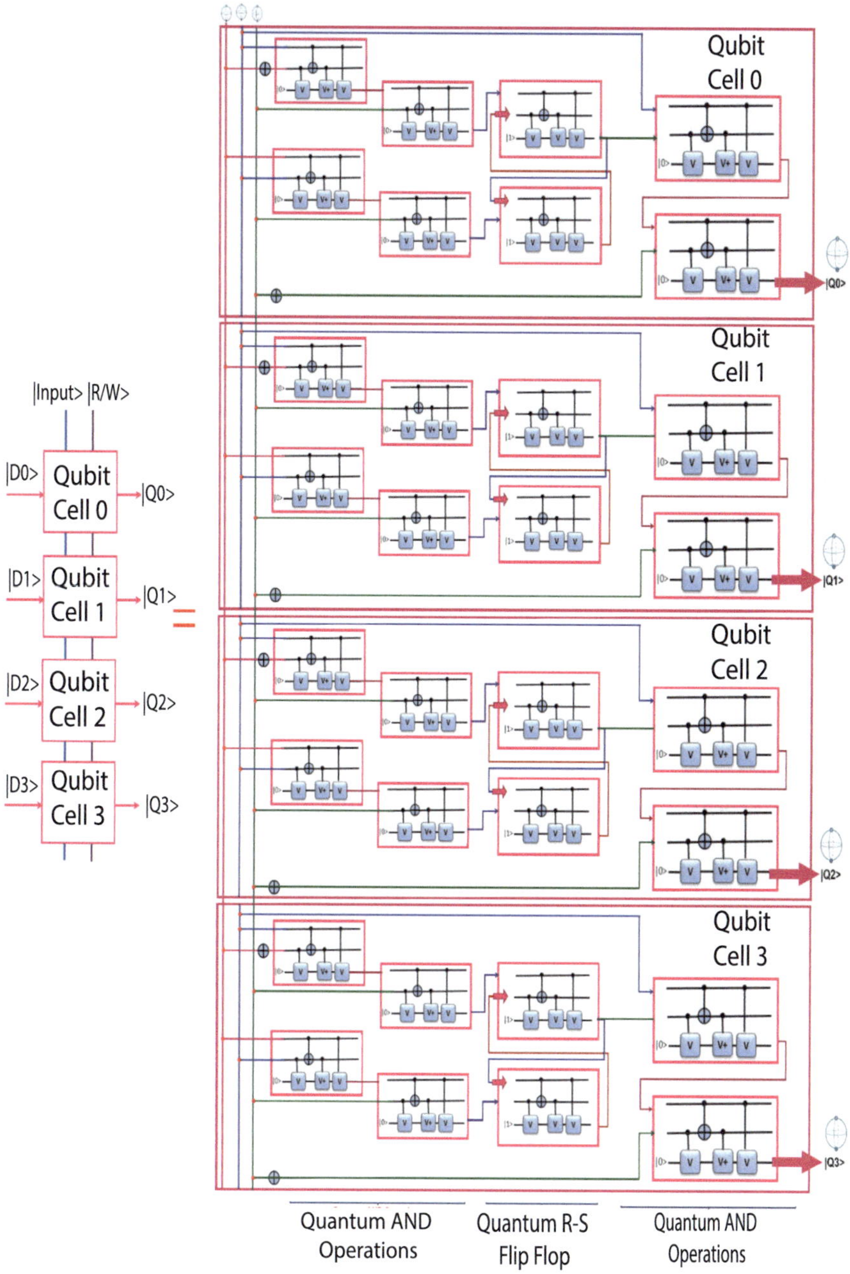

Fig. 2.4 Quantum qubit cells

Table 2.1 Control input of a memory chip

|R/W>	Memory operation
X	*None*
|0>	*Write to the selected word*
|1>	*Read from the selected word*

OR gates to the output |Z1 > terminals. But during the write operation, the data which is available in the input lines are transferred into the four quantum cells of the selected word. The quantum RAM cells that are not selected become disabled and their previous qubit never changes. But when the memory-enabled input that passes into the decoder is equal to |**0** >, none of the words are selected, and all quantum cells remain unchanged regardless of the value of the read/write input. The control input to memory chip in quantum RAM is shown in Table 2.1.

2.2.9 Applications

In addition to serving as a temporary storage for the operating system and applications, RAM is used in numerous other ways:

1. Operand stacks.
2. Register files.
3. Instruction caches.
4. DMA buffers.
5. Instruction memories.
6. Logic functions.
7. Message buffers.
8. Virtual channels.
9. Digital delay lines.
10. Sequential machines.

Virtual Memory

Most modern operating systems employ a method of extending RAM capacity, known as "virtual memory." A portion of the computer's hard drive is set aside for a paging file or a scratch partition, and the combination of physical RAM and the paging file forms the system's total memory. (For example, if a computer has 2 GB of RAM and a 1 GB page file, the operating system has 3 GB total memory available to it.) When the system runs low on physical memory, it can "swap" portions of RAM to the paging file to make room for new data, as well as to read previously swapped information back into RAM. The excessive use of this mechanism results in thrashing and generally hampers overall system performance, mainly because hard drives are far slower than RAM.

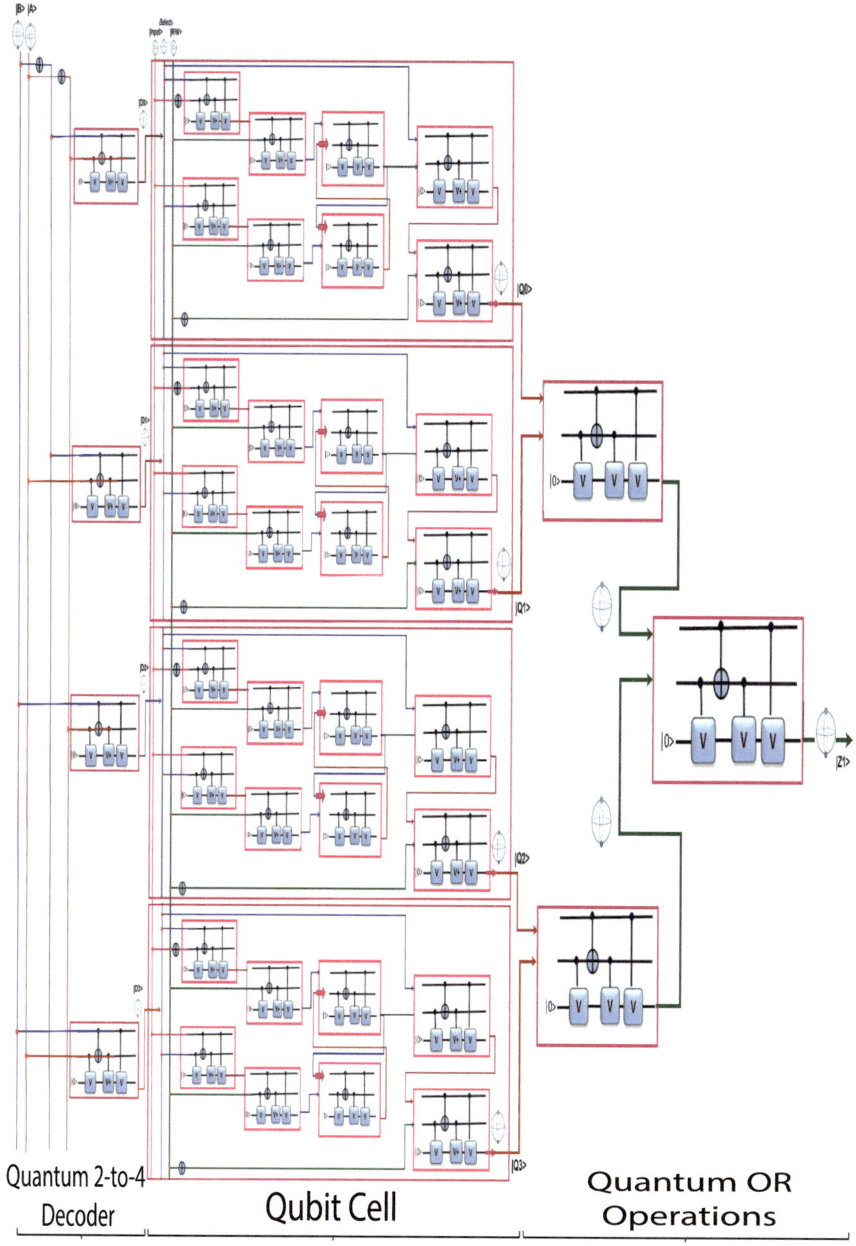

Fig. 2.5 Circuit architecture of a quantum 4-to-1 RAM

RAM Disk
The software can "partition" a portion of a computer's RAM, allowing it to act as a much faster hard drive that is called a RAM disk. A RAM disk loses the stored data when the computer is shut down unless memory is arranged to have a standby battery source.

2.3 Quantum Read-Only Memory

Continuous technological advancement—driven by Moore's law—electronic gadgets and their linked memory systems become simultaneously smaller and more efficient. Current storage capacity and other cognitive mechanisms, on the other hand, consume more energy. Furthermore, internal thermal resistance and other restrictions may hinder further memory storage advancements, even as worldwide demand for enhanced data storage and retrieval capabilities continues to expand. So, the main concern is to provide low-cost, robust, high-density, reliable, and energy-efficient memory technologies through designing quantum-based ROM. This technology can be written on, read from, and erased rapidly, and not deteriorate with time. Though traditional ROM is a slower memory so quantum computing enables the creation of new types of computers capable of operating with qubits as input states, therefore increasing the processing capacity. This section covers the details of quantum ROM (QROM) or quantum read-only memory. In quantum computing, QROM is a type of quantum memory used to store pre-defined quantum states, similar to how classical ROM stores data. It's essentially an operator that loads classical data into a quantum computer by mapping bitstrings to quantum states. While quantum memories hold promise, including QROM, they are currently challenging to implement in noisy intermediate-scale quantum (NISQ) computers due to scaling issues with the number of address lines. The QROM operator might be defined as $\mathrm{QROM}|\mathrm{i}\rangle|0 \otimes \mathrm{m}\rangle = |\mathrm{i}\rangle|\mathrm{bi}\rangle$, where $|\mathrm{i}\rangle$ represents an input bitstring, $|0 \otimes \mathrm{m}\rangle$ represents an initial quantum state, and $|\mathrm{bi}\rangle$ represents the corresponding quantum state stored in memory. Quantum memories like QROM hold the potential to enable quantum advantage in certain applications, such as quantum machine learning. QROM provides a way to bridge classical and quantum computing by allowing the transfer of classical data to the quantum realm.

2.3.1 *History*

Scientists have been more interested in the interaction of quantum radiation with a variety of particles during the last decade. One such field is quantum memory, which maps the quantum state of light onto a collection of atoms before returning it to its original shape. Quantum memory is a key component in information processing, such as optical quantum computing and quantum communication, and it also paves the way for the potential of light-atom interaction. Reconstructing the quantum state of light, on the other hand, is a challenging task. Photon qubits may be stored in quantum memory based on quantum exchange. In 1993, Kessel and Moiseev proposed quantum storage in the single-photon state. In 1998, the experiment was examined, and in 2003, it was proven. To summarize, the research of quantum storage in the single-photon state is a byproduct of the classical optical data storage technique developed in 1979 and 1982. Furthermore, the notion was influenced by the increasing density of data storage in the mid-1970s. Optical data storage may be accomplished by employing absorbers to absorb various frequencies of light, which are then guided to and stored at beam space locations. After 1948, most early stored-program computers, such as the ENIAC, used forms of read-only memory as non-volatile storage for programs. (Previously, it was not a stored-program computer since each program had to be manually wired into the system, which might take days or weeks.) Read-only memory was easier to build since it just required a method to read stored data rather than alter them in place, and so could be done with extremely rudimentary electromechanical devices. When integrated circuits were first introduced in the 1960s, ROM was implemented as arrays of transistors on silicon chips. An ROM memory cell, on the other hand, it might be created with fewer transistors and may consist of the absence (logical qubit $|0>$) or presence (logical qubit $|1>$) of one transistor linking a bit line to a word line.

2.3.2 *Basic Definition*

QROM is an abbreviation for quantum read-only memory. It refers to computer memory chips that hold permanent or semi-permanent data and incorporate both the quantum decoder and quantum OR operations onto a single integrated circuit (IC). The contents of quantum ROM are non-volatile; even if the computer is turned off, the contents of quantum ROM persist. It is used to hold a computer's boot-up instructions. Most of the computers have a tiny bit of quantum ROM that contains the boot software. This is made up of a few kilobytes of code that instructs the computer on what to do when it boots up, such as conducting hardware diagnostics and loading the operating system into quantum RAM. The BIOS is the boot firmware on a computer. To update the programming in quantum ROM, the quantum ROM chips have to be physically removed and replaced. Data saved in quantum ROM cannot be electrically changed once the memory device is manufactured. A block diagram of an ROM is shown in Fig. 2.6. It consists of n input lines and m output lines. Each bit combination of the input variables is called an address. Each bit combination that comes out of the output lines is called a word. The number of bits per word is equal to the number of output lines m. An address is essentially a binary number that denotes one of the minterms of n variables.

Initially, the quantum ROM is a combinational circuit with quantum AND gates connected as a quantum decoder and many quantum OR gates equal to the outputs in the unit. Therefore, it is a two-level implementation in the sum of minterms form.

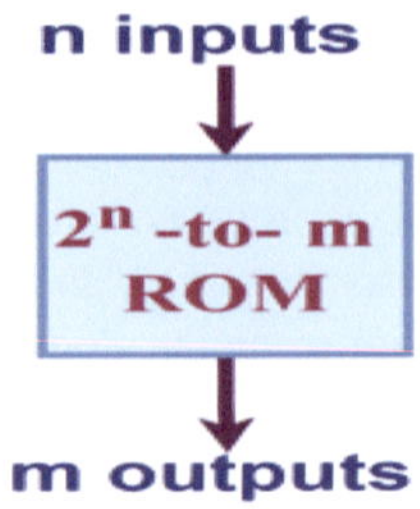

Fig. 2.6 Block diagram of a 2^n-to-m ROM

With n input lines and m output lines in ROM, the output functions will calculate through the sum of minterms form. The number of distinct addresses possible with n input variables is 2^n. An output word can be selected by a unique address, and since there are 2^n distinct addresses in a ROM, there are 2^n distinct words that are said to be stored in the unit. The word available at the output lines at any given time depends on the address value applied to the input lines. Therefore, an ROM is characterized by the number of words 2^n and the number of bits per word m. For input, $n = 2$ and output, $m = 2$ the ROM circuit will be called 4-to-2 ROM and the function output F_1 and F_2 in the sum of minterms form, $\sum$ (0, 1, 2, 3). It does not have to be a quantum AND-OR implementation but it can be any other possible two-level minterms implementation. Thus the second level is usually a wired logic connection to facilitate the fusing of links.

2.3.3 *Advantages*

Quantum ROM stores the instructions required for communication between various hardware components. As previously stated, it is required for the storage and functioning of the BIOS, but it may also be used for basic data management, to store software for basic utility functions, and to read and write to peripheral devices. Other benefits of a quantum ROM include the following:

1. Its static nature means it does not need to be refreshed.
2. It is simple to put to the test.
3. More dependable than RAM since it is non-volatile and cannot be altered or modified unintentionally.
4. Contents are always known and verifiable.
5. It is less costly than RAM.
6. Lower production costs since IC costs are lower.
7. It is a highly compact device.
8. Circuits are simple and quantum ROM is more reliable than RAM.
9. Data can be stored permanently.
10. It helps to start the computer and load the OS.

2.3.4 *Disadvantages*

There are several drawbacks of quantum ROM, which are listed below:

1. ROM is read only and cannot be altered. In other words, ROM memory can only read data and cannot modify the ROM data.
2. It is not feasible to make adjustments if they are necessary.
3. Quantum ROM is a slower form of memory.
4. Improperly deleting the data of the quantum ROM memory would brick the memory.
5. Some ROM memory types allow the user to rewrite the contents.

2.3.5 *Basic Functions*

On the CPU and other computer equipment, memory devices are used to store various types of information such as data, programs, addresses, textual files, and status information. Data structures can be split into bits, bytes, words, blocks, segments, pages, and other identifiers in memory devices. Information is stored in memory cells or memory locations in primary memory. Information that can be accessed is stored in memory locations. A single memory access operation must be performed to read or write data in a memory location, which necessitates the supply of independent control signals to the memory. Memory devices may be classified into two groups based on the information of the addressing method:

1. Memories in which access to places is controlled by addresses and
2. Memories in which access to locations is controlled by the contents of the memory.

Such memories belong to the first category, in which each accessible place has an address that may be used to choose a memory location and conduct needed operations. Addresses are used to address these memories. Hardware circuits perform address decoding and choose the desired place for a memory access operation in such memories. The collection of all addresses available in memory is referred to as the memory's address space. When a computer executes a program of instructions, the CPU constantly retrieves (reads) data from memory regions containing (1) the program codes indicating the operations to be done and (2) the data to be acted on.

2.3.5.1 Read Operation

A random memory enables unrestricted space and time access to any location at any address in the address space. The access is possible independently of the order of all previous accesses. The access can take place to an address in any order. Each location in a random-access memory has independent hardware circuits that provide

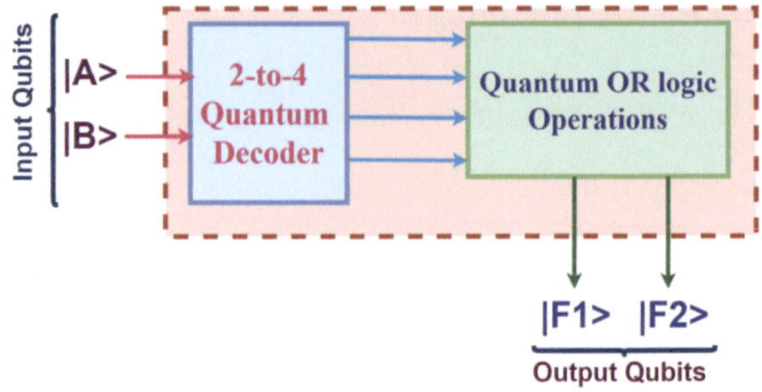

Fig. 2.7 Block diagram of quantum 4-to-2 ROM

the access. These circuits are activated as a result of address decoding. To such memories belong semiconductor memories of the ROM types. The most important parameters are discussed below:

1. **Memory Capacity** or **Memory Volume** is the number of locations that exist in a given memory. Memory capacity is measured in bits, bytes, or words. When words are used, the length of a word in qubits has to be given.
2. **Memory access time** is the time that separates sending a memory access request and the reception of the requested information. The access time determines the unitary speed of memory (the reception time of unitary data). The access time is small for fast memories.
3. **Memory cycle time** is the shortest time that has to elapse between consecutive access requests to the same memory location. The memory cycle time is another parameter that characterizes the overall speed of the memory. The speed is big when the cycle time is small.
4. **Memory transfer rate** is the speed of reading or writing data in the given memory, measured in qu**bits/sec**.

2.3.6 Block Diagram

Consider the block diagram of quantum 4-to-2 ROM shown in Fig. 2.7, the unit consists of four words of two input qubits ($|A>$ and $|B>$) each. This implies there are two output lines ($|F1>$ and $|F2>$) and four distinct words stored in the unit, each of which may be applied to the output lines. The particular word selected that is presently available on the output line is determined from the two input lines. For two input qubits in quantum 4-to-2 ROM, four addresses are specified because $2^2 = 4$. To perform minterms of four addresses, quantum 2-to-4 decoder and quantum OR operations are required. For each address input qubit, there is a unique selected word. Thus, the input qubit address is $|0>|0>$, word number 0 is selected and it appears on the output qubit lines. If the input qubit address is $|1>|1>$, word number 3 is selected and it appears on the output lines. In between, two other input qubit addresses can select the other two words.

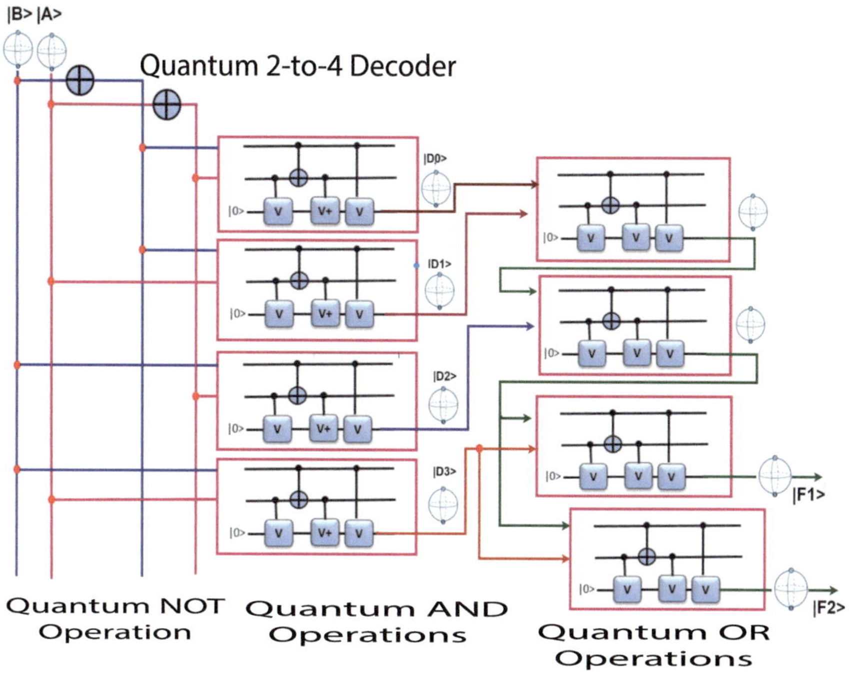

Fig. 2.8 Quantum 4-to-2 ROM

2.3.7 *Circuit Architecture*

In quantum 4-to-2 ROM architecture shown in Fig. 2.8, there are three quantum operations (NOT, AND, and OR) with three quantum gates (NOT, V, and V+). The output of the decoder is performed together with four quantum OR operations and produces desired output qubits $|F1>$ and $|F2>$ of quantum 4-to-2 ROM.

2.3.8 *Working Principle*

According to the truth table of quantum 4-to-2 ROM as shown in Table 2.2 and also considering Table 2.1, it is necessary to do the following operations to perform desired output qubits:

[i] For input qubits $|A>$, $|B> = |0>$, $|0>$, $|D0>$ line will be open. So, the value of $|D0>$ will be $|1>$ and $|D1>$ to $|D3>$ will be $|0>$. For the output qubits of $|F1>$ and $|F2>$, perform OR operations among $|D0> = |1>$, $|D1> = |0>$, $|D2> = |0>$, and $|D3> = |0>$ to generate $|1>$.

Table 2.2 Truth table of quantum 4-to-2 ROM

|**B** >	|**A** >	|**F1** >	|**F2** >
|**0** >	|**0** >	|**1** >	|**1** >
|**0** >	|**1** >	|**1** >	|**1** >
|**1** >	|**0** >	|**1** >	|**1** >
|**1** >	|**1** >	|**1** >	|**1** >

[ii] For input qubits |A >, |B > =|1 >, |0 >, |D1 > line will be open. So, the value of |D1 > will be |1 > and |D0 >, |D2 > and |D3 > will be |0 >. For the output of |F1 > and |F2 >, perform OR operations among |D0 >=|0 >, |D1 > = |1 >, |D2 > = |0 > and |D3 > = |0 > to generate |1 >.
[iii] For input qubits |A >, |B > =|0 >, |1 >, |D2 > line will be open. So, the value of |D2 > will be |1 > and |D0 >, |D1 > and |D3 > will be |0 >. For the output of |F1 > and |F2 >, perform OR operations among |D0 >=|0 >, |D1 > = |0 >, |D2 > = |1 >, and |D3 > = |0 > to generate |1 >.
[iv] For input qubits |A >, |B > =|1 >, |1 >, |D3 > line will be open. So, the value of |D3 > will be |1 > and |D0 > to |D2 > will be |0 >. For the output of |F1 > and |F2 >, perform OR operations among |D0 >=|0 >, |D1 > = |0 >, |D2 > = |0 >, and |D3 > = |1 > to generate |1 >.

2.3.9 Applications

The main concern of quantum ROM application is data storage and memory technology. Quantum ROMs are taken to include all semiconductor, non-volatile memory devices, and they are used in applications where non-volatile storage of information, data, or program codes is needed and where the stored data rarely or never changes. There are some of the most common application areas as listed below.

1. BIOS chip in computers.
2. Network operating systems.
3. Storing sound data in electronic musical instruments.
4. Storage for in-built self-learning functionality in remote-operated transmitters.

2.4 Quantum Programmable Read-Only Memory

PROM is programmable read-only memory (PROM) that can be modified once or can only be programmed once by a user. This is because PROM chips are manufactured with a series of fuses. It is first prepared as blank memory, and then it is programmed to store the information. PROM is manufactured as blank memory and programmed after manufacturing. To program the PROM, a PROM programmer or PROM burner is used. The process of programming the PROM is called burning the PROM. The chip is programmed by burning fuses, which is an irreversible process. The open fuses are read as ones, while the burned fuses are read as zeros. By burning specific fuses, a binary pattern of ones and zeros is imprinted on the chip. This pattern represents the program applied to the ROM. A Quantum Programmable Read Only Memory (QPROM) is a theoretical concept exploring the possibility of storing and manipulating quantum information in a memory that can be programmed once, similar to a standard PROM. It differs from classical PROMs by storing and accessing quantum states (qubits) instead of classical bits.

2.4.1 History

The PROM was invented in 1956 by Wen Tsing Chow, working for the Arma Division of the American Bosch Arma Corporation in Garden City, New York. The invention was conceived at the request of the United States Air Force to come up with a more flexible and secure way of storing the targeting constants in the Atlas E/F ICBM's airborne digital computer. The patent and associated technology were held under a secrecy order for several years while the Atlas E/F was the main operational missile of the United States ICBM force. The term burn, referring to the process of programming a PROM, is also in the original patent, as one of the original implementations was to literally burn the internal whiskers of diodes with a current overload to produce a circuit discontinuity. The first PROM programming machines were also developed by Arma engineers under Mr. Chow's direction and were located in Arma's Garden City lab and Air Force Strategic Air Command (SAC) headquarters. Commercially available semiconductor antifuse-based OTP memory arrays have been around at least since 1969, with initial antifuse bit cells dependent on blowing a capacitor between crossing conductive lines. Texas Instruments developed an MOS gate oxide breakdown antifuse in 1979. A dual-gate-oxide two-transistor (2T) MOS antifuse was introduced in 1982. Early oxide breakdown technologies exhibited a variety of scaling, programming, size, and manufacturing problems that prevented the volume production of memory devices based on these technologies. Although antifuse-based PROM has been available for decades, it wasn't available in standard CMOS until 2001 when Kilopass Technology Inc. patented 1T, 2T, and 3.5T antifuse bit cell technologies using a standard CMOS process, enabling integration of PROM into logic CMOS chips.

2.4.2 Basic Definition

QPROM stands for quantum programmable read-only memory, a programmable semiconductor memory that can be programmed once by the user and only read out.

It is a Programmable Logic Array (PLA) of a Non-Volatile Memory (NVM), which retains its stored contents even after the supply voltage is switched off. PROMs consist of OR gates that are joined together to form an array. In addition to memory elements, PROMs have electronic circuitry that enables bit-by-bit programming. They are used for the permanent storage of programs. The basic architecture of PROMs consists of an input decoder of AND gates and an output matrix of OR gates that are combined. This allows each output to be individually linked to any possible input. The individual links are created by merging the connections. 2^n-to-m PROM is shown in Fig. 2.9.

With n input lines and m output lines in programmable ROM, the output functions will be calculated through the programmable sum of minterms form. For input, $n = 2$ and output, $m = 2$, the PROM circuit will be called 4-to-2 PROM and the function outputs are $F_1 = \sum(0, 2)$ and $F_2 = \sum(1, 3)$.

2.4.3 *Advantages*

The advantages of Quantum Programmable ROM (QPROM) are as follows:

1. The programming can be done using many types of software and does not rely on the hard wiring of the program to the chip.
2. It is easy to configure any programmable device.
3. Since it is not possible to un-blow the fuse, the authenticity of the data remains intact and it is impossible to remove or alter the contents.

2.4.4 *Disadvantages*

The biggest disadvantage of quantum PROM is that the data once burnt cannot be erased or changed when detected with errors. There are several drawbacks of QPROM, which are listed below:

1. The static power consumption is high as the transistors used have higher resistance.
2. It is not possible for a particular byte to be erased, instead the entire content is erased.
3. The UV-based PROM takes time to program the content.

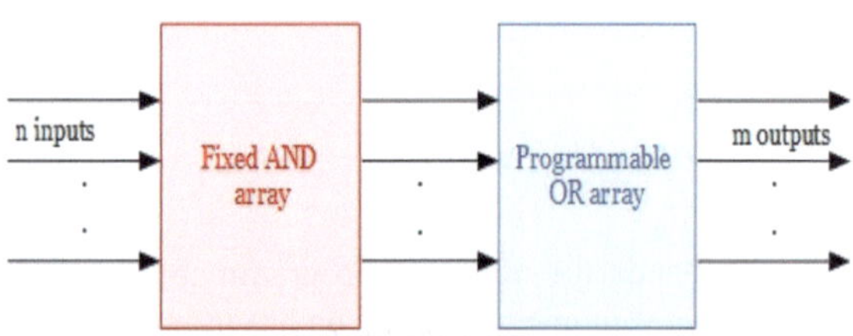

Fig. 2.9 2^n-to-m PROM

4. Data of PROM cannot be modified or rewritten if any error has occurred.

2.4.5 Basic Functions

PROM is sometimes considered in the same category of the circuit as programmable logic, PROM is discussed only in the memory category. Since the PROM architecture reaches its limits when many inputs are linked to many outputs, the PLD architecture, Programmable Logic Device (PLD), provides a more flexible concept.

1. It is possible to write data or program only once. However, once it is written, it can be read any number of times
2. A PROM chip is used mainly in the start-up process of a modern computer
3. A PROM, non-volatile memory stores only several megabytes (MB) of data, up to 4 MB or more per chip.

A random memory enables unrestricted in space and time access to any location at any address in the address space. The access is possible independently on the order of all previous accesses. The access can take place to an address in any order. Each location in a random-access memory has independent hardware circuits that provide the access. These circuits are activated as a result of address decoding. To such memories belong semiconductor memories of the ROM types. The most important parameters are memory capacity, memory access time, memory cycle time, and memory transfer rate.

BOOT PROM

1. Each system has a boot prom chip and a 1mbyte chip is typically located on the same board as the CPU.

2.4.6 Block Diagram

Consider a quantum 4-to-2 PROM with the general organization of block diagram as shown in Fig. 2.10, where the unit consists of four words of two qubits ($|A>$ and $|B>$) each. This implies there are two output lines ($|F1>$ and $|F2>$) and four distinct words stored in the unit, each of which may be applied to the output lines. The particular word selected that is presently available on the output line is determined from the two input lines. There are only two input qubits in quantum 4-to-2 PROM because $2^2 = 4$ and with two-qubit variables can specify four addresses or minterms. To perform minterms of four addresses, a quantum 2-to-4 decoder and some quantum OR operations for minterms $|F1>=\sum(0, 2)$ and $|F2> = \sum(1, 3)$ are required.

Table 2.3 Truth table of quantum 4-to-2 PROM

|B >	|A >	|F1 >	|F2 >
|0 >	|0 >	|1 >	|0 >
|0 >	|1 >	|0 >	|1 >
|1 >	|0 >	|1 >	|0 >
|1 >	|1 >	|0 >	|1 >

2.4.7 Circuit Architecture

In quantum 4-to-2 PROM architecture as shown in Fig. 2.11, there are two quantum NOT and four quantum AND operations that perform quantum 4-to-2 decoder. The output of the decoder then performs two quantum OR operations to produce the desired output of quantum 4-to-2 PROM.

2.4.8 Working Principle

Table 2.3 shows the truth table of quantum 4-to-2 PROM. Following operations are performed to get the desired output in quantum 4-to-2 PROM.
[i] For input qubits |A >, |B > =|0 >, |0 >, |D0 > line will be open. So, the value of |D0 > will be 1 and |D1 > to |D3 > will be |0 >. For the output of |F1 >, the quantum OR operation will be performed between |D0 > and |D2 > that produces |1 > and for |F2 >, quantum OR operations will be performed between |D1 > and |D3 > and generate |0 >.
[ii] For input qubits |A >, |B > =|1 >, |0 >, |D1 > line will be open. So, the value of |D1 > will be 1 and |D0 >, |D2 > and |D3 > will be |0 >. For the output of |F1 >, the quantum OR operation will be performed between |D0 > and |D2 > that produces |0 > and for |F2 > quantum OR operations will be performed between |D1 > and |D3 > and generate |1 >.

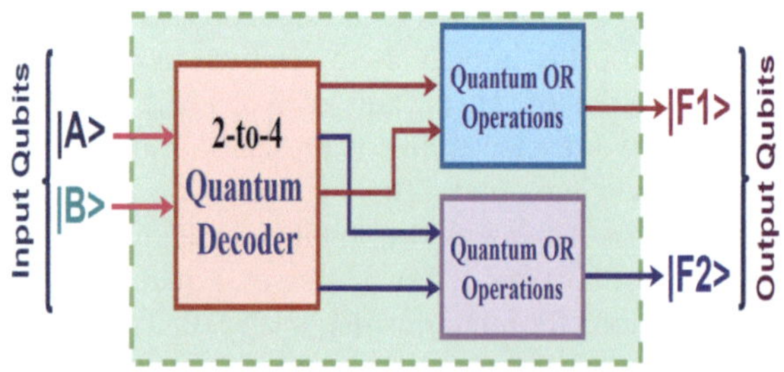

Fig. 2.10 Block diagram of quantum 4-to-2 PROM

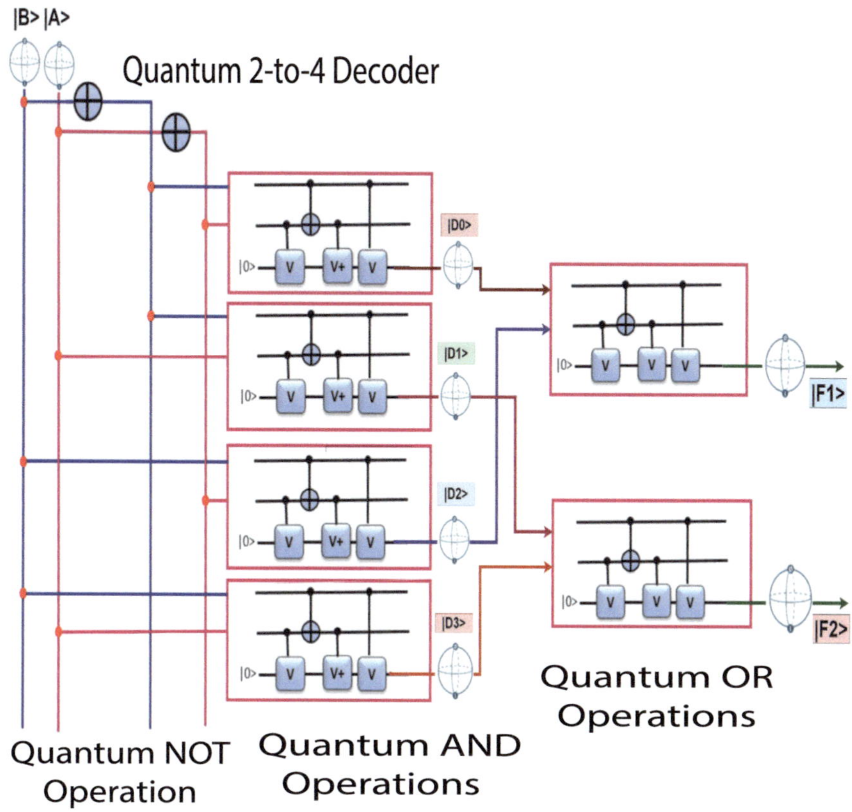

Fig. 2.11 Circuit architecture of quantum 4-to-2 PROM

[iii] For input qubits |A >, |B > =|0 >, |1 >, |D2 > line will be open. So, the value of |D2 > will be 1 and |D0 >, |D1 > and |D3 > will be |0 >. For the output of |F1 >, the quantum OR operation will be performed between |D0 > and |D2 > that produces |1 > and for |F2 > quantum OR operations will be performed between |D1 > and |D3 > and generate |0 >.
[iv] For input qubits |A >, |B > =|1 >, |1 >, |D3 > line will be open. So, the value of |D3 > will be 1 and |D0 > to |D2 > will be |0 >. For the output of |F1 >, the quantum OR operation will be performed between |D0 > and |D2 > that produces |0 > and for |F2 > OR operations will be performed between |D1 > and |D3 > and generate |1 >.

2.4.9 Applications

These types of memories are frequently used in embedded systems or microcontrollers and also in many other consumer and automotive electronics products. They have several different applications such as

1. Mobile Phones for providing User Specific Selections.
2. The Video game consoles.
3. Implantable medical devices.
4. Radio-Frequency Identification (RFID) tags.
5. High definition Multimedia Interfaces (HDMI).

2.5 Quantum Cache Memory

Cache memory is one of the quickest memory types. It serves as a buffer between the CPU and the main memory. Furthermore, it saves the data and instructions that the CPU utilizes the most. Quantum Cache Memory (QCM) is an element for preparing quantum data structures for subsequent quantum processing. The QCM has two primary sections: the addressing encoding and information encoding sections. The addressing section, referred to in our development as the Address qubits, comprises a group dedicated to managing the addresses of data points within the QCM. Concurrently, the information section, the information qubits, consists of qubits responsible for storing the data points ready for quantum processing. QCM has dynamic adaptability to fulfill the specific requirements of varying algorithms, with the total count of qubits within the QCM being dynamically determined based on the data size and the algorithmic demands. Memory is a component or system in computing that stores data for immediate use in a computer or other computer hardware and digital electrical devices. The terms memory and major storage or main memory are frequently used interchangeably. The word store is an old term for memory. Computer memory operates at a high speed compared to storage that is slower but offers higher capacities. If needed, contents of the computer memory can be transferred to storage; a common way of doing this is through a memory management technique called virtual memory. Modern memory is implemented as a semiconductor memory, where data is stored within memory cells built from MOS transistors and other components on an integrated circuit. There are two main kinds of semiconductor memory, volatile and non-volatile. Examples of non-volatile memory are flash memory and ROM, PROM, EPROM, and EEPROM memory. Examples of volatile memory are dynamic random-access memory (DCache) used for primary storage and static random-access memory (SCache) used for CPU cache. Most semiconductor memory is organized into memory cells each storing one bit (0 or 1). The flash memory organization includes both one bit per memory cell and a multi-level cell capable of storing multiple bits per cell. The memory cells are grouped into words of fixed word length, for example, 1, 2, 4, 8, 16, 32, 64, or 128 bits. Each word can be accessed by a binary address of N bits, making it possible to store 2N words in the memory. A CPU cache is hardware. A cache is used by the central processing unit (CPU) of a computer to reduce the average cost (time or energy) to access data from the main memory. A cache is a smaller, faster memory, located closer to a processor core, which stores copies of the data from frequently used main memory locations. Most CPUs have a hierarchy of multiple cache levels (L1, L2, often L3, and rarely even L4), with separate instruction-specific and data-specific caches at level 1. Other types of caches exist (that are not counted toward the "Cache size" of the most important caches mentioned above), such as the translation lookaside buffer (TLB) which is a part of the memory management unit (MMU) as well as it is also the part of the most of the CPU.

2.5.1 Operations

It is placed between the main memory and the CPU. Moreover, for any data, the CPU first checks the cache and then the main memory.

2.5.1.1 Levels of Cache Memory

There can be various levels of cache memory, they are as follows:

Level 1 (L1) or Registers

It stores and accepts the data which is immediately stored in the CPU such as in instruction register, programmable cache memory, cache counter, accumulator, address register, etc.

Level 2 (L2) or Cache Memory

It is the fastest memory that stores data temporarily for fast access by the CPU. Moreover, it has the fastest access time.

Level 3 (L3) or Main Memory

It is the main memory where the computer stores all the current data. It is a volatile memory which means that it loses data on power OFF.

Level 4 (L4) or Secondary Memory

It is slow in terms of access time. But, the data stays permanently in this memory.

Cache is implemented in hardware as a block of memory for storing data that is likely to be used again. Caches are commonly used by CPUs and hard disk drives (HDDs), as well as web browsers and web servers. A cache is a collection of entries. Each item contains related data, which is a duplicate of the identical data stored in a backup storage. Each entry additionally includes a tag that identifies the data in the backup store from which the entry is a copy. Tagging enables stacked cache-oriented algorithms to operate simultaneously without differential relay interference.

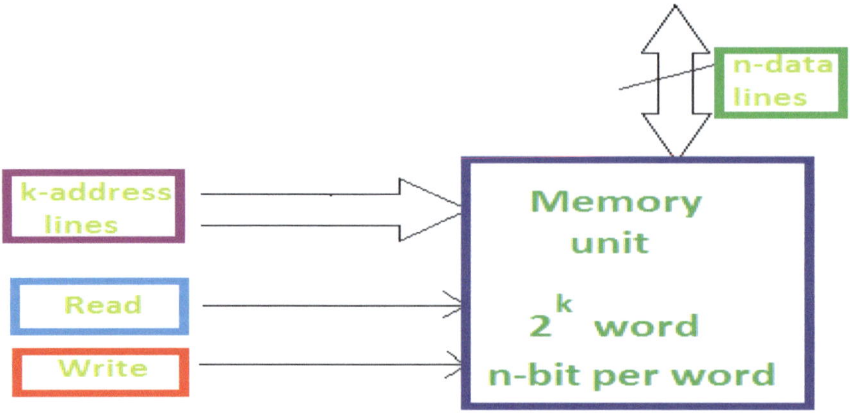

Fig. 2.12 2^k-to-n Cache memory

2.5.2 *Basic Definition*

The steps to access the data from cache memory are

1. A request is made by the CPU.
2. The cache is checked for data.
3. If the data is found in the cache it is returned to the CPU (this is called a cache hit).
4. If the data is not found in the cache then the data will be returned from the main memory.

Cache memory is fast because

1. In the case of a CPU cache, it is faster because it's on the same die as the processor. In other words, the requested data doesn't have to be bussed over to the processor; it's already there.
2. In the case of the cache on a hard drive, it's faster because it's in solid-state memory, and not still on the rotating platters.
3. In the case of the cache on a website, it's faster because the data has already been retrieved from the database (which, in some cases, could be located anywhere in the world).

So it's about locality, mostly. Cache eliminates the data transfer step. Structure of 2^k-to-n cache memory is shown in Fig. 2.12.

Communication between memory and its environment is achieved through data input and output lines, address selection lines, and control lines that specify the direction of transfer. The n data input lines provide the information to be stored in memory, and the n data output lines supply the information coming out of a particular word chosen among the 2^k available inside the memory. The two control inputs specify the direction of transfer desired.

2.5.3 *Advantages*

The advantages of quantum cache memory are as follows:

1. It is faster than the main memory.
2. The access time is quite less in comparison to the main memory.
3. The speed of accessing data increases, hence the CPU works faster.
4. Moreover, the performance of the CPU also becomes better.
5. The recent data is stored in the cache, and therefore the outputs are faster.

2.5.4 *Disadvantages*

Some disadvantages of quantum cache memory are as follows:

1. It is quite expensive.
2. The storage capacity is limited.
3. It is expensive.

2.5.5 *Basic Functions*

1. The CPU first checks any required data in the cache. Furthermore, it does not access the main memory if that data is present in the cache.
2. On the other hand, if the data is not present in the cache then it accesses the main memory.
3. The block of words that the CPU accesses currently is transferred from the main memory to the cache for quick access in the future.
4. The hit ratio defines the performance of the cache memory.

2.5.5.1 Cache Performance

The performance of the cache is in terms of the **hit ratio**. It is the ratio of the number of hits to the total number of accesses.

The CPU searches the data in the cache when it requires writing or reading any data from the main memory. In this case, two cases may occur as follows:

1. If the CPU finds that data in the cache, a **Cache hit** occurs and it reads the data from the cache.
2. On the other hand, if it does not find that data in the cache, a **Cache miss** occurs. Furthermore, during cache miss, the cache allows the entry of data and then reads data from the main memory.

3. Therefore, the hit ratio can be defined as the number of hits divided by the sum of hits and misses.

Hit ratio = hit / (hit + miss) = number of hits/total accesses.

Also, the cache performance can be improved by the following ways:

1. Using a higher cache block size.
2. Higher associativity.
3. Reducing the miss rate.
4. Reducing the time to hit in the cache.

2.5.6 Block Diagram

A 4-to-1 cache memory chip has a memory capacity of four words of one qubit per word. This requires a 2-qubit address and a 1-bit bidirectional data bus. The 1-bit bidirectional data bus allows the transfer of data either from memory to CPU during a read operation or from CPU to memory during a write operation. The read-and-write inputs specify the memory operation, and the two chip select (CS) control inputs are for enabling the chip only when the microprocessor selects it.

Figure 2.13 represents the quantum 4-to-1 cache general organization of block diagram cache memory. This quantum cache memory consists of four separate "Words" of memory and each is single qubits wide. The quantum RAM cell has three inputs and one output. Block diagram of quantum 4-to-1 cache memory is shown in Fig. 2.13.

2.5.7 Design Architecture of Quantum RAM

Quantum cache memory consists of three basic components. Circuit architecture of quantum RAM cell is shown in Fig. 2.14. To execute quantum 4-to-1 cache memory, following operations are required.

1. A quantum 2-to-4 decoder.
2. Quantum RAM cells.
3. Quantum OR operations for corresponding minterms.

Quantum decoder and quantum OR operations are discussed before.

2.5.7.1 Circuit Design of Quantum RAM Cell

The fundamental design of this qubit cell is based on the D flip-flop (Fig. 2.14). To begin with, the cell has three inputs and a single output. The inputs are labeled "|Select >," "|R/W >," and "|Input >." The output line is labeled "|output >." To

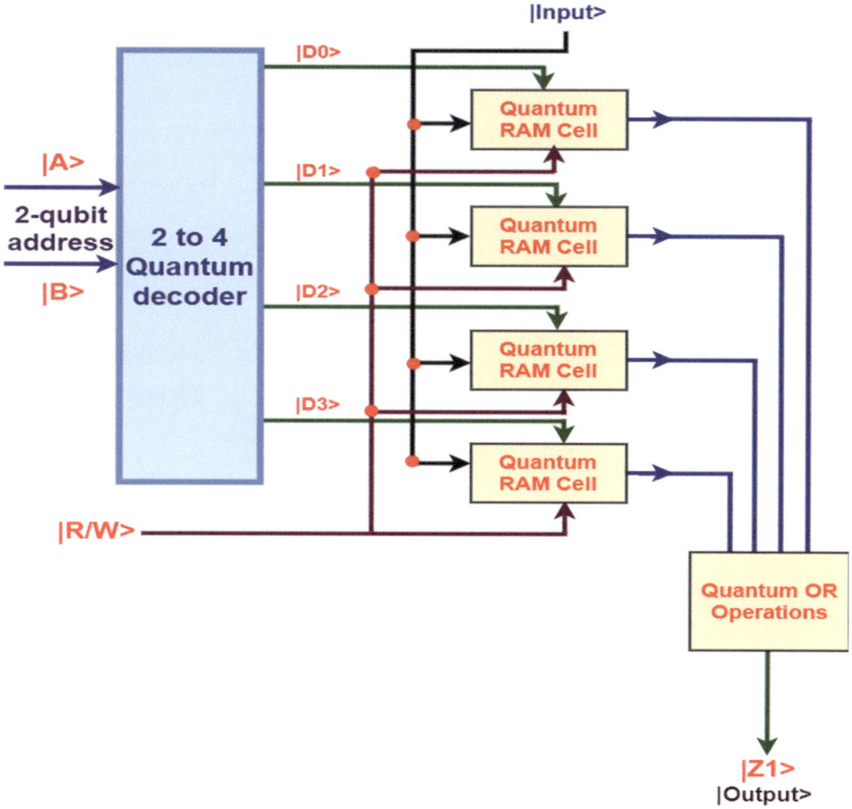

Fig. 2.13 Block diagram of quantum 4-to-1 cache memory

perform the quantum cache memory cell output, two quantum NOT, three quantum AND, and four quantum NAND operations are required.

Step 1:
First draw three input qubits |Input >, |R/W > and |Select >. Two possible states for a qubit are the states |0 > false, and |1 > true.

Step 2:
Draw quantum NOT operation with the |Input > and |R/W > qubits.

Step 3:
Two input qubits (|R/W > and |select >) will go through quantum AND operations.

Step 4:
Again, Two input qubits (NOT of |R/W > and |select >) will go through quantum AND operations.

Step 5:
The outputs of Step 4 and NOT of qubit |input > will go to quantum NAND operations. Also, the outputs of Step 4 and qubit |input > will go to another quantum NAND operation.

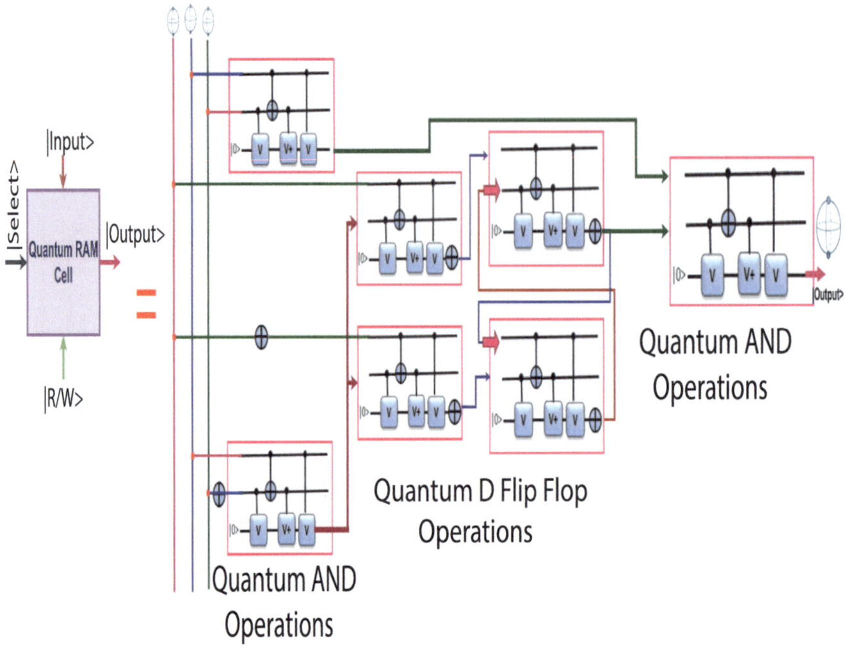

Fig. 2.14 Circuit architecture of quantum RAM cell

Step 6:
The outputs of Step 5 will go to the D flip-flop as input.
Step 7:
Finally, D flip-flop output and Step 3 output qubit will go to a quantum AND operation, then this output with NOT of |R/W will go through another quantum AND operation to produce the desired quantum cache memory |output > qubit.

2.5.7.2 Working Procedure of Quantum RAM Cell

A sequential device as simple as a D flip-flop could be used to remember one bit of data. To develop a complete memory cell, called a qubit cell, based on the flip-flop. The number of total quantum cells per word will be m × n where m represents words with n bits. The "|select >" input is used to access the cell, either for reading or writing also used to access anyone quantum RAM cell when there is more than one quantum RAM cell. When the select line is high or |1 >, the cell performs the memory operation. But when the select line of the quantum RAM cell is low or |0 > the cell is not interested to perform a read from or written to. The next input qubit is "|R/W >" where a system clock will conduct this input. If the clock value on the read/write line is |0 >, this will signify "read" and when it is |1 >, it will perform the "write" phase. When such a cell is selected and in "read" mode, the current value of its underlying flip-flop will be transferred to the cell's output line. When the cell is selected and in "write" mode, an input data signal will determine the value remembered by the flip-flop.

To perform the quantum 4-to-1 cache memory, four selection lines are needed to design four quantum RAM cells. The output of quantum 2-to-4 decoder with four output qubits will perform as selection input qubit of quantum RAM cells. Figure 2.15 shows the 4 quantum RAM cells to perform |Q0 > to |Q3 > for further minterms operation.

2.5.8 Circuit Architecture of Quantum Cache Memory

In quantum 4-to-1 cache memory architecture as shown in Fig. 2.16, two address lines with one ancilla bit are needed for simulating 4-to-1 quantum cache memory and each address line needs to be in CNOT form as well. These address line combinations will be the input of 2-to-4 decoders which consists of four quantum AND gates and this decoder has one enable input. Four select lines from this decoder are obtained and each select line will go through each cache memory cell. Note that word calculation of cache memory will be 2^k where k is the address line and 2^k is the total words of n-bit and decoder combination will be $k \times 2^k$. This two-qubit cache memory consists of four separate cache memory cells and each cell has three inputs- |D0 >, anyone selects a line and read/write inputs. The obtained output from four quantum cache memory cells will be the input of a quantum OR gate which produces the final output. This is the whole design procedure of 4-to-1 quantum cache memory.

2.5.9 Working Principle

Figure 2.16 represents the implementation of 4-to-1 quantum cache memory. This quantum cache memory consists of four separate "Words" of memory and each is one qubits wide. The quantum cache memory cell has three inputs and one output. The complete circuit of a quantum cache memory cell is described in Fig. 2.16 with proper explanation. A word consists of two quantum cache memory cells arranged in such a way so that both qubits can be accessed simultaneously. Four words of memory need two address lines. |A > and |B > are the two-qubit address line inputs that goes through a 2-to-4 decoder that selects one of the four words. The memory-enabled input enables the decoder. If the memory enabled is |0 >, all output of the decoder will be |0 > and in that case, none of the memory addresses will be selected. But when the memory enabled is |1 >, one of the four words is selected. The word is selected by the value in the two address lines. When a word has been selected, the read/write input determines the operation. During the read operation, the four qubits of the selected word pass to the quantum OR gates to the output |Z1 > terminals. But during the write operation, the data which is available in the input lines are transferred

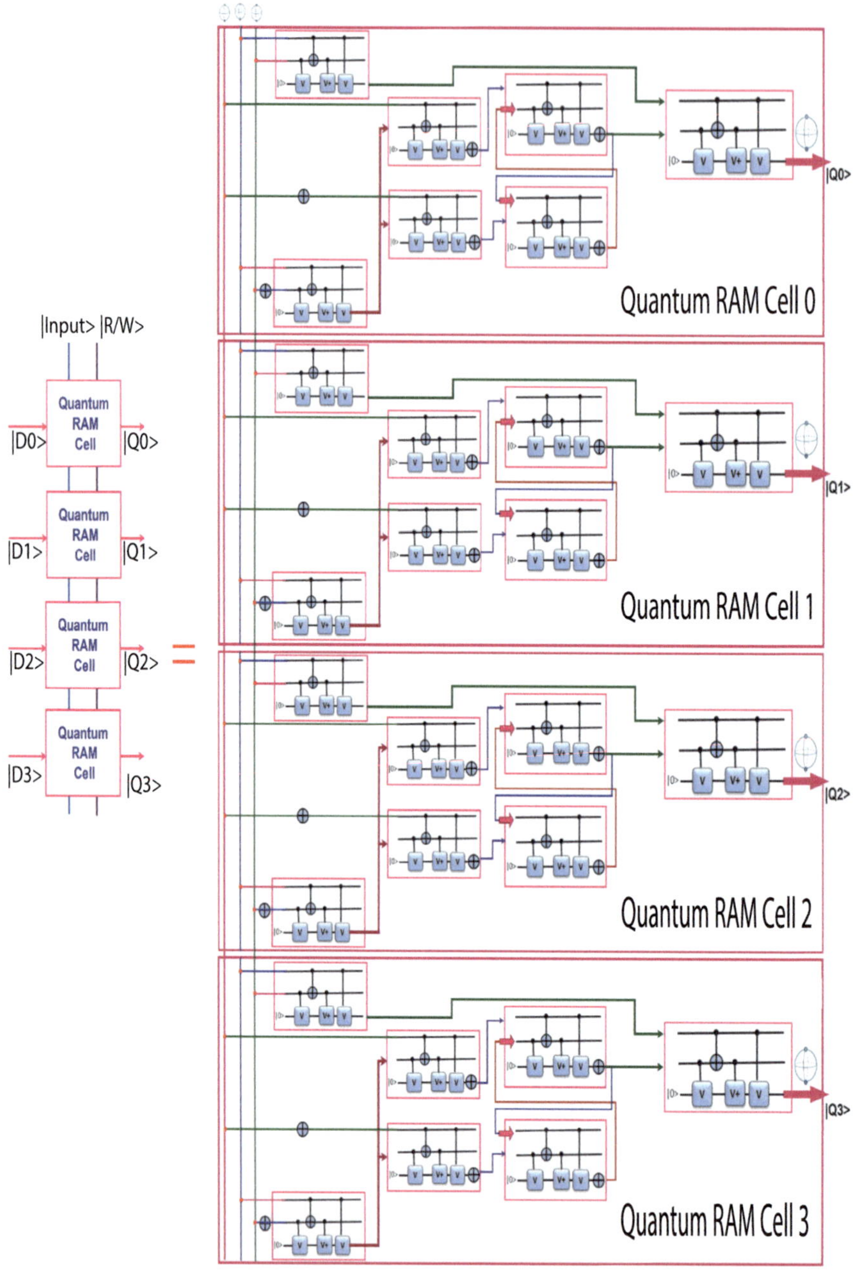

Fig. 2.15 Quantum RAM cells

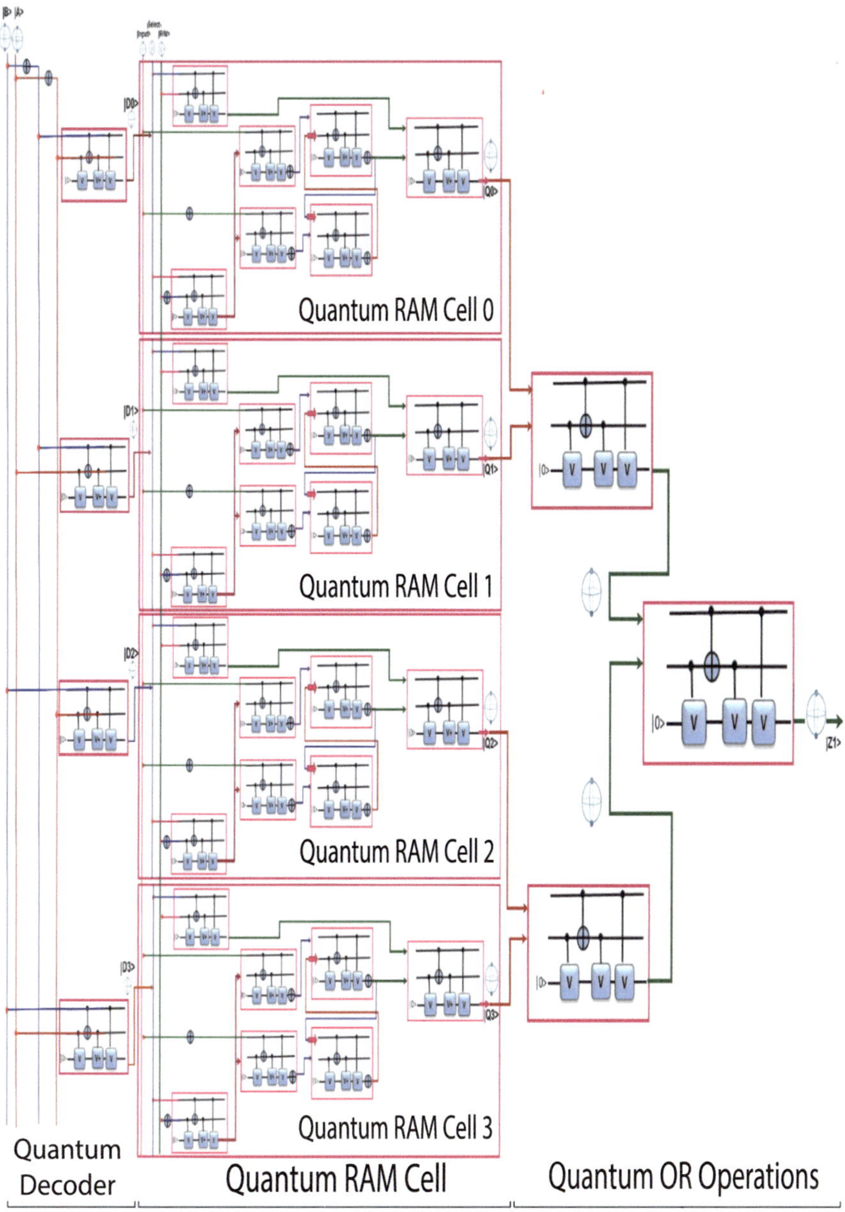

Fig. 2.16 Circuit architecture of quantum 4-to-1 cache memory

into the four quantum cells of the selected word. The quantum cache memory cells that are not selected become disable and their previous qubit never changes. But when the memory-enable input that passes into the decoder is equal to $|0>$, none

Table 2.4 Control input to memory chip

$\|R/W >$	Memory Operation
X	*None*
$\|0 >$	*Write to the selected word*
$\|1 >$	*Read from the selected word*

of the words are selected, and all quantum cells remain unchanged regardless of the value of the read/write input as shown in Table 2.4.

2.5.10 *Applications*

The software can "partition" a portion of a computer's cache memory, allowing it to act as a much faster hard drive that is called a cache memory disk. A cache memory disk loses the stored data when the computer is shut down unless memory is arranged to have a standby battery source. Most modern operating systems employ a method of extending cache memory capacity, known as "virtual memory". A portion of the computer's hard drive is set aside for a paging file or a scratch partition, and the combination of physical cache memory and the paging file forms the system's total memory (For example, if a computer has 2 GB of cache memory and a 1 GB page file, the operating system has 3 GB total memory available to it). When the system runs low on physical memory, it can "swap" portions of cache memory to the paging file to make room for new data, as well as to read previously swapped information back into cache memory. Excessive use of this mechanism results in thrashing and generally hampers overall system performance, mainly because hard drives are far slower than cache memory.

2.6 Summary

This chapter has presented the details of quantum memory devices. The history, basic definition and functions, block diagram and circuit diagram and working principle, applications and more information on RAM, ROM, PROM, and cache memory in quantum computing. With the advancement of modern science, these quantum memory devices will be developed and easier to implement everywhere. The quantum computing is an exciting field for researchers, so in the upcoming future, it will overcome all current difficulties very soon.

Chapter 3
Memory Devices in Quantum-DNA Computing

3.1 Introduction

Quantum computing and DNA computing are gaining attention as alternatives to classical computing systems. There exist some quantum algorithms which are significantly faster than their conventional counterparts and these processes establish quantum computing as a superior future technology that involves quantum circuits and quantum gates. Traditional silicon computers consume much more power as compared to the computing systems, based on Deoxyribonucleic Acid (DNA), whereas DNA-based logic gates are stable and reusable. DNA computation might save a billion times the energy of an electrical computer while storing data in a trillion times less space. DNA computing mechanism for accurately storing and retrieving information inside a DNA sequence. A novel logic gate design based on chemical reactions is presented in which observance of double-stranded sequences indicates a truth evaluation. Both of these properties might be captured by integrating quantum-DNA computing. In this chapter, combined quantum and DNA computing are discussed, a novel theoretical methods for designing different memory devices considering a binary logic system. This is an entirely new way of merging these two technologies and representing memory devices in a computer architecture. In quantum biocomputing or quantum-DNA computing, this chapter will cover all four memory devices.

3.2 Quantum-DNA Random-Access Memory

A Quantum Random Access Memory (QRAM) is a quantum computing memory technology that leverages quantum principles to store and retrieve data efficiently. In analogy to classical RAM, QRAM allows for accessing any memory location in a superposition of states. This means it can hold and access multiple pieces of

H. Md. Hasan Babu, *Quantum Biocomputing in Quantum Biology Volume II*,
https://doi.org/10.1007/978-981-97-5349-9_3

data simultaneously, unlike classical RAM which can only hold one value at a time. When in the biological computing or DNA computing or biocomputing, QRAM could be used to model and simulate complex biological systems, such as protein folding or DNA sequencing, by storing and accessing vast amounts of data related to these systems in a superposition of states, this QRAM is called quantum-DNA RAM or quantum biological computing RAM or quantum biocomputing RAM. This memory could lead to faster and more efficient simulations and potentially new insights into biological processes. Its design consists of various devices such as quantum and DNA logic gates and decoder for the quantum-DNA 4-to-1 random-access memory. Quantum-DNA RAM is a basic but it is very important thing in the computer system. So, the quantum-DNA RAM is a mandatory part of quantum-DNA computing system. Therefore, integrated computing system as quantum-DNA for RAM is the main concern of this section.

3.2.1 Block Diagram

Figure 3.1 represents the block diagram of quantum-DNA 4-to-1 RAM with NMR relaxation in normal room temperature, consisting of four separate “Words” of memory and each is single sequence wide.

3.2.2 Working Principle

Circuit architecture of quantum-DNA RAM memory with NMR relaxation in normal room temperature is shown in Fig. 3.2. This quantum RAM consists of four separate “Words” of memory and each is 1 qubit wide. The quantum RAM cell has three inputs and one output. The complete circuit of a quantum RAM cell is described in Fig. 8 with proper explanation. A word consisting of two quantum RAM cells is arranged in such a way so that both qubits can be accessed simultaneously. Four words of memory need two address lines. |A> and |B> are the two-qubit address lines input that goes through a 2-to-4 decoder that selects one of the four words. The memory-enabled input enables the decoder. If the memory enables is |0>, all output of the decoder will be |0> and in that case, none of the memory addresses will be selected. But when the memory enabled is |1>, one of the four words is selected. The word is selected by the value in the two address lines. When a word has been selected, the read/write input determines the operation. During the read operation, the four qubits of the selected word pass to the DNA OR gates to the output Z1 terminals. But during the write operation, the data which is available in the input lines are transferred into the four quantum cells of the selected word. The quantum RAM cells that are not selected are become disabled and their previous qubit never changes. But when the memory-enabled input that passes into the decoder is equal to |0>, none of the words are selected, and all quantum cells remain unchanged regardless of the value of the read/write input.

3.3 Quantum-DNA Read-Only Memory

DNA computing can save money, billions of energy in the electrical computer when storing data in spatial space. DNA computing, precanic coloring mechanism, and pick-up information within the DNA sequence. In addition, computing calculations of DNA nanotubes are very important as essentially billion DNA molecules of chemical reactions for calculations can be performed simultaneously. In other way, a Quantum Read-Only Memory (QROM) is a type of quantum memory that allows classical data to be loaded into a quantum computer. QROM is essential for storing and accessing data within a quantum computing environment. It works by representing data as bitstrings, which are then loaded into the quantum memory as qubits. While QROM is primarily associated with quantum computing, it is also relevant to the emerging field of biological computing, where researchers explore using biological systems to perform computations. In this context, QROM could potentially

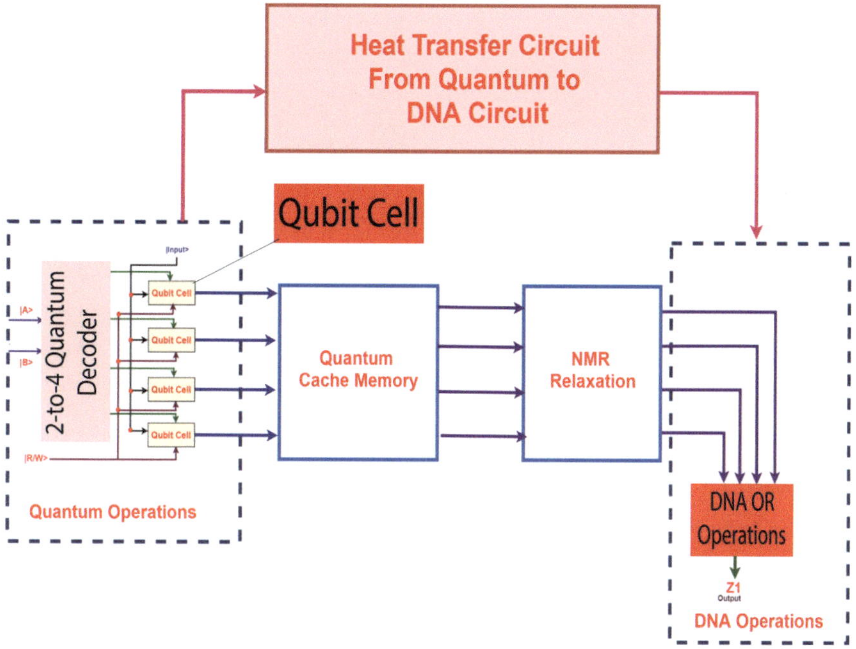

Fig. 3.1 Block diagram of quantum-DNA 4-to-1 RAM

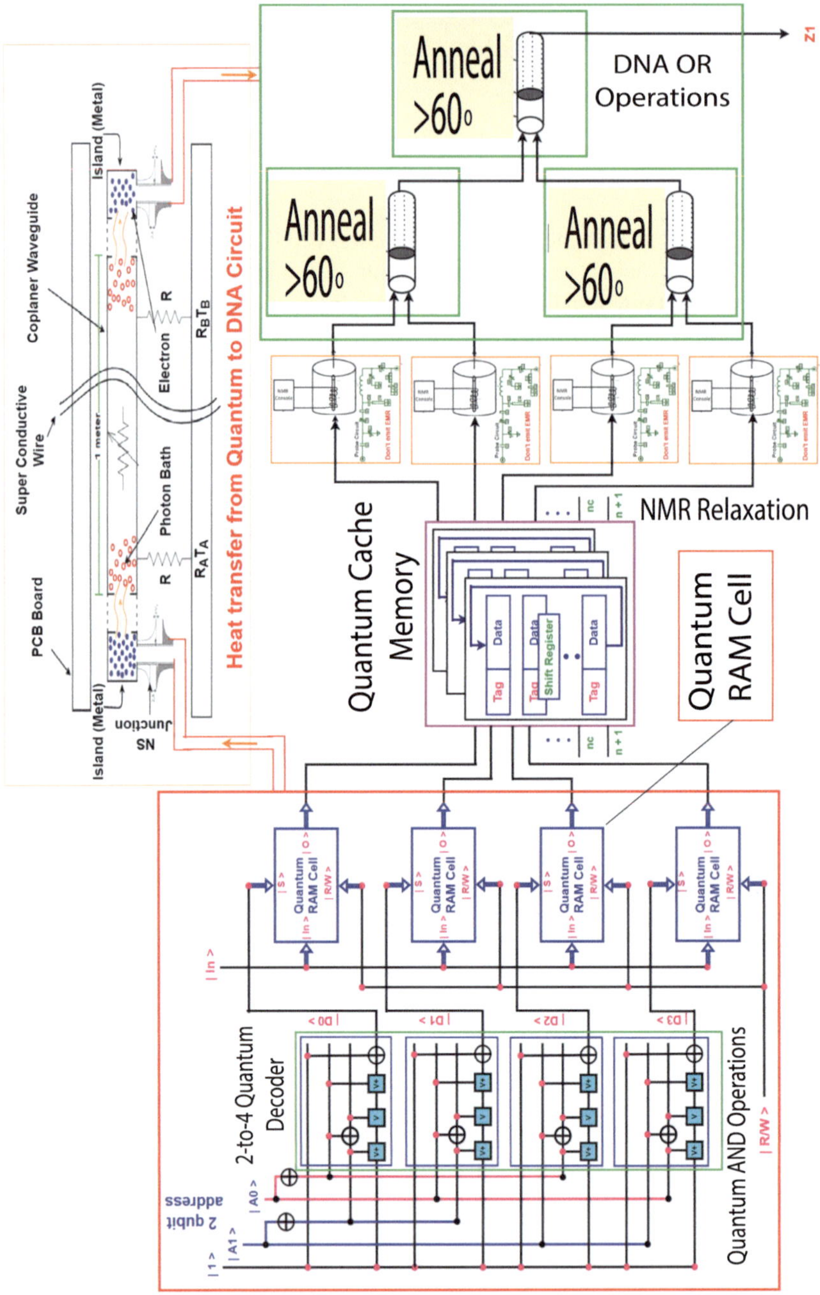

Fig. 3.2 Circuit architecture of quantum-DNA 4-to-1 RAM

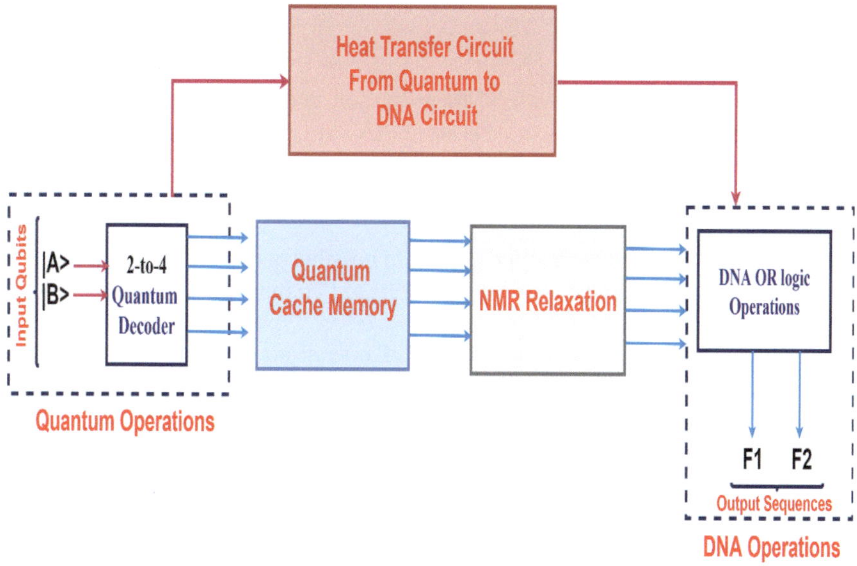

Fig. 3.3 Block diagram of quantum-DNA 4-to-2 ROM

be used to store and retrieve biological data, such as DNA sequences, or to implement biological algorithms. In this case, the QROM is called quantum-DNA ROM or quantum biocomputing ROM or quantum biological computing ROM. Therefore, integrated computing systems as quantum-to-DNA for ROM memory are the main concern of this section.

3.3.1 Block Diagram

Quantum-DNA 4-to-2 ROM general organization of block diagram (Fig. 3.3), the unit consists of four words of two input qubits (|A> and |B>) each. This implies there are 2-DNA sequence output lines (F1 and F2) and four distinct words stored in the unit, each of which may be applied to the output lines. According to the block diagram, the operations of 4-to-2 ROM have been divided into two distinct computing systems as the input section will perform in quantum and in the output DNA sequences will be produced.

So, input qubits |A> and |B> in quantum computing operations store the qubits in cache memory and transform those qubits to DNA sequences to form outputs F1 and F2 in DNA computing operations.

3.3.2 Design Procedure

Quantum-DNA 4-to-2 ROM with NMR relaxation in normal room temperature is depicted in Fig. 3.4. To design a quantum-DNA 4-to-2 ROM circuit (considering NMR relaxation), the input values are in qubit (quantum computing) and the output will be DNA sequence (DNA computing) at normal room temperature.

Step 1:

First, draw two input qubits |A> and |B>. Two possible qubits are |0> false and |1> true. These two will produce four combinations of two input qubits.

Step 2:

To design a decoder, draw two quantum NOT operations and four quantum AND operations.

Step 3:

After getting qubits of all the corresponding values from four quantum AND operations of the 2-to-4 decoder, it needs to store all the quantum bits into the cache memory. Quantum cache memory is actually used for storing logical qubits and error correction. Quantum cache also increases the logical qubit collection.

Step 4:

After storing qubits of all the corresponding values from four quantum AND operations in cache memory, NMR relaxation is performed to transform qubits to DNA sequences. Here, NMR relaxation is done at room temperature, where it doesn't need to emit EMR.

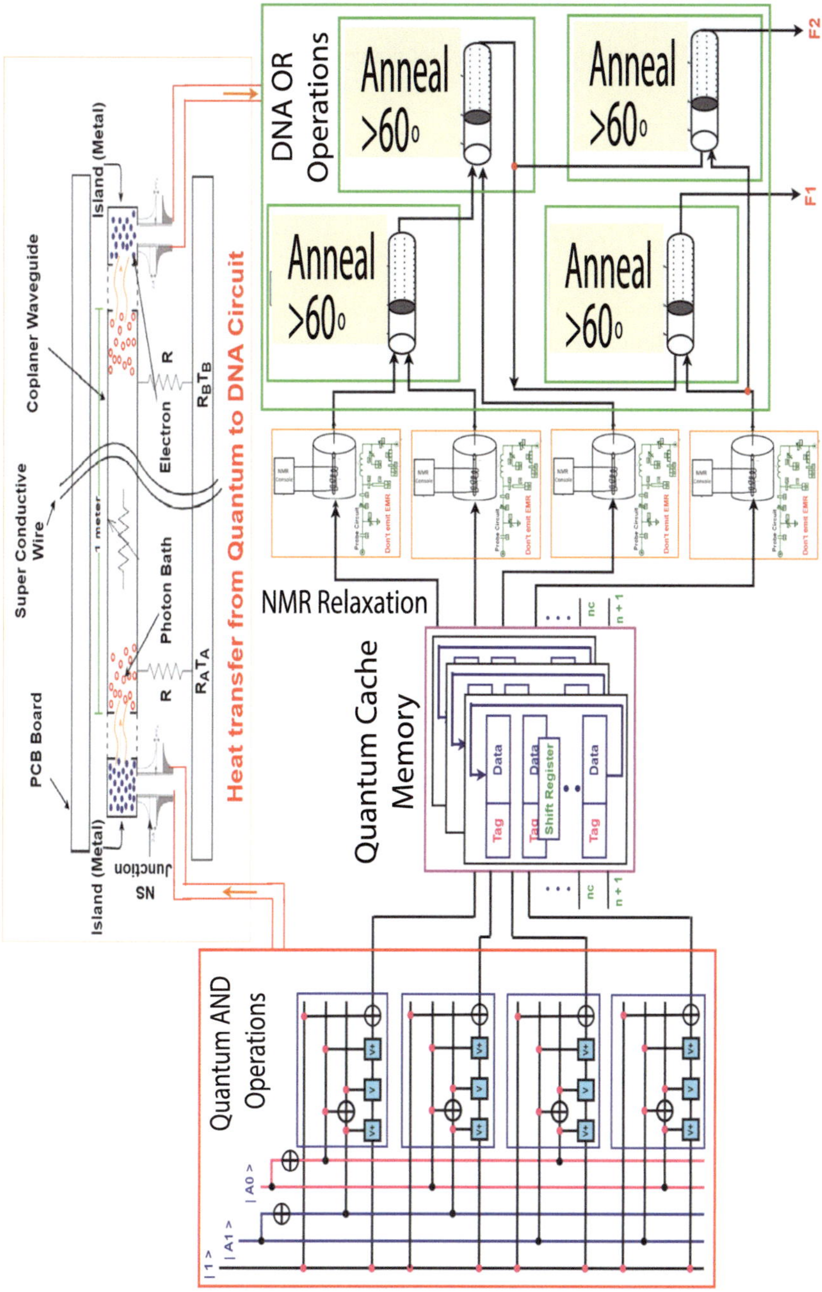

Fig. 3.4 Circuit architecture of quantum-DNA 4-to-2 ROM

Step 5:
Finally, after NMR relaxation, the DNA sequences are found from qubits and these will go through two DNA OR gates to produce the desired quantum-DNA 4-to-2 ROM output sequences.

3.3.3 *Working Principle*

Truth table of quantum-DNA 4-to-2 ROM is shown in Table 3.1. The operational procedures of quantum-DNA 4-to-2 ROM are discussed as follows:
[i] For input combination A, B = **|0>**, **|0>**, |D0> line will be open and |D1> to |D3> will be closed. So, the value of |D0> = |1>, |D1> to |D3> = |0> will be stored in cache memory and after NMR relaxation all of these will transform into DNA sequence as D0 = **ACCTAG** and D1 to D3 = **TGGATC**. For the outputs of F1 and F2, perform DNA OR operations among D0 to D3 and generate **ACCTAG** as the desired output.
[ii] For input combination A, B = **|1>**, **|0>**, |D1> line will be open and |D0>, |D2> and |D3> will be closed. So, the value of |D1> = |1>, |D0> |D2>, |D3> =|0> will be stored in cache memory and after NMR relaxation all of these will transform into DNA sequence as D1 = **ACCTAG** and D0, D2, D3 = **TGGATC**. For the outputs of F1 and F2, perform DNA OR operations among D0 to D3 and generate **ACCTAG** as the desired output.
[iii] For input combination A, B = **|0>**, **|1>**, |D2> line will be open and |D0>, |D1> and |D3> will be closed. So, the value of |D2> = |1>, |D0> |D2>, |D3> = |0> will be stored in cache memory and after NMR relaxation all of these will transform into DNA sequence as D2 = **ACCTAG** and D0, D1, D3 = **TGGATC**. For the outputs of F1 and F2, perform DNA OR operations among D0 to D3 and generate **ACCTAG** as the desired output.
[iv] For input combination A, B = **|1>**, **|1>**, |D3> line will be open and |D0>, |D1> and |D2> will be closed. So, the value of |D3> = |1>, |D0> |D1>, |D2> = |0> will be stored in cache memory and after NMR relaxation all of these will transform into DNA sequence as D3 = **ACCTAG** and D0, D1, D2 = **TGGATC**. For the outputs of F1 and F2, perform DNA OR operations among D0 to D3 and generate **ACCTAG** as the desired output.

Table 3.1 Truth table of a quantum-DNA 4-to-2 ROM

<table>
<tr><th>|B></th><th>|A></th><th>F1</th><th>F2</th></tr>
<tr><td>|0></td><td>|0></td><td>ACCTAG</td><td>ACCTAG</td></tr>
<tr><td>|0></td><td>|1></td><td>ACCTAG</td><td>ACCTAG</td></tr>
<tr><td>|1></td><td>|0></td><td>ACCTAG</td><td>ACCTAG</td></tr>
<tr><td>|1></td><td>|1></td><td>ACCTAG</td><td>ACCTAG</td></tr>
</table>

3.4 Quantum-DNA Programmable Read-Only Memory

As an alternative to traditional digital computing systems, quantum computing and DNA computing are gaining traction. In this chapter, unique theoretical approaches for creating alternative memory devices using a binary logic system are described using a combination of quantum and DNA computing. Quantum and DNA logic gates and decoders for quantum-DNA 4-to-2 read-only memory are among the memory devices designed. In terms of processing power and storage, both DNA and quantum computers have the potential to have a considerable impact on standard digital computers. Because of its coherent superposition of states and biomedical technology, quantum computers are more powerful than regular Turing processors. Both of these properties might be captured by integrating DNA and quantum computing. Quantum biological computing explores the possibility of using biological systems or biological materials to perform quantum computations, and it also includes the concept of quantum read-only memory (QROM). Quantum Programmable Read Only Memory (QPROM) in this context refers to memory devices where the content can be changed once after manufacture, a type of read-only memory (ROM). In summary, quantum biological computing is a broad field exploring the intersection of quantum physics, biology, and computer science. QPROM is a specific type of quantum memory that allows for loading data into quantum systems, while PROM is a classical memory that can be programmed once. This type of QPROM memory which is used in quantum biological computing systems, is called quantum-DNA PROM memory or quantum biocomputing PROM memory or quantum biological PROM memory. Therefore, the integrated computing system such as quantum-to-DNA PROM memory is the main concern of this section.

3.4.1 Block Diagram

Quantum-DNA 4-to-2 PROM block diagram with NMR relaxation in normal room temperature is shown in Fig. 3.5. So, input qubits |A> and |B> in quantum computing operations, store the qubits in cache memory and transform those qubits to DNA sequences to form outputs F1 and F2 in DNA computing operations.

3.4.2 Design Procedure

The design architecture of quantum-DNA 4-to-2 PROM with NMR relaxation in normal room temperature is shown in Fig. 3.6. To perform PROM in normal room temperature, a 2-to-4 decoder is needed and minterms of decoder output as the OR operations input to produce the desired PROM output function. In quantum-DNA computing (considering NMR relaxation), the input values are in qubit (quantum

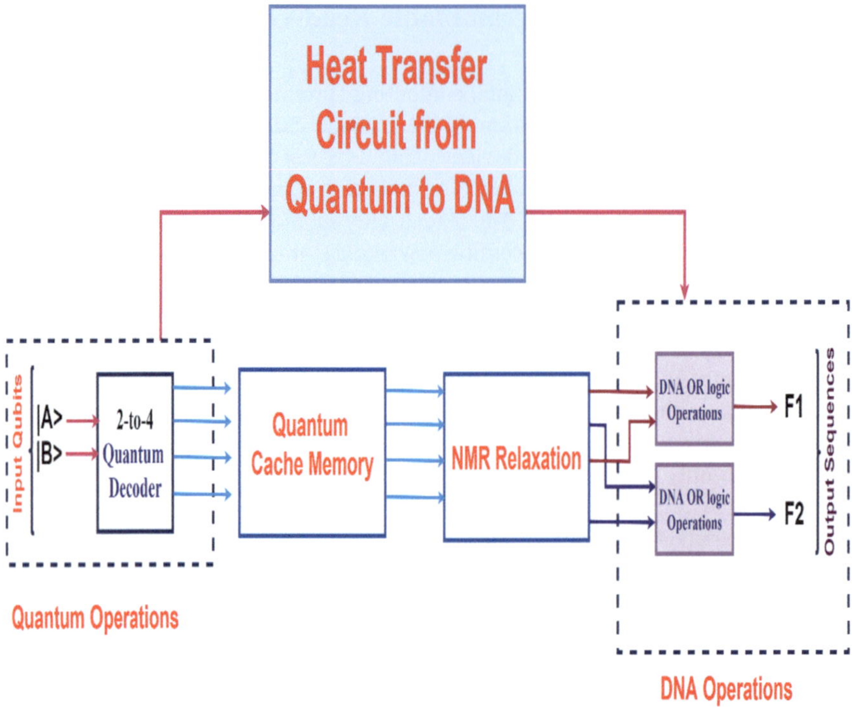

Fig. 3.5 Block diagram of quantum-DNA 4-to-2 PROM

computing) and the output will be DNA sequence (DNA computing) at normal room temperature.

Step 1:
First, draw two input qubits |A> and |B>. Two possible qubits are |0> false and |1> true. These two will produce four combinations of two input qubits.
Step 2:
To design a decoder, draw two quantum NOT operations and four quantum AND operations.

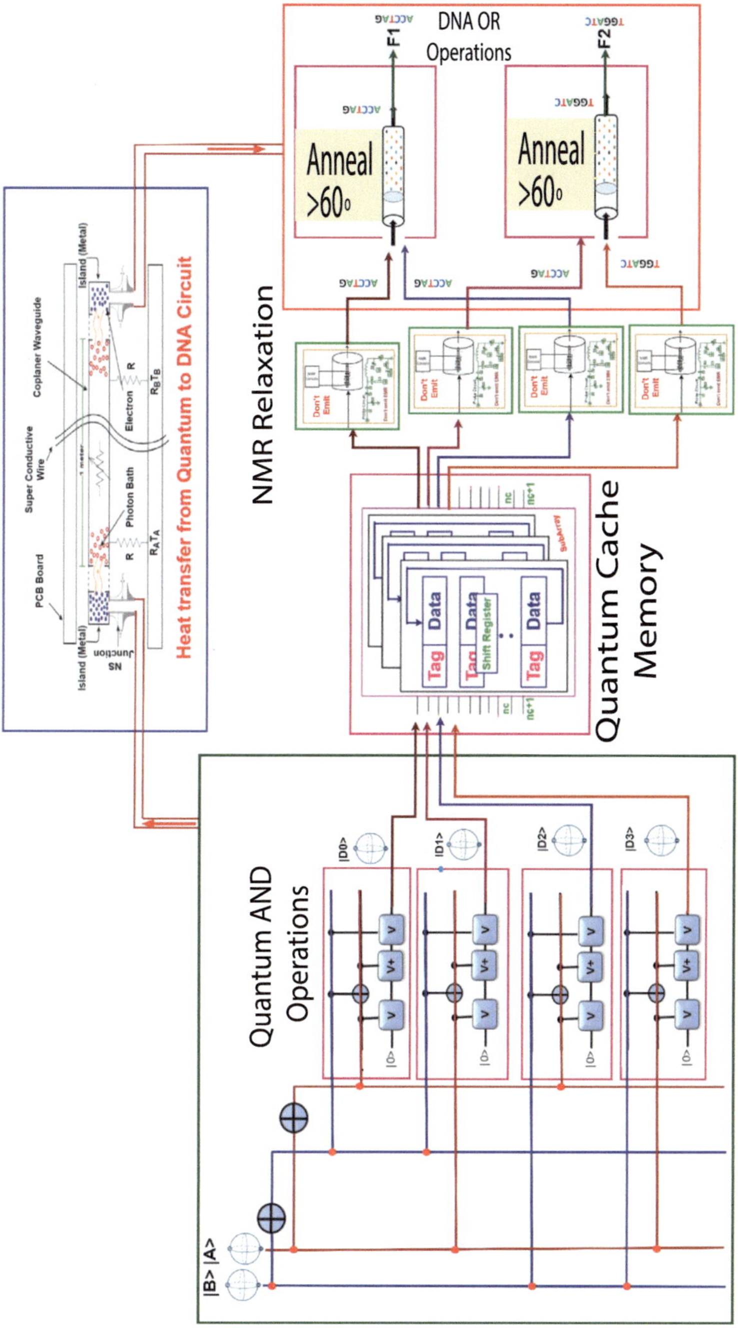

Fig. 3.6 Circuit architecture of quantum–DNA 4-to-2 PROM

Step 3:
After getting qubits of all the corresponding values from four quantum AND operations of the 2-to-4 decoder, it needs to store all the quantum bits into a quantum cache memory.
Step 4:
After storing qubits of all the corresponding values from four quantum AND operations in cache memory, we need to perform NMR relaxation to transform the qubit to the DNA sequence. Here, NMR relaxation is done at room temperature, where there is no need to emit EMR.
Step 5:
Finally, after NMR relaxation, the DNA sequences will be found from qubits and these will go through two DNA OR gates to produce desired quantum-DNA 4-to-2 PROM output sequences.

3.4.3 Working Principle

The truth table of a quantum-DNA 4-to-2 ROM is given in Table 3.2. The operational procedures of quantum-DNA 4-to-2 ROM are explained as follows:
[i] For input combination A, B = **|0>**, **|0>**, |D0> line will be open and |D1> to |D3> will be closed. So, the value of |D0> =|1>, |D1> to |D3> = |0> will be stored in cache memory and after NMR relaxation all of these will transform into DNA sequence as D0 = **ACCTAG** and D1 to D3= **TGGATC**.
For the output F1 performs DNA OR operations between D0 and D2 that generate **ACCTAG**, and for F2 performs DNA OR operations between D1 and D3 that generate **TGGATC** as desired output.
[ii] For input combination A, B = **|1>**, **|0>**, |D1> line will be open and |D0>, |D2> and |D3> will be closed. So, the value of |D1>=|1>, |D0> |D2>, |D3> =|0> will store in cache memory and after NMR relaxation all of these will transform into DNA sequence as D1 = **ACCTAG** and D0, D2, D3= **TGGATC**.
For the output F1, the DNA OR operation will perform between D0 and D2 that produces **TGGATC** and for F2 performs DNA OR Operations between D1 and D3 and generates **ACCTAG**.

Table 3.2 Truth table of quantum-DNA 4-to-2 PROM

|B>	|A>	F1	F2
|0>	**|0>**	**ACCTAG**	**TGGATC**
|0>	**|1>**	**TGGATC**	**ACCTAG**
|1>	**|0>**	**ACCTAG**	**TGGATC**
|1>	**|1>**	**TGGATC**	**ACCTAG**

[iii] For input combination A, B = **|0>**, **|1>**, |D2> line will be open and |D0>, |D1> and |D3> will be closed. So, the value of |D2> = |1>, |D0> |D2>, |D3> = |0> will be stored in cache memory and after NMR relaxation all of these will transform into DNA sequence as D2 = **ACCTAG** and D0, D1, D3= **TGGATC**.
For the output F1 performs DNA OR operations between D0 and D2 that generate **ACCTAG**, and for F2 performs DNA OR operations between D1 and D3 that generate **TGGATC** as the desired output.
[iv] For input combination A, B = **|1>**, **|1>**, |D3> line will be open and |D0>, |D1> and |D2> will be closed. So, the value of |D3> = |1>, |D0> |D1>, |D2> = |0> will be stored in cache memory and after NMR relaxation all of these will transform into DNA sequence as D3 = **ACCTAG** and D0, D1, D2= **TGGATC**.
For the output F1, the DNA OR operation will be performed between D0 and D2 that produces **TGGATC** and for F2, DNA OR operations will be performed between D1 and D3 and generate **ACCTAG**.

3.5 Quantum-DNA Cache Memory

The design of various memory devices includes quantum and DNA logic gates and decoder for quantum-DNA 4-to-1 cache memory. Both DNA and quantum computers have the potential to significantly affect traditional digital computers in terms of processing power and depot. Quantum computers are more powerful than traditional Turing machines computing because of their coherent superposition of states and biotechnology techniques can be used to evolve DNA computers. Both of these properties might be captured by integrating DNA and quantum computing. In other way, quantum biological computing explores the potential of quantum mechanics in biological systems and the use of quantum information for biological applications. The concept of a "quantum cache memory (QCM)" in this context, as developed in research like the QCM framework, focuses on efficient and reliable storage and manipulation of genetic information within a quantum computing environment. This framework aims to preserve the integrity of DNA sequences during quantum processing, allowing for tasks like SNP detection and pattern searching. In essence, the concept of a "quantum cache memory" in the context of quantum biological computing refers to a method for efficiently storing and manipulating genetic information using quantum states and algorithms, with the goal of enabling powerful biological analyses like SNP detection and pattern search. In quantum biological computing, this cache memory is called quantum-DNA cache memory or quantum biocomputing cache memory or quantum biological computing cache memory. Therefore, the integrated computing systems such as the quantum-DNA systems for cache memory are the main concern of this section.

3.5.1 Block Diagram

Figure 3.7 represents the general organization of the quantum-DNA 4-to-1 cache memory block diagram, which consists of four separate "Words" of memory and each is single sequence wide.

3.5.2 Circuit Architecture and Working Principle

Circuit architecture of quantum-DNA cache memory is shown in Fig. 3.7. This figure represents the implementation of 4-to-1 quantum-DNA cache memory. This quantum RAM consists of four separate "Words" of memory and each is one qubit wide. The quantum RAM cell has three inputs and one output. A word consisting of two quantum RAM cells is arranged in such a way so that both qubits can be accessed simultaneously. Four words of memory need two address lines. |A> and |B> are the two-qubit address line inputs that go through a 2-to-4 decoder that selects one of the four words. The memory-enabled input enables the decoder. If the memory is enabled |0>, all outputs of the decoder will be |0> and in that case, none of the

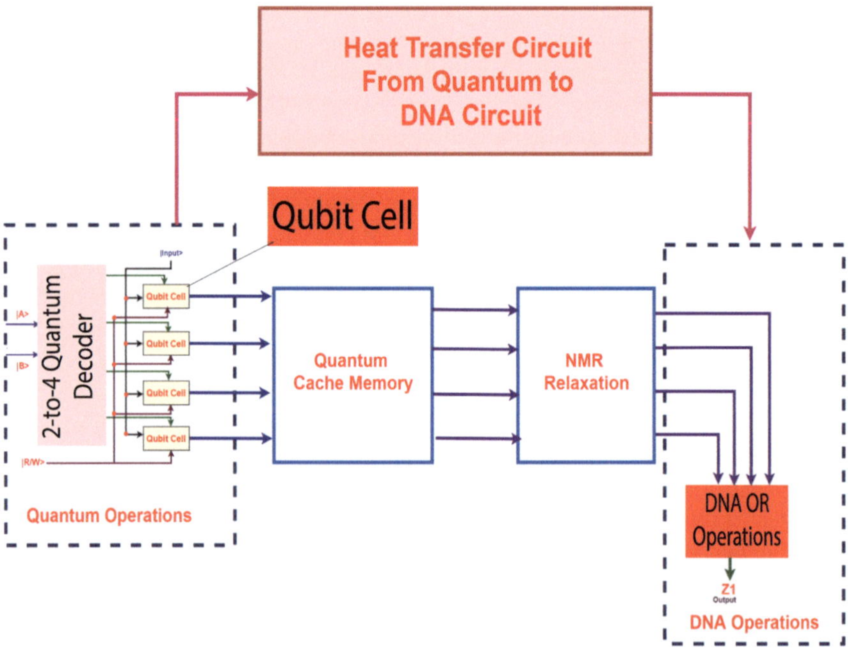

Fig. 3.7 Block diagram of quantum-DNA 4-to-1 cache memory

Table 3.3 Control input to memory chip

$\vert R/W>$	Memory operation
X	*None*
$\vert 0>$	*Write to the selected word*
$\vert 1>$	*Read from the selected word*

memory addresses will be selected. But when the memory enabled is |1>, one of the four words is selected. The word is selected by the value in the two address lines.

When a word has been selected, the read/write input determines the operation. During the read operation, the four qubits of the selected word pass to the DNA OR gates to the output Z1 terminals. But during the write operation, the data which is available in the input lines are transferred into the four quantum cells of the selected word. The quantum RAM cells that are not selected have become disabled and their previous qubit never changes. But when the memory-enabled input that passes into the decoder is equal to |0>, none of the words are selected, and then all quantum cells remain unchanged regardless of the value of the read/write input. The control input to memory chip of quantum-DNA cache memory is given in Table 3.3.

3.6 Applications

It might be possible to capture the benefits of quantum computing and DNA computing together after combining these two. The applications of memory devices in quantum computing are described separately in Chap. 2 and the applications of memory devices in DNA computing are also described separately in Chap. 4. So, both applications can be achieved in all memory devices in quantum-DNA computing.

3.7 Summary

This chapter has presented the details of memory devices in quantum-DNA computing. Necessary figures with their working principle have also been shown. Quantum-DNA computing is a new era of modern technology where two exciting technologies are combined together. The book is introducing this technology for the first time. The upcoming future will be benefited with the new technology. In quantum-DNA circuit, the excessive heat is needed to be transferred to the DNA part where the extra heat is needed to perform calculations.

Chapter 4
Memory Devices in DNA-Quantum Computing

4.1 Introduction

DNA-quantum computing is the opposite of quantum-DNA computing. Both are called quantum biocomputing. Both DNA and quantum computers have the potential to significantly affect traditional digital computers in terms of processing power and depot. Furthermore, nanotech DNA computing is highly parallel: In principle, billions upon trillions of DNA molecules might be undergoing chemical reactions or doing calculations, at the same time. In creating DNA-based approaches to simulate digital data manipulation, demonstrating how data may be digested using a circuit made up of DNA-based dynamic logic gates. Quantum computers are also much quicker than conventional supercomputers. A DNA-quantum computing system combines DNA with quantum computing. In DNA-quantum computing, this chapter will go through all of the intricacies of the same four memory devices.

4.2 DNA-Quantum Random-Access Memory

DNA-quantum computing for Random-Access Memory (RAM) is the main concern of this chapter. Biological quantum computing, in the context of Random Access Memory (RAM), refers to exploring biological systems for potential use in quantum memory, specifically for quantum random access memory (QRAM). This involves investigating how biological components could be engineered or manipulated to store and retrieve quantum information efficiently and reliably. QRAM is a key component of quantum computers, enabling them to access and process quantum data stored in a quantum format. Some researchers are investigating the possibility of using biological systems as quantum memory devices. These systems could be engineered to exhibit quantum phenomena and store quantum data in a stable and controllable

H. Md. Hasan Babu, *Quantum Biocomputing in Quantum Biology Volume II*,
https://doi.org/10.1007/978-981-97-5349-9_4

manner. This type of memory is called biological quantum RAM memory or bio-quantum RAM memory or DNA-quantum RAM memory. DNA-quantum RAM is a volatile memory. Quantum RAM is for quantum computer and DNA RAM is for DNA computer. So, to process data in DNA-quantum computing system, DNA-quantum RAM is necessary where the first part would be DNA and the last part would be quantum as well.

4.2.1 Block Diagram

A quadrupole ion trap traps charged particles using dynamic electric fields. In honor of Wolfgang Paul, these are also known as radio-frequency (RF) traps or Paul traps. The process of catching and cooling ions using quadrupole ion traps has advanced dramatically during the last two decades. These trapping techniques allow researchers to confine charged particles of a single species to the trap center, allowing them to study these ions in a well-controlled environment. Because of the lengthy storage durations of the ions, it is feasible in these traps, the transit-time broadening is eliminated, allowing for precision spectroscopic observations of these ions. Several significant experiments with a single electron or ion have been carried out to address challenges in basic physics, including the determination of the electron radius, precise measurements of fundamental characteristics, and testing of quantum mechanics predictions. The block diagram of quantum-DNA 4-to-1 RAM is shown in Fig. 4.1.

4.2.2 Working Principle and Circuit Architecture

Figure 4.2 presents the implementation of 4-to-1 DNA-quantum RAM. This DNA-quantum RAM consists of four separate "Words" of memory and each is one sequence wide. The DNA RAM cell has three inputs and one output. The complete circuit of a DNA RAM cell is described in Fig. 4.2 with proper explanation. A word consisting of two DNA RAM cells is arranged in such a way so that both sequences can be accessed simultaneously. Four words of memory need two address lines. A and B are the two-sequence address lines inputs that go through a DNA 2-to-4 decoder that selects one of the four words. The memory-enabled input enables the decoder. If the memory enabled is **TGGATC**, all outputs of the decoder will be **TGGATC** and in that case, none of the memory addresses will be selected. But when the memory enabled is **ACCTAG**, one of the four words is selected. The word is selected by the value in the two address lines. When a word has been selected, the read/write input determines the operation. During the read operation, the four sequences of the selected word pass to the quantum OR operations to the output |Z1> terminals. But during the write operation, the data which is available in the input lines are transferred into the four DNA cells of the selected word. The DNA RAM cells that are not selected become disabled and their previous sequence never changes. But when the memory-enabled input that passes into the decoder is equal to **TGGATC**, none of the words are selected, and then all DNA cells remain unchanged regardless of the value of the read/write input. This is the working procedure of 4-to-1 DNA-quantum RAM.

4.3 DNA-Quantum Read-Only Memory

Both DNA and quantum computers have the potential to significantly affect traditional digital computers in terms of processing power and depot. Quantum computers are more powerful than traditional Turing machines computing because of their coherent superposition of states and biotechnology techniques can be used to evolve DNA computers. The general idea of quantum computing is that some quantum algorithms are faster than those associated with time, and these systems make quantum computing one of the best technologies of the future. The DNA-quantum is the opposite of quantum-DNA computing. In other way, In biological quantum computing, a "read-only memory" (ROM) would be a mechanism for storing and retrieving quantum information, similar to a classical ROM, but using biological components and principles. While the concept of "biological ROM" is still speculative, it explores

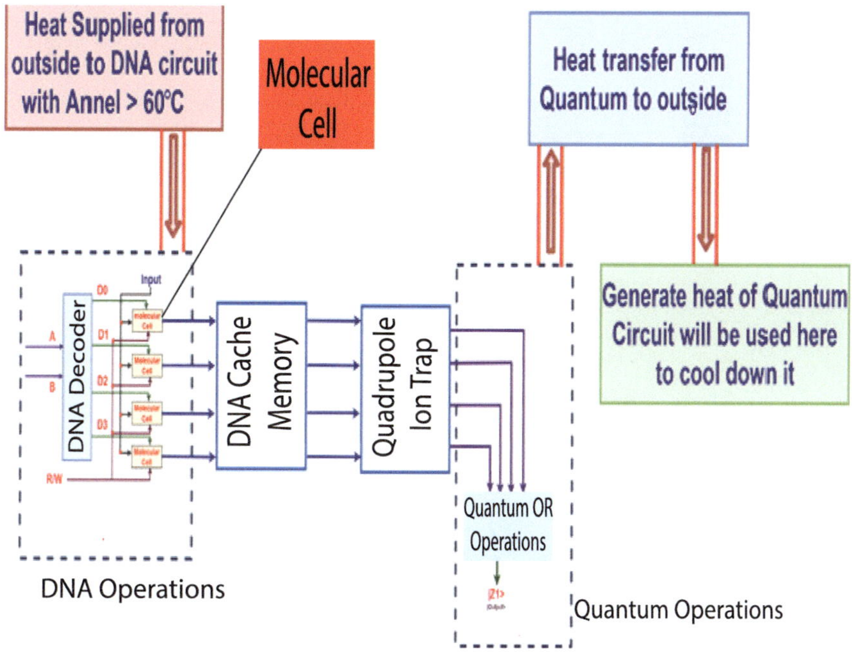

Fig. 4.1 Block diagram of quantum-DNA 4-to-1 RAM

the possibility of using biological systems or biological materials to build quantum memories. If a biological quantum ROM could be developed, it could offer unique advantages, such as the potential for building quantum computers with biological

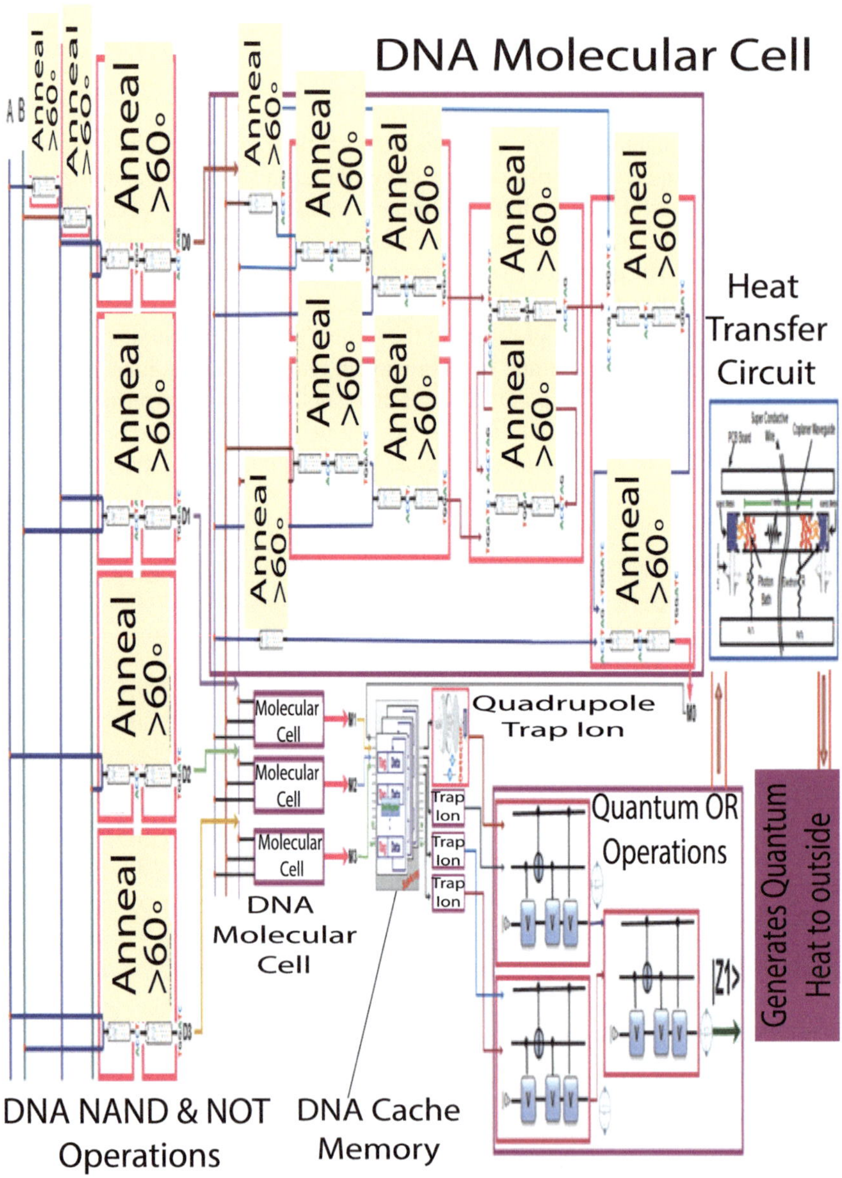

Fig. 4.2 Circuit architecture of DNA-quantum 4-to-1 RAM

materials and potentially harnessing biological processes for quantum computation. This section will describe about the details of DNA-quantum read-only memory.

4.3.1 Design Procedure

To perform ROM at normal room temperature, a 2-to-4 decoder and minterms of decoder output as the OR operations input are needed to produce the desired ROM output function. In the DNA-quantum computing (considering quadrupole ion trap), the input values are in DNA sequences (DNA computing) and the output will be qubits (quantum computing) at normal room temperature. Figure 4.3 shows the block diagram of DNA-quantum 4-to-2 ROM memory and the circuit architecture of DNA-quantum 4-to-2 ROM with quadrupole ion trap in normal room temperature is shown in Fig. 4.4. The design procedure of DNA-quantum 4-to-2 ROM is explained as follows:

Step 1 First, draw two input DNA sequences A and B. Two possible sequences are **ACCTAG** true and **TGGATC** false. These two will produce four combinations of two input sequences.

Step 2
To design a decoder, draw two DNA NOT operations and four DNA AND operations.

Step 3
After getting sequences of all the corresponding values from four DNA AND operations of 2-to-4 decoder, it needs to store all the DNA sequences into the cache memory. DNA cache memory is used for storing logical qubits and error correction. DNA cache also increases the logical qubit collection.

Step 4
After storing sequences of all the corresponding values from four DNA AND operations in cache memory, quadrupole ion trap is used to transform DNA sequences to qubits.

Step 5
Finally, the qubits from DNA sequences can be found and these will go through quantum OR operations to produce desired DNA-quantum 4-to-2 ROM output qubits.

4.3.2 Working Principle

The truth table of DNA-quantum 4-to-2 ROM is given in Table 4.1 and the working principle is explained as follows:

[i] For input combination A, B = **TGGATC**, **TGGATC**, D0 line will be open and D1 to D3 will be closed. So, the value of D0 = **ACCTAG**, D1 to D3 = **TGGATC**

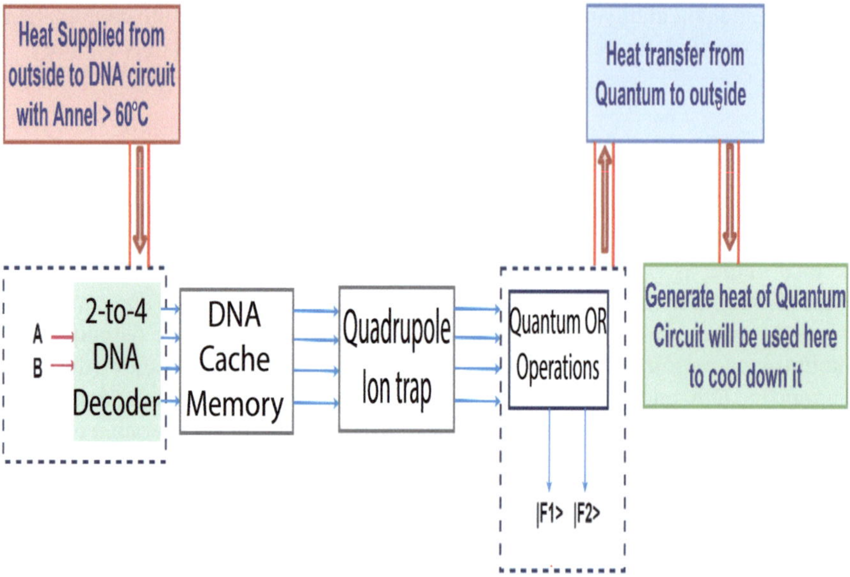

Fig. 4.3 DNA-quantum 4-to-2 ROM memory

will be stored in DNA cache memory and after quadrupole ion trap all of these will be transformed into quantum qubits as |D0< = **|1**> and |D1> to |D3>**= |0>**. For the output qubits of |F1> and |F2>, it performs quantum OR operations among |D0> to |D3> and generates **|1>** as the desired output qubits.

[ii] For input combination A, B = **ACCTAG**, **TGGATC**, D1 line will be open and D0, D2 and D3 will be closed. So, the value of D1= **ACCTAG**, D1, D2, D3 = **TGGATC** will be stored in DNA cache memory and after Quadrupole Ion Trap all of these will be transformed into quantum qubits as |D1> = **|1>** and |D0>, |D2>, |D3>**= |0>**. For the output qubits of |F1> and |F2>, it performs quantum OR operations among |D0> to |D3> and generates **|1>** as the desired output qubits.

[iii] For input combination A, B = **TGGATC, ACCTAG**, D2 line will be open and D0, D1 and D3 will be closed. So, the value of D2= **ACCTAG**, D0, D1, D3 = **TGGATC** will be stored in DNA cache memory and after quadrupole ion trap all of these will

Table 4.1 Truth table of quantum-DNA 4-to-2 ROM

B	A	|F1>	|F2>
TGGATC	TGGATC	|1>	|1>
TGGATC	ACCTAG	|1>	|1>
ACCTAG	TGGATC	|1>	|1>
ACCTAG	ACCTAG	|1>	|1>

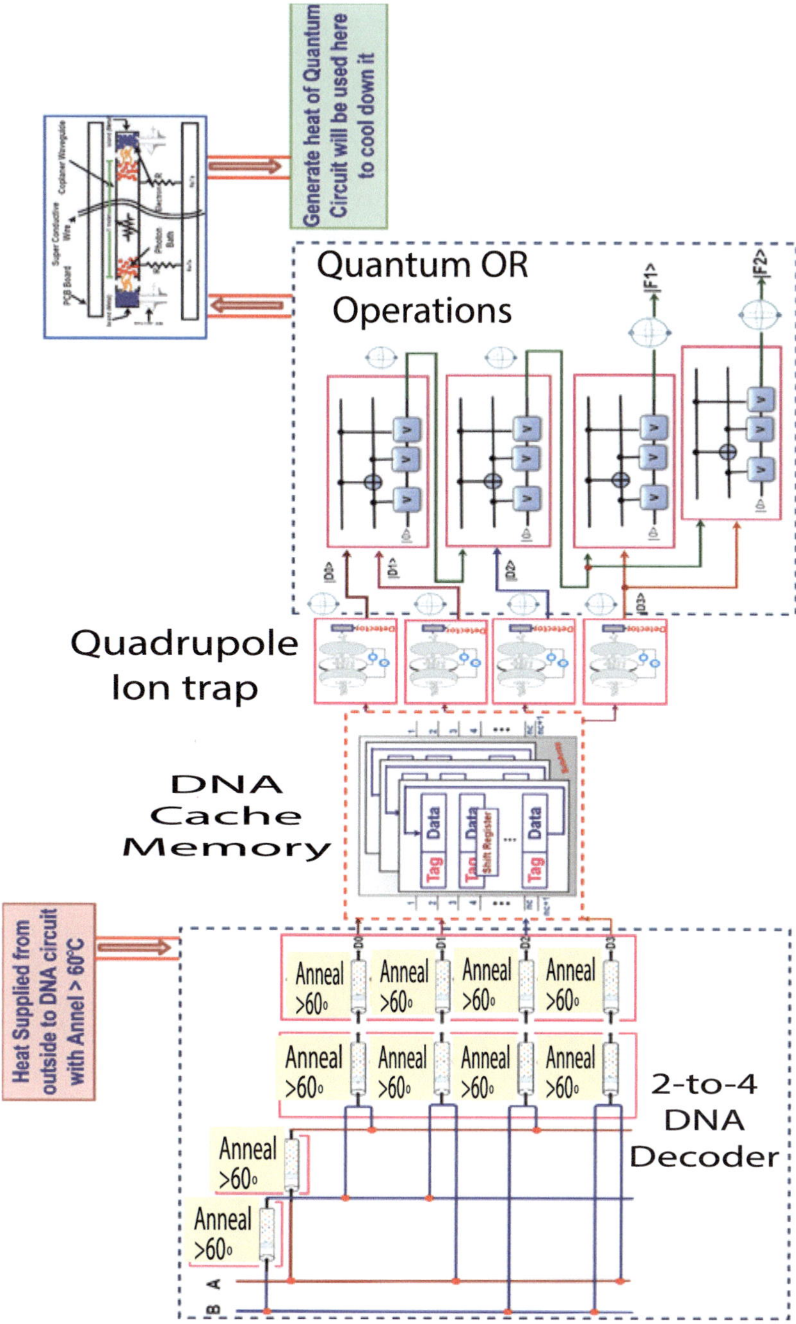

Fig. 4.4 Circuit architecture of DNA-quantum 4-to-2 ROM

be transformed into quantum qubits as |D2> = **|1>** and |D0>, |D1>, |D3>**= |0>**. For the output qubits of |F1> and |F2>, it performs quantum OR operations among |D0> to |D3> and generates **|1>** as the desired output qubits.

[iv] For input combination A, B = **ACCTAG, ACCTAG**, D3 line will be open and D0 to D2 will be closed. So, the value of D3= **ACCTAG**, D0, D1, D2 = **TGGATC** will be stored in DNA cache memory and after quadrupole ion trap all of these will be transformed into quantum qubits as |D3> = **|1>** and |D0>, |D1>, |D2>**= |0>**. For the output qubits of |F1> and |F2>, it performs quantum OR operations among |D0> to |D3> and generates **|1>** as the desired output qubits.

4.4 DNA-Quantum Programmable Read-Only Memory

The chapter focuses on DNA-quantum programmable read-only memory. This chapter presents a novel theoretical strategy for creating alternative storage devices based on a binary logical system defined by a combination of all compensation, which includes DNA and quantum computing. When it comes to feeding and storing DNA or quantum computers, the capacity to make a significant influence on ordinary computers is quite important. In other way, Biological quantum computing explores the idea of using or mimicking biological systems for quantum computation, potentially incorporating biological materials or processes to build or interface with quantum devices. Programmable read-only memory (PROM) is a type of memory where data can be written once and then read repeatedly, offering potential applications in quantum systems. Biological quantum computing could leverage PROM-like mechanisms within biological systems, or use biological materials to create programmable quantum memories. This type of memory is called the biological quantum ROM memory or bio-quantum ROM memory or DNA-quantum ROM memory. The greatest and most recent technology introduced in this chapter is the combination of these two.

4.4.1 Design Procedure

Block diagram of DNA-quantum 4-to-2 PROM is shown in Fig. 4.5 and the circuit architecture of DNA-quantum 4-to-2 PROM is depicted in Fig. 4.6. To perform PROM at normal room temperature, a 2-to-4 decoder and minterms of decoder output as the OR operations input are needed to produce the desired PROM output function. In the DNA-quantum computing (considering Quadrupole Ion Trap), the input values are in DNA sequences (DNA computing) and the output will be qubits (quantum computing) at normal room temperature.

Step 1
First, draw two input DNA sequences A and B. Two possible sequences are **ACCTAG** true and **TGGATC** false. These two will produce four combinations of two input sequences.

Step 2
To design a decoder, draw two DNA NOT operations and four DNA AND operations.

Step 3
After getting sequences of all the corresponding values from four DNA AND operations of 2-to-4 decoder, it needs to store all the DNA sequences into the cache memory.

Step 4
After storing sequences of all the corresponding values from four DNA AND operations in cache memory, the quadruple ion trap helps to transform DNA sequences to qubits.

Step 5
Finally, qubits from DNA sequences are found and these will go through quantum OR operations to produce the desired DNA-quantum 4-to-2 PROM output qubits.

4.4.2 Working Principle

The truth table of quantum-DNA 4-to-2 PROM is given in Table 4.2 and the working

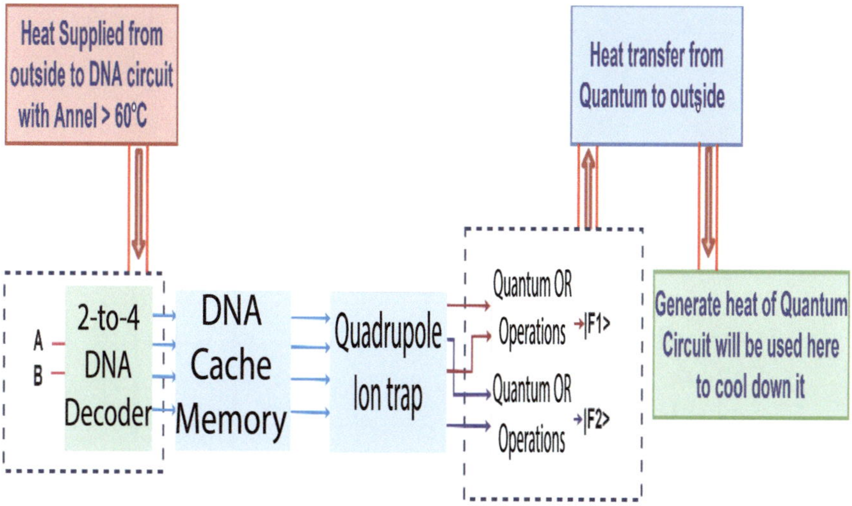

Fig. 4.5 Block diagram of DNA-quantum 4-to-2 PROM

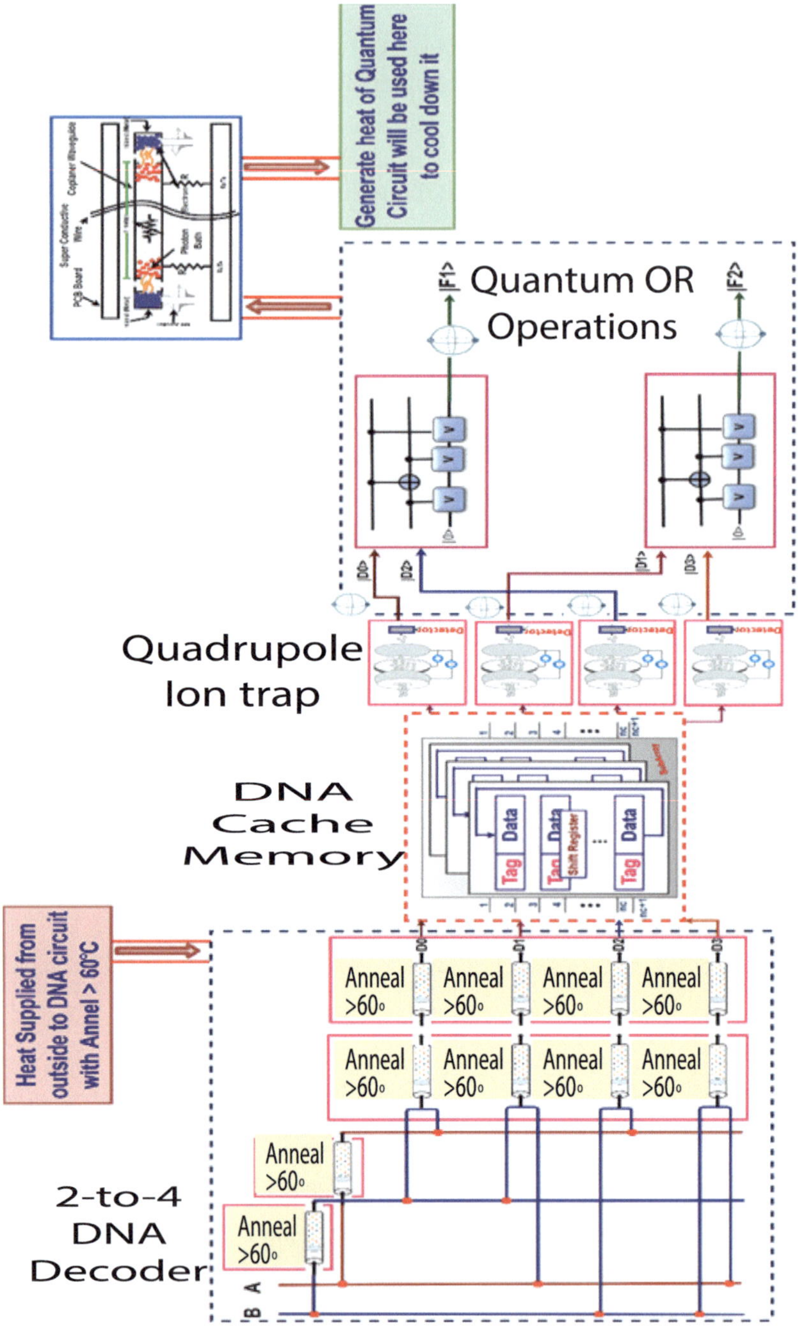

Fig. 4.6 Circuit architecture of DNA-quantum 4-to-2 PROM

principle is given as follows:

[i] For input combination A, B = **TGGATC, TGGATC**, D0 line will be open and D1 to D3 will be closed. So, the value of D0= **ACCTAG**, D1 to D3 = **TGGATC** will be stored in DNA cache memory and after quadrupole ion trap all of these will be transformed into quantum qubits as |D0> = **|1>** and |D1> to |D3>= **|0>**.
For the output of |F1>, the quantum OR operation will be performed between |D0> and |D2> that produces **|1>** and for |F2>, quantum OR operations will be performed between |D1> and |D3> and generates **|0>**.

[ii] For input combination A, B = **ACCTAG, TGGATC**, D1 line will be open and D0, D2 and D3 will be closed. So, the value of D1= **ACCTAG**, D1, D2, D3 = **TGGATC** will be stored in DNA cache memory and after quadrupole ion trap all of these will be transformed into quantum qubits as |D1> = **|1>** and |D0>, |D2>, |D3>= **|0>**.
For the output of |F1>, the quantum OR operation will be performed between |D0> and |D2> that produces **|0>** and for |F2>, quantum OR operations will be performed between |D1> and |D3> and generates **|1>**.

[iii] For input combination A, B = **TGGATC, ACCTAG**, D2 line will be open and D0, D1, and D3 will be closed. So, the value of D2= **ACCTAG**, D0, D1, D3 = **TGGATC** will be stored in DNA cache memory and after quadrupole ion trap all of these will be transformed into quantum qubits as |D2> = **|1>** and |D0>, |D1>, |D3>= **|0>**.
For the output of |F1>, the quantum OR operation will be performed between |D0> and |D2> that produces **|1>** and for |F2>, quantum OR operations will be performed between |D1> and |D3> and generates **|0>**.

[iv] For input combination A, B = **ACCTAG, ACCTAG**, D3 line will be open and D0 to D2 will be closed. So, the value of D3= **ACCTAG**, D0, D1, D2 = **TGGATC** will be stored in DNA cache memory and after quadrupole ion trap all of these will be transformed into quantum qubits as |D3> = **|1>** and |D0>, |D1>, |D2> = **|0>**.
For the output of |F1>, the quantum OR operation will be performed between |D0> and |D2> that produces **|0>** and for |F2>, OR operations will be performed between |D1> and |D3>, generates **|1>**.

Table 4.2 Truth table of quantum-DNA 4-to-2 PROM

B	A	$\vert F1\rangle$	$\vert F2\rangle$
TGGATC	TGGATC	$\vert 1\rangle$	$\vert 0\rangle$
TGGATC	ACCTAG	$\vert 0\rangle$	$\vert 1\rangle$
ACCTAG	TGGATC	$\vert 1\rangle$	$\vert 0\rangle$
ACCTAG	ACCTAG	$\vert 0\rangle$	$\vert 1\rangle$

4.5 DNA-Quantum Cache Memory

The cache is a type of high-performance memory. It's utilized for high-speed CPU acceleration and synchronization. Cache memory costs more than main memory or disk memory, although it is less expensive than CPU registers. Cache is a form of memory that works as a buffer between the RAM and the processor. The relevant data and instructions are frequently provided so that they are instantly available to the CPU when it is required. Caching is a technique for reducing the time it takes to retrieve critical memory data. The cache is a smaller and, faster memory that stores copies of data from memory regions that are often accessed. The CPU instructions and data store in a variety of independent caches. Bioquantum computing, a field exploring the integration of quantum computing with biological systems, leverages the inherent quantum properties of biological molecules to perform computations. This approach aims to harness the massive parallelism and information processing capabilities of biological systems while also benefiting from the power of quantum computing. Cache memory, in this context, would likely refer to a temporary storage area for frequently accessed data within a bioquantum computing system, potentially using biological materials or quantum states to store and retrieve information efficiently. To develop a stable and reliable quantum cache memory is a major challenge in quantum computing, and the same applies to bioquantum computing. Researchers are exploring different approaches to build quantum cache memory, including using biomolecules or quantum states. Further research is also needed to understand how to best utilize biological materials and quantum states for creating effective cache memory in bioquantum computing systems. The main concern of this chapter is to discuss about cache memory in DNA-quantum computing.

4.5.1 Design Procedure

Block diagram of Quantum-DNA 4-to-1 cache memory is shown in Fig. 4.7. Each of the four "Words" of memory in this DNA-quantum RAM is one sequence wide. There are three inputs and one output on the DNA RAM cell. A word is made up of two DNA RAM cells that are organized so that both sequences may be accessible at the same time. Two address lines are required for four words of memory. The two-sequence address lines A and B are sent to a DNA 2-to-4 decoder, which picks one of the four words. The decoder is activated by the memory-enabled input. During the read operation, the four sequences of the chosen word are sent to the output |Z1> terminals through quantum OR operations.

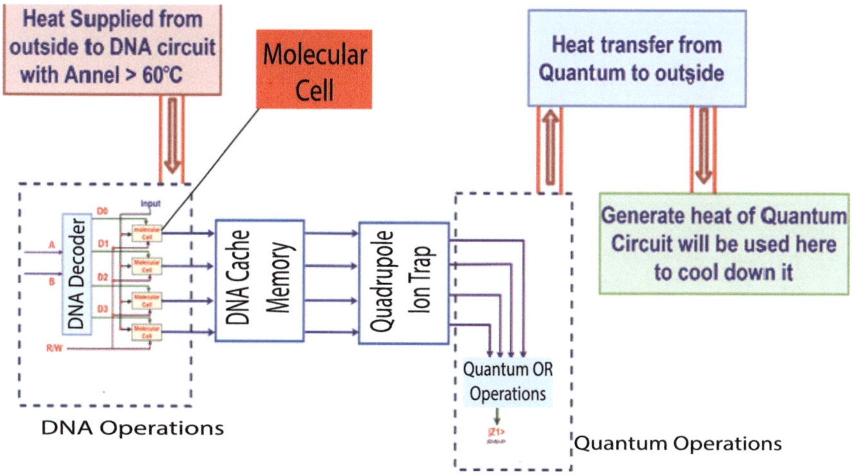

Fig. 4.7 Block diagram of Quantum-DNA 4-to-1 Cache memory

4.5.2 *Working Principle*

Figure 4.8 shows the implementation of 4-to-1 DNA-quantum cache memory. DNA-quantum RAM consists of four separate "Words" of memory and each is one sequence wide. The DNA RAM cell has three inputs and one output. A word consisting of two DNA RAM cells is arranged in such a way so that both sequences can be accessed simultaneously. Four words of memory need two address lines. A and B are the two-sequence address line inputs that go through a DNA 2-to-4 decoder that selects one of the four words. The memory-enabled input activates the decoder. If the memory enabled is **TGGATC**, all outputs of the decoder will be **TGGATC** and in that case, none of the memory addresses will be selected. But when the memory enabled is **ACCTAG**, one of the four words is selected. The word is selected by the value in the two address lines. When a word has been selected, the read/write input determines the operation. During the read operation, the four sequences of the selected word pass to the quantum OR operations to the output |Z1> terminals. But during the write operation, the data which is available in the input lines are transferred into the four DNA cells of the selected word. The DNA RAM cells that are not selected have become disabled and their previous sequence never changes. But when the memory-enabled input that passes into the decoder is equal to **TGGATC**, none of the words are selected, and then all DNA cells remain unchanged regardless of the value of the read/write input.

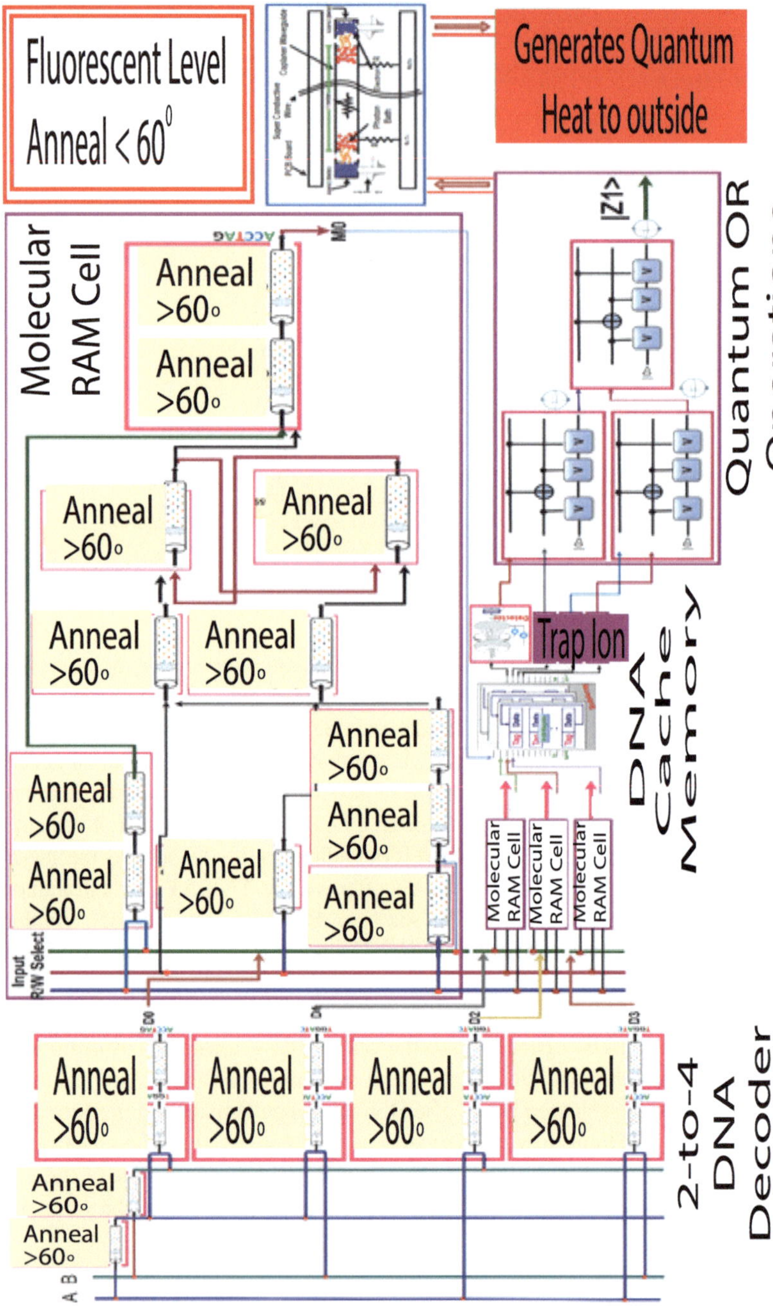

Fig. 4.8 DNA-quantum 4-to-1 cache memory

4.6 Summary

The specifics of DNA-quantum memory devices were provided in this chapter. The initial component of the circuit was always DNA, and the latter part was always quantum. In DNA-quantum computing, the inputs are always DNA sequences and the outputs are qubits, which is the polar opposite of quantum-DNA computing. All of the memory devices are illustrated with diagrams, and their working principles and explanations are also provided. In DNA-quantum computing, DNA cache memory is needed and the extra heat is needed to supply to the DNA part. The quantum part produces excessive heat which is needed to be transferred to the cooler to maintain the temperature of the circuit.

Part II
Programmable Devices in Quantum Biocomputing

Overview

Programmable Logic Devices (PLD) are a generic term for an integrated circuit that can be programmed in a laboratory to perform complex functions. A PLD consists of arrays of AND and OR gates. A system designer implements a logic design with a device programmer that blows fuses on the PLD to control gate operation. A programmable logic array (PLA) has a programmable-AND gate array, which links to a programmable OR gate array, which can then be conditionally complemented to produce an output. PLA is similar to ROM concept. However, a PLA does not provide full decoding of a variable and does not generate all the minterms as in a ROM. Programmable Array Logic (PAL) devices have arrays of transistor cells arranged in a "fixed-OR, programmable-AND" plane which are used to implement "sum-of-products" quantum logic equations for each of the outputs in terms of the inputs and either synchronous or asynchronous feedback from the outputs. System designers can use development software that converts basic code into instructions and a device programmer which needs to implement a design. FPGAs contain an array of programmable logic blocks, and a hierarchy of reconfigurable interconnects allowing blocks to be wired together. Logic blocks can be configured to perform complex combinational functions, or act as simple logic gates like AND and XOR. In most FPGAs, logic blocks also include memory elements, which may be simple flip-flops or more complete blocks of memory. Many FPGAs can be reprogrammed to implement different logic functions, allowing flexible reconfigurable computing as performed in computer software. A complex programmable logic device (CPLD) is a programmable logic device with complexity between that of PALs and FPGAs, and architectural features of both. In this part, these four programmable logic devices, namely, Programmable Logic Array (PLA), Programmable Array Logic (PAL), Field Programmable Gate Arrays (FPGA), Complex Programmable Logic Devices (CPLD) will be implemented in quantum biocomputing. Programmable logic devices

(PLDs) play a crucial role in quantum biocomputing, enabling the design and implementation of complex biological circuits using DNA and other biomolecules. PLDs like FPGAs and CPLDs can be configured to perform specific logic operations, which are then utilized in biomolecular systems. This allows for the development of sensitive and customizable circuits that mimic biological processes and can be used for various applications, including sensing and treatment of diseases. These hybrid systems leverage both quantum mechanics and biomolecules to create more sophisticated circuits. DNA sequences can be used as inputs to quantum-based PLDs, and the system's behavior can be modulated by quantum effects.

Chapter 5
Programmable Devices in Quantum Computing

5.1 Introduction

In quantum computing it usually uses an object's quantum state to generate a qubit. These are the undefined qualities of an item before they have been identified, such as an electron's spin or a photon's polarization. Unmeasured quantum states exist in a mixed "superposition," similar to a coin spinning in the air before landing.

These superpositions can get entangled with those of other things, implying that their eventual outcomes will be mathematically connected. The complicated mathematics underpinning these unsettled states of entangled "spinning coins" may be fed into specific algorithms to quickly solve issues that would take a traditional computer a long time to solve, if it could even solve them at all. Solving complicated mathematical problems, creating difficult-to-crack security codes, and forecasting numerous particle interactions in chemical processes might be benefited from such algorithms.

A programmable logic device (PLD) is an electronic component used to build reconfigurable device which is constructed using discrete logic gates with fixed functions, where a PLD has an undefined function at the time of manufacture. Before the PLD can be used in a circuit, it must be programmed to implement the desired function. Compared to fixed logic devices, programmable logic devices simplify the design of complex logic and may offer superior performance. Unlike microprocessors, programming a PLD changes the connections made between the gates in the device. In other way, Programmable Logic Devices (PLDs), like CPLDs and FPGAs, are not directly used to perform quantum computations in the same way as they are used for classical computing. However, they play a crucial role in controlling and managing the hardware involved in quantum systems. Specifically, they can be used to implement the logic for controlling quantum gates, measurements, and other operations. In essence, PLDs are like the "brains" behind the control and execution of quantum hardware. They translate the instructions from a classical computer into the signals needed to manipulate and measure quantum systems. The quantum logic used

H. Md. Hasan Babu, *Quantum Biocomputing in Quantum Biology Volume II*,
https://doi.org/10.1007/978-981-97-5349-9_5

to design PLD devices makes the device more powerful which is the main concern of this chapter.

5.2 Quantum Programmable Logic Array

The programmable logic array is a type of Programmble Logic Device (PLD), which has both programmable quantum AND array and programmable quantum OR array. In a combinational circuit, because of don't care conditions, not all the minterms are used. Programmable Logic Array (PLA) is a fixed architecture logic device with programmable quantum AND gates followed by programmable quantum OR gates. PLA is basically a type of programmable logic device used to build a reconfigurable digital circuit. PLDs have an undefined function at the time of manufacturing, but they are programmed before being made into use. PLA is a combination of memory and other logic circuits. PLAs can also be implemented using Quantum Dot Cellular Automata (QCA) technology, offering potential advantages in quantum computing and nanotechnology.

5.2.1 *Block Diagram*

Quantum PLA is a programmable logic device that has both programmable quantum AND array and programmable quantum OR array. Hence, it is the most flexible PLD. The block diagram of quantum PLA is shown in Fig. 5.1. Here, the inputs of quantum AND operations are programmable. That means each quantum AND operation has both normal and complemented inputs of variables. So, based on the requirement, any of those inputs can be programmed. So, the required product terms can be generated by using these quantum AND operations.

Here, the inputs of quantum OR operations are also programmable. So, it can be possible to program any number of required product terms, since all the outputs of quantum AND operations are applied as inputs to each quantum OR operation. Therefore, the outputs of quantum PLA will be in the form of the sum of product forms.

5.2.2 *Circuit Architecture*

The circuit diagram of PLA for functions F1, F2, and F3 is shown in Fig. 5.2 and the truth table of quantum PLA for functions F1, F2, and F3 is given in Table 5.1. In Fig. 5.2, a number of AND gates and a number of OR gates are used to get the function values.

Table 5.1 Truth table of quantum PLA for functions F1, F2, and F3

$\|A>$	$\|B>$	$\|C>$	$\|F1>$	$\|F2>$	$\|F3>$
$\|0>$	$\|\mathbf{0}>$	$\|\mathbf{0}>$	$\|\mathbf{0}>$	$\|\mathbf{1}>$	$\|\mathbf{0}>$
$\|0>$	$\|\mathbf{0}>$	$\|\mathbf{1}>$	$\|\mathbf{1}>$	$\|\mathbf{0}>$	$\|\mathbf{0}>$
$\|0>$	$\|\mathbf{1}>$	$\|\mathbf{0}>$	$\|\mathbf{0}>$	$\|\mathbf{1}>$	$\|\mathbf{1}>$
$\|0>$	$\|\mathbf{1}>$	$\|\mathbf{1}>$	$\|\mathbf{1}>$	$\|\mathbf{0}>$	$\|\mathbf{1}>$
$\|1>$	$\|\mathbf{0}>$	$\|\mathbf{0}>$	$\|\mathbf{0}>$	$\|\mathbf{1}>$	$\|\mathbf{1}>$
$\|1>$	$\|\mathbf{0}>$	$\|\mathbf{1}>$	$\|\mathbf{0}>$	$\|\mathbf{1}>$	$\|\mathbf{1}>$
$\|1>$	$\|\mathbf{1}>$	$\|\mathbf{0}>$	$\|\mathbf{1}>$	$\|\mathbf{0}>$	$\|\mathbf{0}>$
$\|1>$	$\|\mathbf{1}>$	$\|\mathbf{1}>$	$\|\mathbf{1}>$	$\|\mathbf{0}>$	$\|\mathbf{0}>$

F1 (A, B, C) = m (1, 3, 6, 7)
F2 (A, B, C) = m (0, 2, 4, 5)
F3 (A, B, C) = m (2, 3, 4, 5)

K-map to reduce the function:

A\BC	$\|\mathbf{0}>\|\mathbf{0}>$	$\|\mathbf{0}>\|\mathbf{1}>$	$\|\mathbf{1}>\|1>$	$\|\mathbf{1}>\|0>$
$\|\mathbf{0}>$	$\|\mathbf{0}>$	$\|\mathbf{1}>$	$\|\mathbf{1}>$	$\|\mathbf{0}>$
$\|\mathbf{1}>$	$\|\mathbf{0}>$	$\|\mathbf{0}>$	$\|\mathbf{1}>$	$\|\mathbf{1}>$

$$|F1>=|A>.|B>+|A'>.|C>$$

A\BC	$\|\mathbf{0}>\|\mathbf{0}>$	$\|\mathbf{0}>\|\mathbf{1}>$	$\|\mathbf{1}>\|1>$	$\|\mathbf{1}>\|0>$
$\|\mathbf{0}>$	$\|\mathbf{1}>$	$\|\mathbf{0}>$	$\|\mathbf{0}>$	$\|\mathbf{1}>$
$\|\mathbf{1}>$	$\|\mathbf{1}>$	$\|\mathbf{1}>$	$\|\mathbf{0}>$	$\|\mathbf{0}>$

$$|F2>=|A'>.|C'>+|A>.|B'>$$

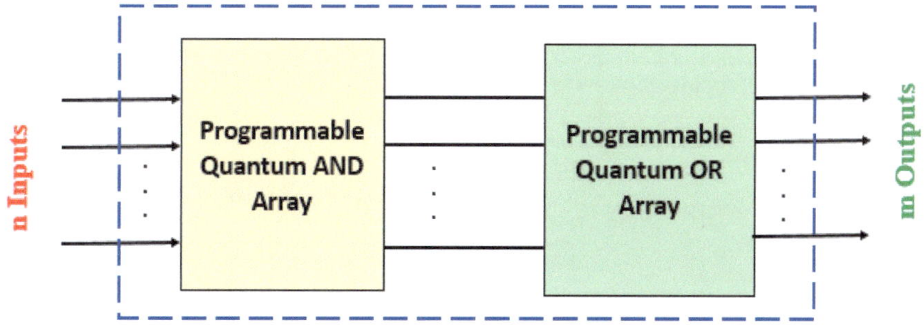

Fig. 5.1 Block diagram of quantum PLA

A\BC	$\|0>\|0>$	$\|0>\|1>$	$\|1>\|1>$	$\|1>\|0>$
$\|0>$	$\|0>$	$\|0>$	$\|1>$	$\|1>$
$\|1>$	$\|1>$	$\|1>$	$\|0>$	$\|0>$

$$|F3>=|A'>.|B>+|A>.|B'>$$

Consider the implementation of the following **quantum logic functions** using Quantum PLA:

$$|F1>=|A>.|B>+|A'>.|C>$$
$$|F2>=|A'>.|C'>+|A>.|B'>$$
$$|F3>=|A'>.|B>+|A>.|B'>$$

The given three functions are in sum of product forms. The number of product terms present in the given quantum logic functions $|F1>$, $|F2>$, and $|F3>$ are two.

$|A>.\ |B'>$ product are use in both functions $|F2>$ and $|F3>$. So, five programmable quantum AND gates and three programmable quantum OR gates are required for producing those three functions. The corresponding quantum **PLA** is shown in Fig. 5.3.

Consider the realization of the quantum logic expression $|F1>=|A>.\ |B>+|A'>.\ |C>$, $|F2>=|A'>.\ |C'>+|A>.\ |B'>$, and $|F3>=|A'>.\ |B>+|A>.\ |B'>$ using quantum programmable logic array.

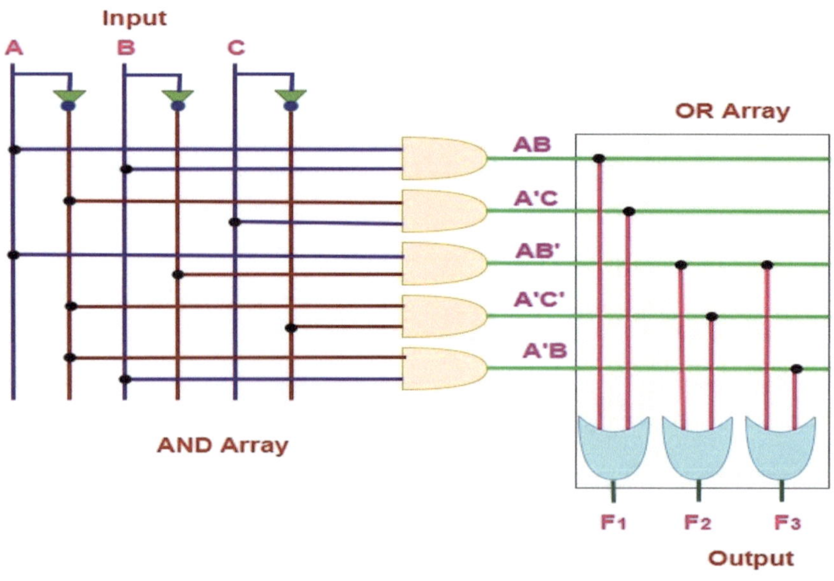

Fig. 5.2 Circuit diagram of PLA for functions F1, F2, and F3

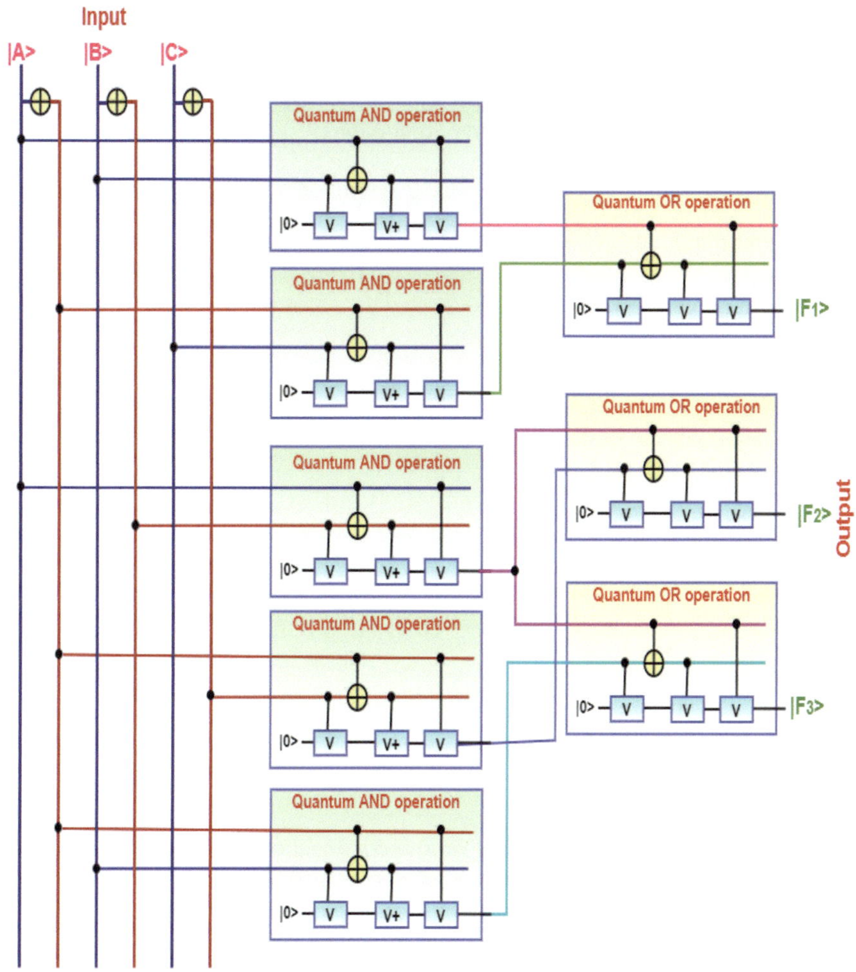

Fig. 5.3 Circuit architecture of quantum PLA

For the given problem, there are three inputs (|A >, |B >, |C >) and three outputs (|F1 >, |F2 >, |F3 >). The complement of three inputs is obtained through quantum NOT operation. Thus the realization has six input lines (input with its complement).

The given expression has five product terms and so the fuses are placed in the corresponding literals to obtain the product terms.

5.2.3 *Working Principle*

According to truth table (Table 5.1) of Quantum PLA, it is necessary to do the following operations to get the desired output qubits:

1. For input qubits $|A>, |B>, |C>= |0>, |0>, |0>$ function $|F2>$ produces $|1>$.
2. For input qubits $|A>, |B>, |C>= |0>, |0>, |1>$ function $|F1>$ produces $|1>$.
3. For input qubits $|A>, |B>, |C>= |0>, |1>, |0>$ functions $|F2>$ and $|F3>$ produce $|1>$.
4. For input qubits $|A>, |B>, |C>= |0>, |1>, |1>$ functions $|F1>$ and $|F3>$ produce $|1>$.
5. For input qubits $|A>, |B>, |C>= |1>, |0>, |0>$ functions $|F2>$ and $|F3>$ produce $|1>$.
6. For input qubits $|A>, |B>, |C>= |1>, |0>, |1>$ functions $|F2>$ and $|F3>$ produce $|1>$.
7. For input qubits $|A>, |B>, |C>= |1>, |1>, |0>$ function $|F1>$ produces $|1>$.
8. For input qubits $|A>, |B>, |C>= |1>, |1>, |1>$ function $|F1>$ produces $|1>$.

5.2.4 *Applications*

PLA is used for the implementation of various combinational circuits using quantum AND operation and quantum OR operation. In quantum PLA, all the minterms are not realized but only required minterms are implemented. As quantum PLA has a programmable quantum AND operation array and programmable quantum OR operation array, it provides more flexibility but the disadvantage is that, it is not easy to use.

Applications of a quantum PLA are as follows:

1. It is used to provide control over datapath.
2. It also is used as a counter.
3. This device may use as a decoder.
4. In many cases, it is used as a BUS interface in programmed I/O.

5.3 Quantum Programmable Array Logic

A Quantum Programmable Array Logic (QPAL) is a logic device, which has a programmable quantum AND array and fixed quantum OR array. It is used to realize a

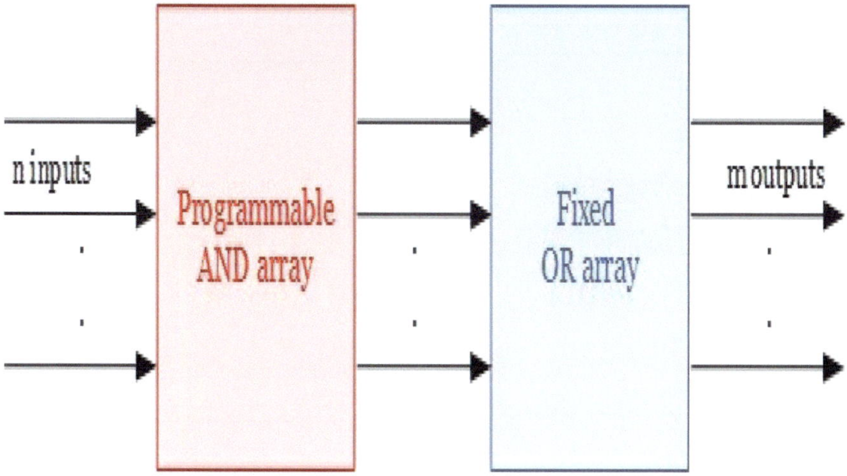

Fig. 5.4 Programmable array logic

logic function. In this QPAL, only quantum AND gates are programmable and hence it is easier to work with QPAL. In other way, a quantum PAL or QPAL refers to the concept of building quantum computers with logic arrays that can be programmed to perform different quantum operations. These arrays are similar to classical programmable logic devices but utilize quantum mechanics for computation. In essence, quantum programmable array logic aims to harness the power of quantum mechanics for computation by using programmable quantum gate arrays, paving the way for more versatile and powerful quantum computers. The goal of quantum PALs is to create a quantum processor where the logic array can be reconfigured to implement different quantum algorithms or circuits.

Figure 5.4 shows the internal structure of programmable array logic. The product terms can be programmed through the fuse link. It means the user can decide the connection between the inputs and the AND gates. If a particular input line is to be connected to the AND gate, then the fuse link must be placed at the interconnection. The AND gate outputs are then fed as an input to the fixed-OR gate. Depending upon the required function, the output line of the AND gate is connected to the corresponding input of the OR gate.

5.3.1 Block Diagram

Quantum Programmable Array Logic (PAL) is a form of Programmable Logic Device (PLD) that may be used to implement a certain logical function. A quantum AND gate array is followed by a quantum OR gate array in quantum PALs. It should be emphasized, however, that only the quantum AND gate array is programmable here,

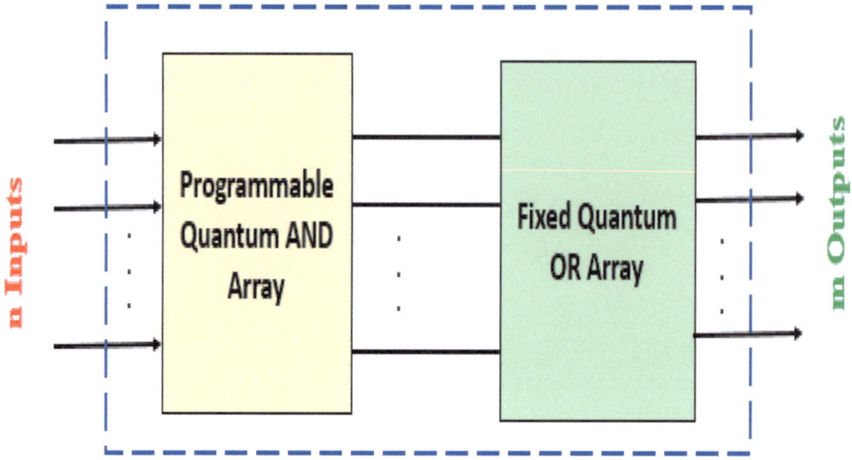

Fig. 5.5 Block diagram of a quantum PAL

as opposed to the quantum OR gate array, which has fixed logic. This is due to the fact that the inputs to the quantum AND gates are routed through fuses that operate as programmable connections. Quantum PALs have less programming flexibility than programmable logic arrays because of their programmable-AND and fixed-OR architectures (PLAs). However, due to the same reason PALs are less expensive than PLAs. Hence, it is the most flexible quantum PLD. The block diagram of quantum PAL is shown in Fig. 5.5.

Here, the inputs of quantum AND operations are programmable. That means each quantum AND operation has both normal and complemented inputs of variables.

5.3.2 *Circuit Architecture*

Consider the PAL for the following functions:

F1 (A, B, C) = m (3, 5, 6, 7)
F2 (A, B, C) = m (0, 2, 4, 5).

Circuit architecture for the corresponding functions (F1, F2) is shown in Fig. 5.6 and the truth table is given in Table 5.2.

A K-map to reduce the function:

A\BC	|0 > |0 >	|0 > |1 >	|1 > |1 >	|1 > |0 >
|0 >	|0 >	|0 >	|1 >	|0 >
|1 >	|0 >	|1 >	|1 >	|1 >

$$|F1>= |A> .|B> +|B> .|C> +|A> .|C>$$

Table 5.2 Truth table of a quantum PAL for functions F1 and F2

|A >	|B >	|C >	|F1 >	|F2 >
|0 >	|**0** >	|**0** >	|**0** >	|**1** >
|0 >	|**0** >	|**1** >	|**0** >	|**0** >
|0 >	|**1** >	|**0** >	|**0** >	|**1** >
|0 >	|**1** >	|**1** >	|**1** >	|**0** >
|1 >	|**0** >	|**0** >	|**0** >	|**1** >
|1 >	|**0** >	|**1** >	|**1** >	|**1** >
|1 >	|**1** >	|**0** >	|**1** >	|**0** >
|1 >	|**1** >	|**1** >	|**1** >	|**0** >

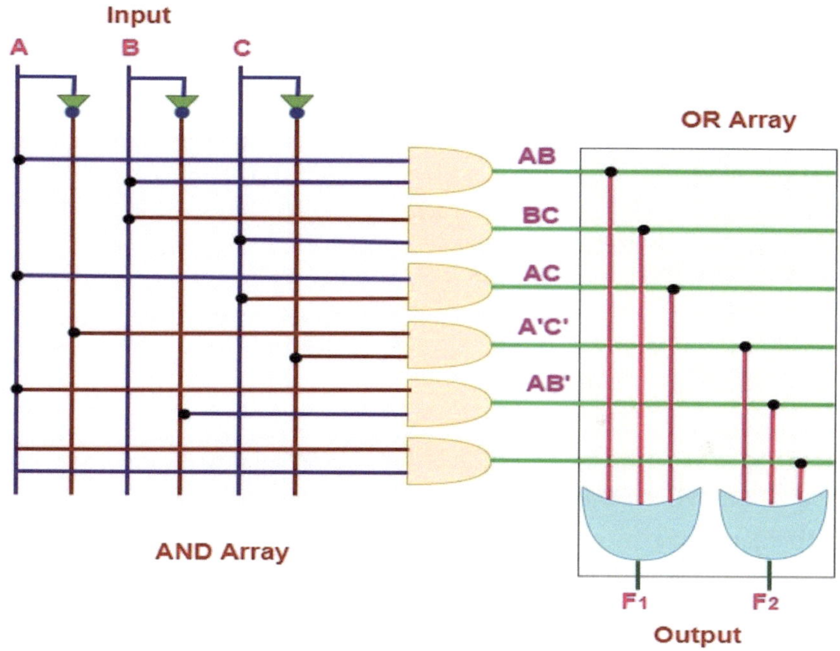

Fig. 5.6 PAL for functions F1 and F2

A\BC	|**0** > |**0** >	|**0** > |**1** >	|**1** > |1 >	|**1** > |0 >
|**0** >	|**1** >	|**0** >	|**0** >	|**1** >
|**1** >	|**1** >	|**1** >	|**0** >	|**0** >

$$|F2>=|A'>.|C'>+|A>.|B'>$$

Consider the following **quantum logic functions** using a quantum PAL:

$$|F1>=|A>.|B>+|B>.|C>+|A>.|C>$$
$$|F2>=|A'>.|C'>+|A>.|B'>$$

The given two functions are in sum-of-products form. The number of product terms present in the given quantum logic functions $|F1>$, $|F2>$ are three and two, respectively.

Six programmable quantum AND operations and two fixed quantum OR operations are required for producing those two functions. But, it is required to perform an extra two quantum OR operations as three product terms for each function are required to do OR operation. The corresponding quantum **PAL** is shown in Fig. 5.7.

Consider the realization of the quantum logic expression $|F1>=|A>.|B>+|B>.|C>+|A>.|C>$; $|F2>=|A'>.|C'>+|A>.|B'>$ using quantum programmable logic array.

For the given problem, there are three inputs ($|A>$, $|B>$, $|C>$) and two outputs ($|F1>$ and $|F2>$). The complement of three inputs is obtained through quantum NOT operation. Thus, the realization has six input lines (input with its complement).

The given first expression has three product terms and the second expression has two product terms. But as the OR operation is fixed the fuses are placed in the corresponding literals to obtain the product terms.

5.3.3 *Working Principle*

According to truth table as shown in Table 5.2 of quantum PAL, it is necessary to do the following operations to perform the desired output qubits:

1. For input qubits $|A>, |B>, |C>=|0>, |0>, |0>$ function $|F2>$ produces $|1>$.
2. For input qubits $|A>, |B>, |C>=|0>, |0>, |1>$ function $|F1>$ produces $|1>$.
3. For input qubits $|A>, |B>, |C>=|0>, |1>, |0>$ function $|F2>$ produces $|1>$.
4. For input qubits $|A>, |B>, |C>=|0>, |1>, |1>$ function $|F1>$ produces $|1>$.
5. For input qubits $|A>, |B>, |C>=|1>, |0>, |0>$ function $|F2>$ produces $|1>$.
6. For input qubits $|A>, |B>, |C>=|1>, |0>, |1>$ functions $|F1>$ and $|F2>$ produce $|1>$.
7. For input qubits $|A>, |B>, |C>=|1>, |1>, |0>$ function $|F1>$ produces $|1>$.
8. For input qubits $|A>, |B>, |C>=|1>, |1>, |1>$ function $|F1>$ produces $|1>$.

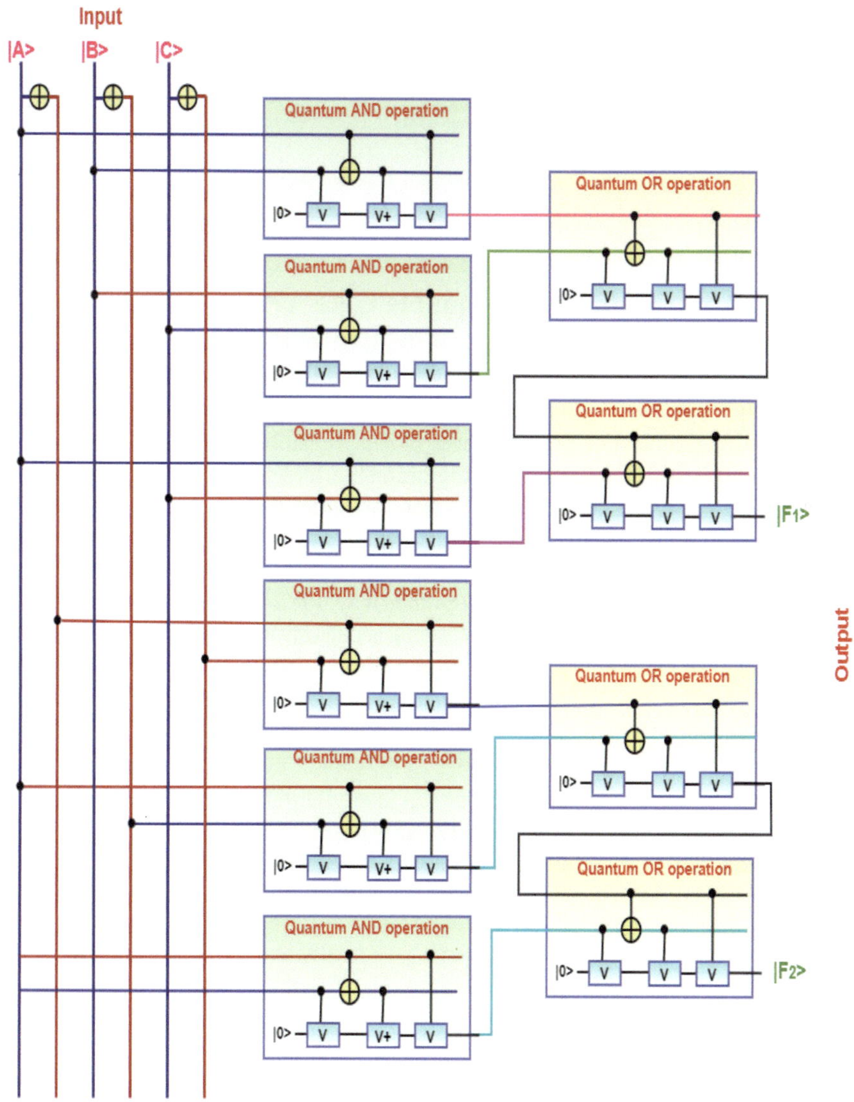

Fig. 5.7 Circuit architecture of quantum PAL

5.3.4 Advantages

The PLA is more flexible than a PAL. PLA is costlier as compared to the PAL. A number of functions provided by PLA are more relatively useful because it enables the programming of the OR plane also. PAL works faster while PLA is slower comparatively. Besides these, the PAL has the following benefits:

1. Highly efficient.
2. Low production cost as compared to PLA.
3. Highly secure.
4. High reliability.
5. Low power is required for working.
6. More flexible to design.

5.4 Quantum Field Programmable Gate Arrays

A Field Programmable Gate Array (FPGA) is a Programmable Logic Device (PLD). It is a digital integrated circuit that contains configurable logic blocks along with programmable interconnection between these blocks. It can be configured by end-users to implement specific applications. Programmable interconnections are to be had for customers or designers to carry out given functions without difficulty. A typical model FPGA chip is shown in Fig. 5.8. There are I/O blocks, which are designed and numbered consistent with function. There are CLBs (configurable logic blocks) for implementing different functions. In other way, FPGAs are playing an increasingly important role in the development and acceleration of quantum computing, primarily through their ability to simulate and emulate quantum circuits. While not a direct quantum processor, FPGAs offer a programmable hardware platform that can be adapted to execute specific quantum algorithms or simulate aspects of quantum behavior. This flexibility allows researchers to quickly prototype and test quantum algorithms before committing to specialized quantum hardware. As quantum computing technology matures, FPGAs are likely to play an even more important role in quantum hardware development and testing. Researchers are exploring new FPGA architectures and programming techniques to further optimize their performance for quantum computing applications. The development of specialized quantum-aware FPGAs is also a promising area of research, potentially leading to more efficient and powerful quantum computing systems.

5.4.1 Block Diagram

A basic FPGA architecture consists of thousands of fundamental elements called configurable logic blocks (CLBs) surrounded by a system of programmable interconnects, called a fabric that routes signals between CLBs. FPGA configurable logic block is shown in Fig. 5.9.

On an FPGA, a customizable logic block (CLB) is the most basic repeating logic block. Hundreds of such logic blocks are accessible on the FPGA, all of which are coupled by routing resources. These logic blocks are used to build sequential and combinational logic.

There are three essential CLB components:

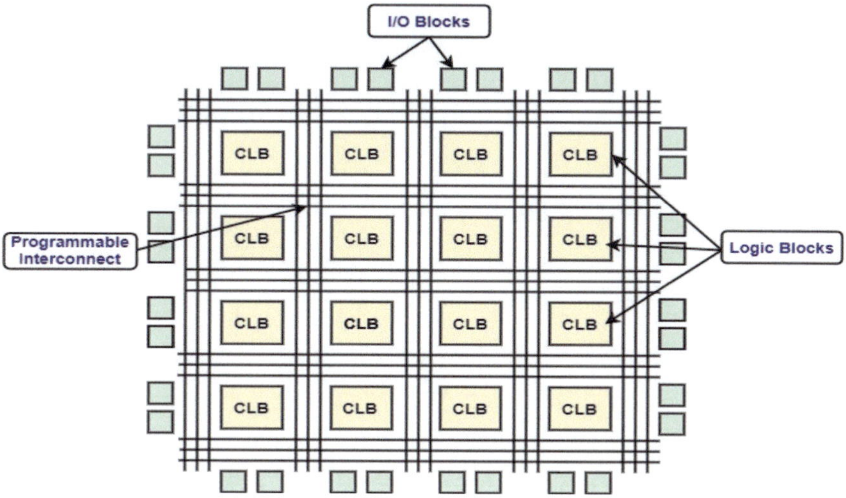

Fig. 5.8 FPGA circuit diagram

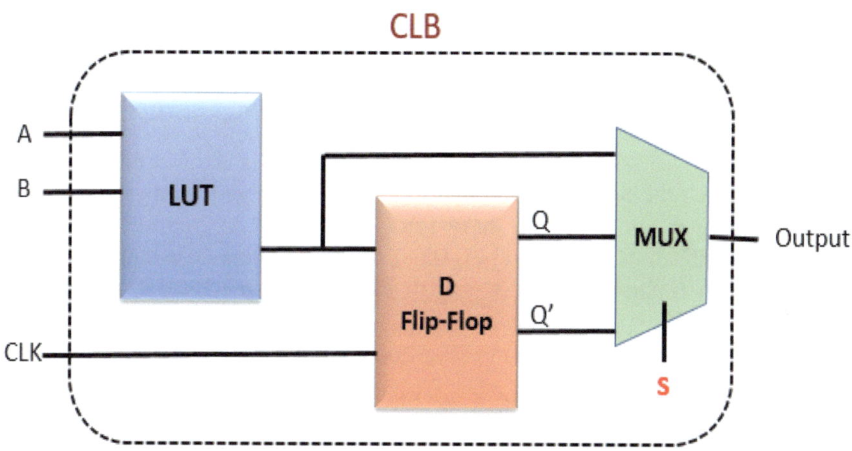

Fig. 5.9 FPGA configurable logic block

1. Flip-flops.
2. Look-up tables (LUTs).
3. Multiplexers.

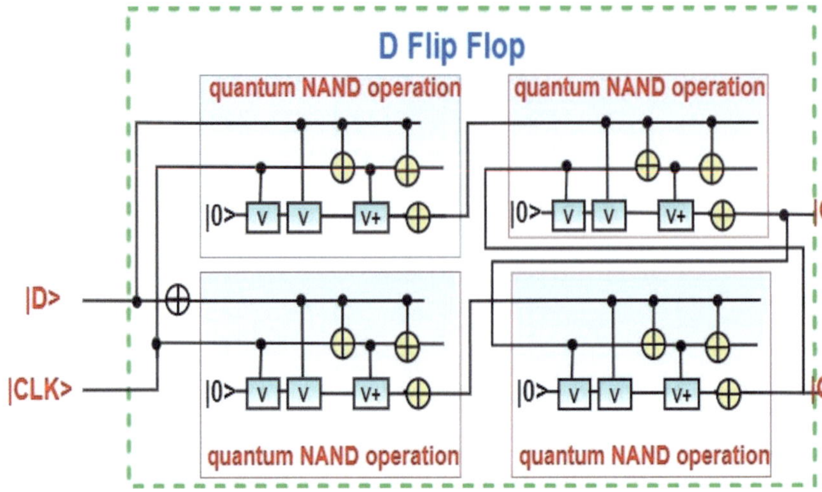

Fig. 5.10 Quantum D flip-flop

5.4.2 Design Architecture of Basic Components

Flip-Flop: The most important component of the timed flip-flops is the D flip-flop. The S and R inputs become complements of each other when an inverter (NOT gate) is added between the Set and Reset inputs, guaranteeing that the two inputs S and R are never equal (0 or 1) at the same time, allowing us to control the toggling action of the flip-flop with only one D (Data) input. Then this Data input is labeled "D" and is used in place of the "Set" signal, and the inverter is used to generate the complementary "Reset" input thereby making a level-sensitive D-type flip-flop from a level-sensitive SR-latch. Quantum D flip-flop is shown in Fig. 5.10.

Look-up Table (LUT): The core of the FPGA is an LUT. It contains all of the design's logically possible outputs. The digital designer programs the LUT to solve a quantum logic equation. The look-up table must be able to store all conceivable combinations of quantum logic expressions. The structure of an LUT is depicted in Fig. 5.11.

Quantum Multiplexer: A multiplexer (MUX) is a device that can accept multiple input signals and synthesize a single output signal for each input signal in a recoverable way. It's also an integrated system with a set of data inputs and a single output. Figure 5.12 depicts a quantum 2-to-1 MUX.

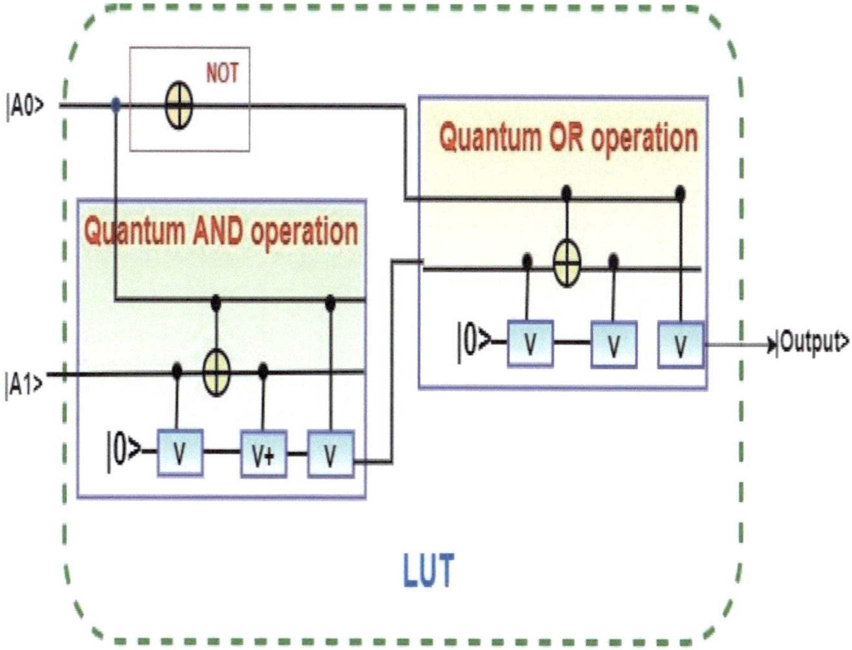

Fig. 5.11 Two inputs look-up table

5.4.3 Circuit Architecture

FPGA logic block is designed by connecting flip-flops, look-up tables (LUT), and multiplexers. Here a simple quantum FPGA logic block is designed by using a D flip-flop, a look-up table (LUT), and a multiplexer. A simple two-input LUT consists of one quantum AND, one quantum NOT, and one quantum OR operation. The output of the LUT will go through the D flip-flop and multiplexer as input. The D flip-flop is a sequential circuit that consists of four quantum NAND operations and one quantum NOT operation. The output of the quantum D flip-flop will go through the multiplexer as input. A quantum 2-to-1 MUX is designed using two quantum AND operations, one quantum NOT operation, and one quantum OR operation. The multiplexer generates the desired output for the FPGA logic block. Circuit architecture of quantum FPGA is shown in Fig. 5.13.

5.4.4 Working Principle

Two inputs $|A0>$ and $|A1>$ firstly go through the look-up tables (LUT). In LUT, then $|A0>$ will go through quantum NOT operation. $|A0>$ and $|A1>$ will go

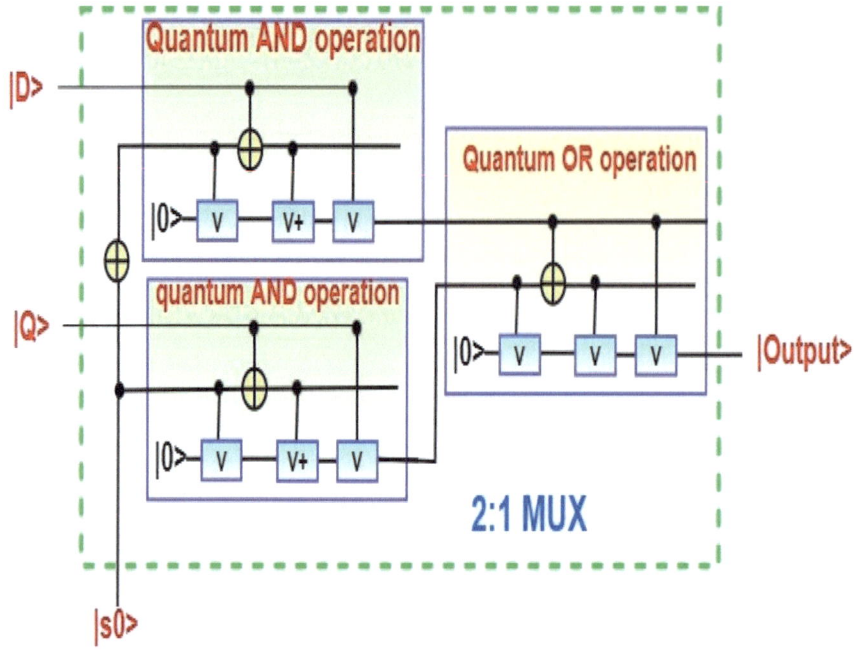

Fig. 5.12 Quantum 2-to-1 MUX

through quantum AND operation. Finally, the outputs of the quantum NOT and AND operations will go through the quantum OR operation and produce the LUT output.

Two inputs, one is the output of LUT and another is clock input, will go through the D flip-flop. The D flip-flop uses one quantum NOT operation and four quantum NAND operations. The D flip-flop transfers the LUT output, if the CLK input sequence is $|1>$. If the CLK input is $|0>$, one of the inputs to each of the last two quantum NAND operations will be $|1>$, thus the output of the D flip-flop remains unchanged regardless of the values of the LUT output.

Two quantum AND operations, one quantum NOT operation, and one quantum OR operation are used in 2-to-1 MUX. S_0' (complement of S_0) is created by applying the quantum NOT operation for input S_0. The NOT gate and LUT output will go through as inputs for the first quantum AND operation. S_0 and D flip-flop (DFF) output will go through as inputs for the second quantum AND operation. The output of these two quantum AND operations will go through the quantum OR operation to generate the FPGA logical block output.

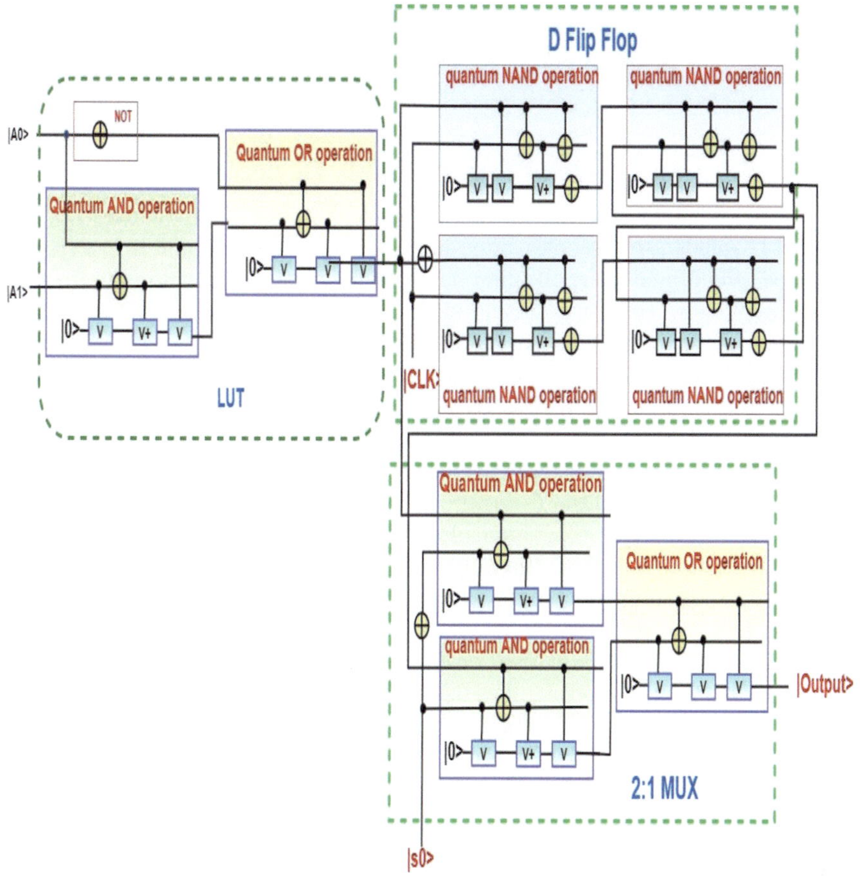

Fig. 5.13 Circuit architecture of quantum FPGA

5.4.5 Applications

A quantum FPGA (QFPGA) architecture is presented by combining the advantages of the measurement-based quantum computation. QFPGA consists of quantum logic blocks (QLBs). The QLB is used to realize a small quantum logic while the QRC is to combine them properly for larger logic realization. There are two types of buses in QFPGA, the local bus in the CLB. QFPGA: There are a two main different types of buses which are as follows:

1. Register reads and writes; and
2. Data transfer or streaming

The register reads and writes is normally for reading or writing a register, often 32 bits. This is normally includes a clock, a ready signal, and a read or write bit.

The data transfer often has a data valid signal when the data bus is valid, and there is feedback if the receiver is not ready for more data.

5.5 Quantum Complex Programmable Devices

The CPLD stands for "Complex programmable logic devices," and it is a type of integrated circuit used by application designers to construct digital hardware such as mobile phones. These can handle more design complexity than SPLDs (simple programmable logic devices), but they have simpler logic than FPGAs (field-programmable gate arrays). CPLDs have a large number of logic blocks, each with 8–16 macrocells. All of the macrocells in a logic block are fully linked because each logic block performs a specific function. These blocks may or may not be linked to one another, depending on the application. Figure 5.14 shows the block diagram of connected logic block PLDs.

Most CPLDs (complex programmable logic devices) have macrocells with a sum of logic functions, a D FF (D flip-flop), and a MUX. Depending on the chip, the combinatorial logic function supports from 4 to 16 product terms with inclusive fan-in. CPLDs also differ in terms of shift registers and logic gates. Due to this reason, CPLDs with a huge number of logic gates may be used instead of FPGAs. Another CPLD specification signifies the number of product terms that a macrocell

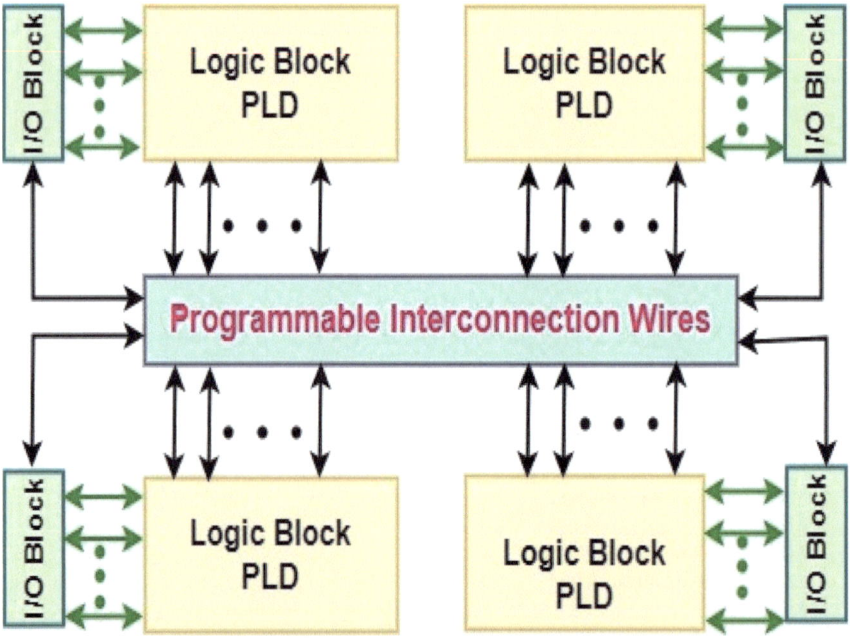

Fig. 5.14 Block diagram of connected logic block PLDs

can accomplish. Product terms are the product of digital signals that execute a specific logic function. CPLDs are available in several IC package forms and logic families. CPLDs also differ in terms of supply voltage, operating current, standby current, and power dissipation. CPLDs are not directly used in quantum computing hardware, but they are crucial for controlling and interacting with quantum hardware at various stages. While not the core quantum processing unit, CPLDs help manage qubits, implement control circuits, and interface with other parts of the system. In essence, CPLDs play a supporting role in quantum computing, enabling the precise control and interaction with quantum hardware, which is crucial for realizing the full potential of quantum algorithms and experiments.

In terms of complexity, CPLD (complex programmable logic device) lies in between SPLD (simple programmable logic device) and FPGA, and thus inherits features from both these devices. CPLDs are more complex than SPLDs but less complex than FPGAs.

5.5.1 Block Diagram

A set of programmable functional blocks (FBs) constructs a sophisticated programmable logic device. Macrocells containing a sum of logic functions, a D FF (D flip-flop), and a MUX which are found in most CPLDs (complex programmable logic devices). Quantum CPLD configurable logic block is shown in Fig. 5.15. Logic block of PLD (Programmable Logic Device) is shown in Fig. 5.16.

On a CPLD, a functional block (FB) is the most basic repeating logic block. The linkages between the function blocks are programmable. The programmable FB resembles a logic gate array, with an array of AND gates that can be programmed and OR gates that are stable. Interconnections between function blocks are made using a switch matrix. Furthermore, a CPLD's switch matrix may or may not be entirely linked. A normal PAL (Programmable Array Logic) device has a few hundred logic gates of complexity, but a CPLD has tens of thousands of logic gates of complexity. The CPLDs have predictable timing characteristics and hence are suitable for critical control applications and other applications where a high-performance level is required. Further, due to low power consumption and low cost, CPLDs are mostly used for battery-operated portable applications such as mobile phones, digital assistants, etc. The functional block is shown in Fig. 5.17.

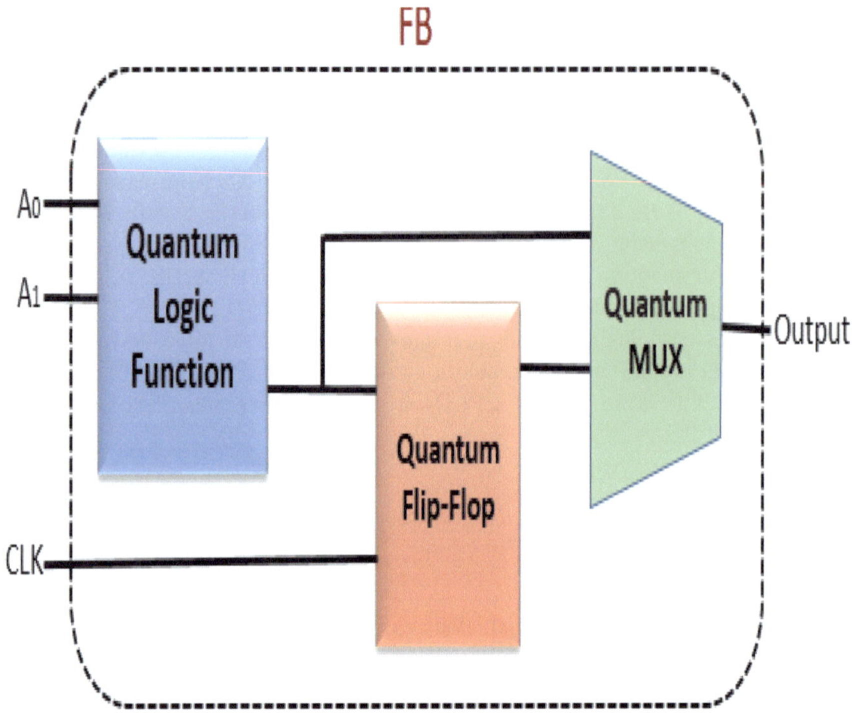

Fig. 5.15 Quantum CPLD configurable logic block

There are three essential components of FBs which are as follows:

1. Quantum flip-flops.
2. Quantum logic function.
3. Quantum multiplexers.

5.5.2 Circuit Architecture

CPLD logic block is designed by connecting flip-flops, PAL (logic function), and multiplexers. Here a simple quantum CPLD logic block is designed by using a D flip-flop, logic function, and a multiplexer. A simple logic function consists of a quantum AND and one quantum OR operation. The output of the logic function will be XOR with zero. The output of the XOR will go through the D flip-flop and multiplexer as input. The D flip-flop is a sequential circuit that consists of four quantum NAND operations and one quantum NOT operation. The output of the quantum D flip-flop will go through the multiplexer as input. A quantum 2-to-1 MUX is designed using two quantum AND operations, one quantum NOT operation, and one quantum OR

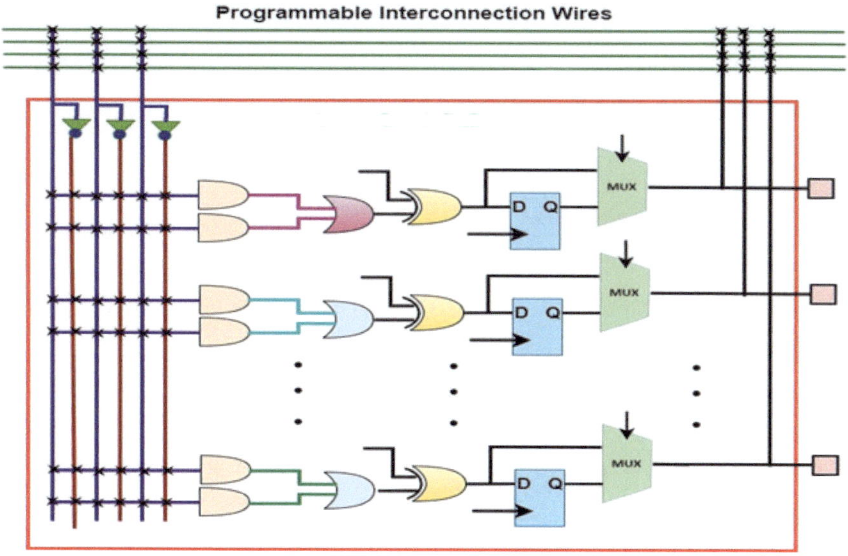

Fig. 5.16 Logic block of a PLD

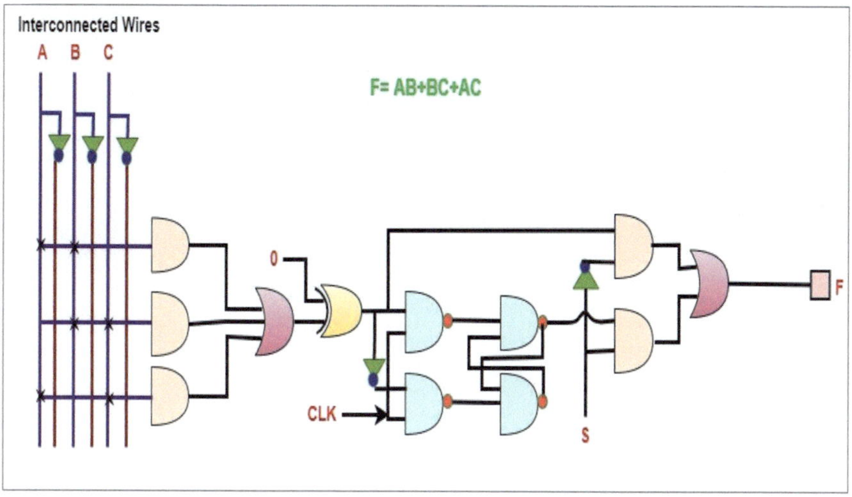

Fig. 5.17 Functional block

operation. The multiplexer generates the desired output for the CPLD functional block. The circuit architecture of a quantum CPLD is shown in Fig. 5.18.

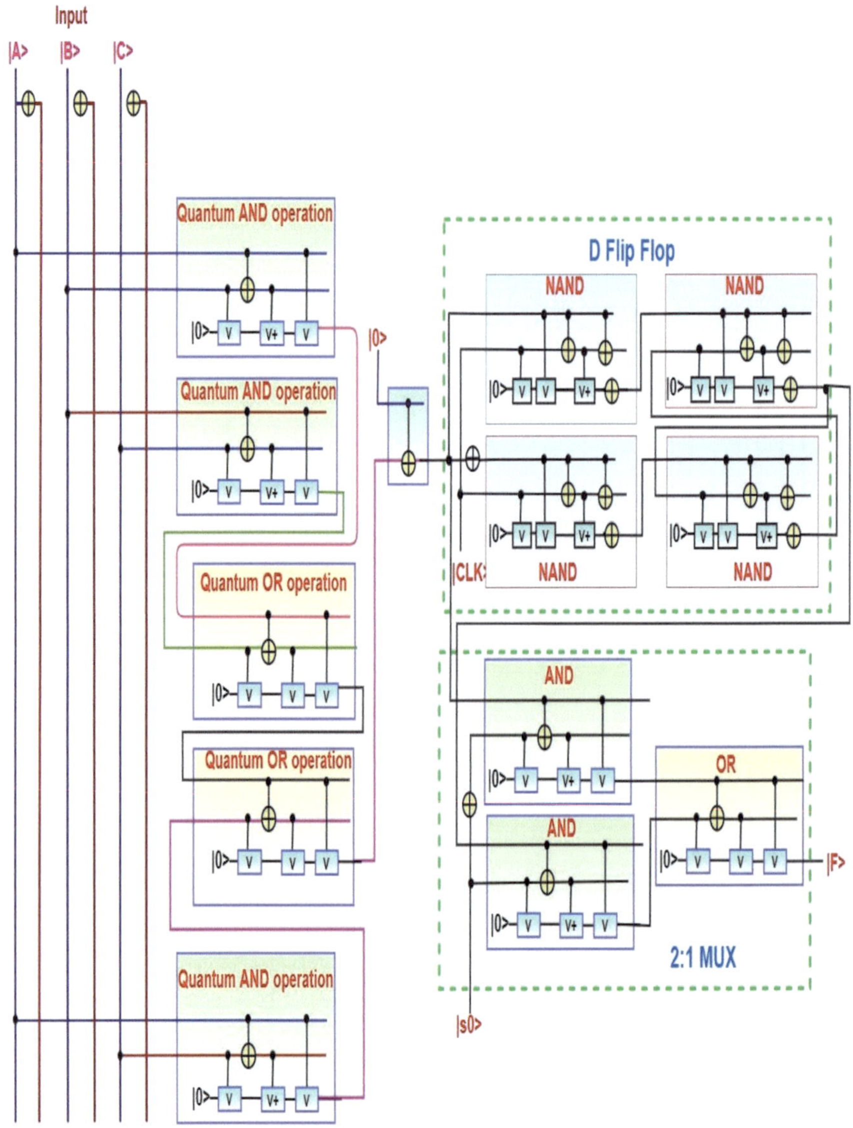

Fig. 5.18 Circuit architecture of quantum CPLD

5.5.3 Working Principle

Three inputs |A >, |B >, and |C > firstly go thought the logic function (F = AB + BC + AC). Then, the logic function is implemented using three quantum AND operations and two quantum OR operations. Outputs of the logic function will go through the quantum XOR (XOR with |0 >) operation and produce the output.

Two inputs, where one is the output of XOR and another is the clock input, which will go through the D flip-flop. The D flip-flop uses one quantum NOT operation and four quantum NAND operations. The D flip-flop transfers the logic function output, if the CLK input sequence is |1 >. If the CLK input is |0 >, one of the inputs to each of the last two quantum NAND operations will be |1 >, thus the output of the D flip-flop remains unchanged regardless of the values of the XOR output.

Two quantum AND operations, one quantum NOT operation, and one quantum OR operation are used in 2-to-1 MUX. S_0' is created by applying the quantum NOT operation for input S_0. The NOT gate and logic function output will go through as inputs for the first quantum AND operation. S_0 and D flip-flop (DFF) output will go through as inputs for the second quantum AND operation. The output of these two quantum AND operations will go through the quantum OR operation to generate the CPLD functional block output.

5.5.4 Applications

CPLDs find their applications in much low-to-medium complexity digital control and signal processing circuits. Some of the important applications are listed below.

1. CPLDs can be used as boot loaders for FPGAs and other programmable systems.
2. CPLDs are often used as address decoders and custom state machines in digital systems.
3. Due to their small size and low power consumption, CPLDs are ideal for use in portable and handheld digital devices.
4. CPLDs are also used in safety-critical control applications.
5. Complex programmable logic devices are ideal for high-performance, critical control applications.
6. CPLD can be used in digital designs to perform the functions of the boot loader.
7. CPLD is used for loading the configuration data of a field-programmable gate array from non-volatile memory.
8. Generally, these are used in small design applications like address decoding.
9. CPLDs are frequently used in many applications like in cost-sensitive, battery-operated portable devices due to their low size and usage of low power.

5.6 Summary

This chapter has presented the overall idea of PLD in quantum computing. Individual applications and working principles of programmable logic devices such as Programmable Logic Array (PLA), Programmable Array Logic (PAL), Field

Programmable Gate Arrays (FPGA), and Complex Programmable Logic Devices (CPLD) in quantum computing are discussed in detail in this chapter. In a quantum computing circuit, the excessive heat is produced from the circuit which can be very harmful and can destroy the whole circuit and ruin the computation. So, there must be a cooler attached to the circuit to consume the extra heat.

Chapter 6
Programmable Devices in Quantum-DNA Computing

6.1 Introduction

This is a new computing mechanism the history where quantum computing and DNA computing combinely work. The programmable logic devices are the same here but the first portion will maintain the quantum computing technology and the last portion will maintain the DNA computing technology. By combining these two, the benefits of both could be captured. The input portion will be quantum part and the output portion will be DNA part. So, the inputs will be qubits and the outputs will be DNA molecular sequences.

Quantum-DNA Programmable logic device needs some quantum operations and some DNA operations. This book is the first move where this new technology (Quantum biocomputing) has been introduced for the first time. In the coming future, more implementation and development will come through research. Researchers are so excited with these quantum and DNA technology. Programmable devices in quantum biological computing or quantum-DNA computing or quantum biocomputing involve using quantum technologies like qubits and quantum gates to design and manipulate biological systems, such as DNA, to perform computations. This field combines quantum computing principles with biological materials to create powerful and potentially transformative computational tools. The programmable logic devices like PLA, PAL, FPGA, and CPLD in Quantum-DNA computing are the main concern of this chapter.

6.2 Quantum-DNA Programmable Logic Array

Programmable Logic Array is a type of PLD, which has both programmable AND array and programmable OR array. Quantum-DNA Programmable Logic Array (PLA) is a fixed architecture logic device with programmable Quantum AND oper-

H. Md. Hasan Babu, *Quantum Biocomputing in Quantum Biology Volume II*,
https://doi.org/10.1007/978-981-97-5349-9_6

ations followed by programmable DNA OR operations. PLA is a type of programmable logic device used to build a reconfigurable digital circuit. In other way, a PLA in quantum biocomputing refers to the application of PLA technology within the context of using quantum mechanics to solve biological problems. PLAs are programmable devices that can implement various combinational logic circuits, offering a flexible and reconfigurable approach to designing digital circuits. In quantum biocomputing, this can be used to model and simulate biological systems, potentially leading to new insights and solutions in fields like drug discovery or genetic engineering. In essence, PLAs offer a powerful and versatile tool for exploring the intersection of quantum mechanics and biology, potentially leading to new discoveries and solutions in various fields.

6.2.1 Block Diagram

Quantum-DNA PLA is a programmable logic device that has both Programmable Quantum AND array and Programmable DNA OR array. Hence, it is the most flexible PLD. The block diagram of Quantum-DNA PLA is shown in Fig. 6.1.

Here, the inputs of Quantum AND operations are programmable. That means each Quantum AND operation has both normal and complemented inputs of variables.

Quantum computing provides faster computation in logic devices than DNA computing. So it is required to use a storage device (CACHE MEMORY) for making a balance between these two computing systems. DNA computing gives more storage capacity, as a result, the output operation of the logic devices will perform in DNA computing. It is required to transform qubit into DNA molecule and, for this, a

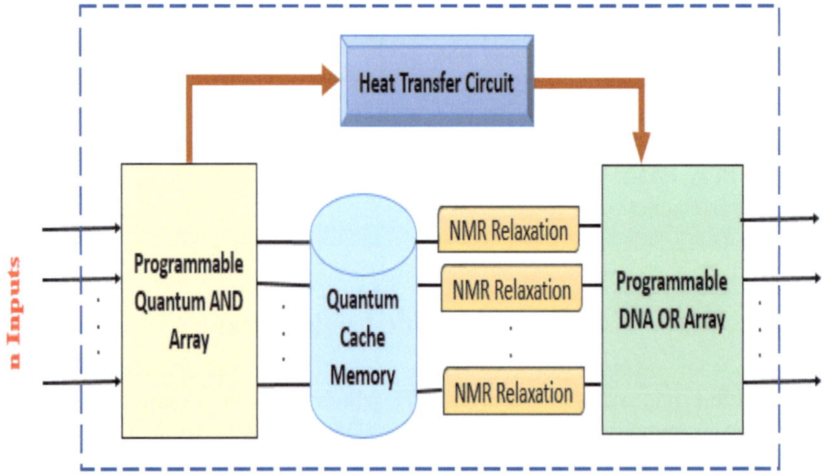

Fig. 6.1 Block diagram of quantum-DNA PLA

transformation process is used by which an excited magnetic state returns to its equilibrium distribution. Here, the inputs of DNA OR operations are also programmable. So, any number of required product terms can be programmed, since all the outputs of DNA AND operations are applied as inputs to each DNA OR operation. Therefore, the outputs of DNA PAL will be in the form of the sum of products form. Quantum Computing produces more heat in circuits which can be used in DNA computing to perform DNA logic operations. For this, it is required to use a heat transfer circuit between quantum operations and DNA operations. The Quantum-DNA PLA implementation methods will improve the result of any quantum and DNA device.

6.2.2 Circuit Architecture

Quantum-DNA PLA is a programmable quantum and DNA device that has both Programmable Quantum AND array and Programmable DNA OR array.
The following logic functions are used to construct a Quantum-DNA PLA:

F1 = AB + A'C
F2 = A'C' + AB'
F3 = A'B + AB'.

The given three functions are in sum of products form. The number of product terms present in the given logic functions F1, F2, and F3 are two.

A.B's products are used in both functions F2 and F3. So, five programmable Quantum AND gates and three programmable DNA OR gates are required for producing those three functions. For the given problem, there are three inputs (A, B, C) and three outputs (F1, F2, F3). The complement of three inputs is obtained through quantum NOT gates. Thus the realization has six input lines (input with its complement). The given expression has six product terms and so the fuses are placed in the corresponding literals to obtain the product terms. Five quantum AND operations are designed using quantum computing**.** Quantum computing provides fast computation in logic devices than DNA computing. So, it is required to use a cache memory is used to store the quantum qubits. It is required to transform qubit into DNA molecule and for this, NMR Relaxation is used by which an excited magnetic state returns to its equilibrium distribution. Here, the inputs of DNA OR operations are also programmable. So, any number can be programmed for required product terms, since all the outputs of Quantum AND operations are applied as inputs to each DNA OR operation. Therefore, the outputs of Quantum-DNA PAL will be in the form of the sum of products form. The circuit architecture of Quantum-DNA PLA for functions F1, F2, and F3 is shown in Fig. 6.2 and the truth table of quantum-DNA PLA for functions F1, F2, and F3 is given in Table 6.1.

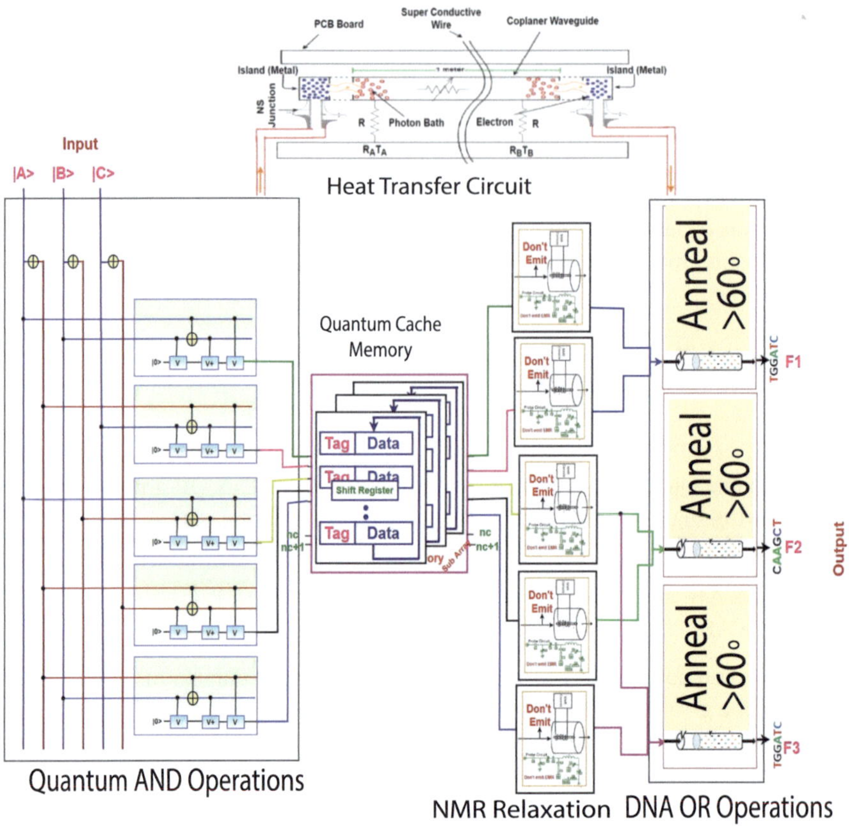

Fig. 6.2 Quantum-DNA PLA for functions F1, F2, and F3

Table 6.1 Truth table of quantum-DNA PLA for functions F1, F2, and F3

|A>	|B>	|C>	|F1>	|F2>	|F3>
|0>	**|0>**	**|0>**	TGGATC	ACCTAG	TGGATC
|0>	**|0>**	**|1>**	ACCTAG	TGGATC	TGGATC
|0>	**|1>**	**|0>**	TGGATC	ACCTAG	ACCTAG
|0>	**|1>**	**|1>**	ACCTAG	TGGATC	ACCTAG
|1>	**|0>**	**|0>**	TGGATC	ACCTAG	ACCTAG
|1>	**|0>**	**|1>**	TGGATC	ACCTAG	ACCTAG
|1>	**|1>**	**|0>**	ACCTAG	TGGATC	TGGATC
|1>	**|1>**	**|1>**	ACCTAG	TGGATC	TGGATC

6.2.3 *Working Principle*

According to truth table (Table 6.1) of Quantum-DNA PLA, it is necessary to do the following operations to get desired output qubits:

1. For input qubits |A>, |B>, |C> =|0>, |0>, |0> function F2 produces ACCTAG.
2. For input qubits |A>, |B>, |C> =|0>, |0>, |1> function F1 produces ACCTAG.
3. For input qubits |A>, |B>, |C> =|0>, |1>, |0> functions F2 and F3 produce ACCTAG.
4. For input qubits |A>, |B>, |C> =|0>, |1>, |1> functions F1 and F3 produce ACCTAG.
5. For input qubits |A>, |B>, |C> =|1>, |0>, |0> functions F2 and F3 produce ACCTAG.
6. For input qubits |A>, |B>, |C> =|1>, |0>, |1> functions F2 and F3 produce ACCTAG.
7. For input qubits |A>, |B>, |C> =|1>, |1>, |0> function F1 produces ACCTAG.
8. For input qubits |A>, |B>, |C> =|1>, |1>, |1> function F1 produces ACCTAG.

6.3 Quantum-DNA Programmable Array Logic

Programmable Array Logic (PAL) is a logic device, which has a programmable AND array and a fixed OR array. It is used to realize a logic function. In this PLD, only AND gates are programmable. Hence it is easier to work with PAL. The product terms can be programmed through the fuse link. It means the user can decide the connection between the inputs and the AND gates. If a particular input line is to be connected to the AND gate, then the fuse link must be placed at the interconnection. The AND gate outputs are then fed as an input to the fixed OR gate. Depending upon the required function, the output line of the AND gate is connected to the corresponding input of the OR gate. On other way, quantum biocomputing explores the use of quantum systems to model and simulate biological processes, while programmable array logic (PAL) is a type of programmable logic device used in classical computing. While both involve manipulating data to solve problems, quantum biocomputing leverages quantum phenomena, like superposition and entanglement, whereas PALs implement logic circuits with programmable AND and OR gates. Here, superposition, entanglement, and other quantum phenomena are exploited to represent and manipulate biological data. In essence, quantum biocomputing focuses on using quantum mechanics to understand and solve biological problems, while programmable array logic provides a tool for building custom logic circuits in classical computing.

6.3.1 *Block Diagram*

Quantum-DNA PAL is a programmable logic device that has both Programmable Quantum AND array and fixed DNA OR array. Hence, it is the most flexible PLD. The block diagram of Quantum-DNA PAL is shown in Fig. 6.3.

Here, the inputs of quantum AND operations are programmable. That means each quantum AND operation has both normal and complemented inputs of variables. So, based on the requirement, any of those inputs can be programmed. So, the required product terms can be generated by using these quantum AND operations.

Quantum computing provides faster computation in logic devices than DNA computing. So, it is required to use a storage device (CACHE MEMORY) to make a balance between these two computing systems. DNA computing gives more storage capacity, as a result, the output operation of the logic devices will perform in DNA computing. It is required to transform qubit into DNA molecule and for this, it is required to use a transformation process by which an excited magnetic state returns to its equilibrium distribution. Here, the inputs of DNA OR operations are fixed. Quantum computing produces more heat in circuits which can be used in DNA computing to perform DNA logic operations. For this, it is required to use a heat transfer circuit between quantum operations and DNA operations. The quantum-DNA PAL implementation methods will improve the result of any logic device.

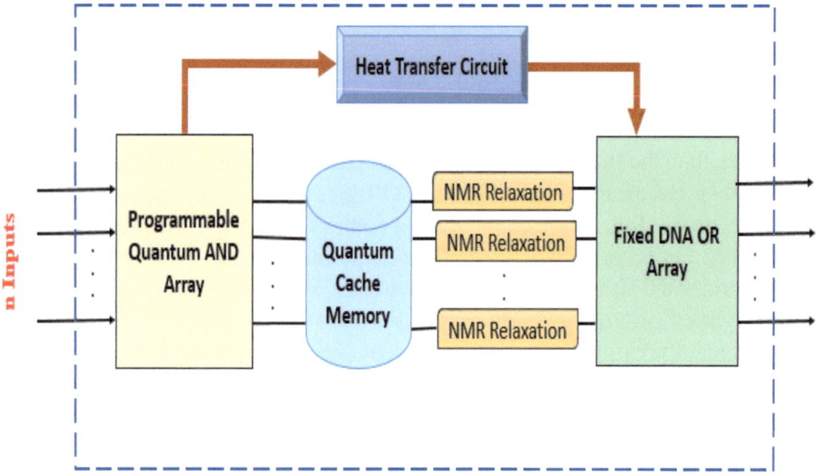

Fig. 6.3 Block diagram of quantum-DNA PAL

6.3.2 Circuit Architecture

Quantum-DNA PAL is a programmable logic device that has Programmable Quantum AND array & fixed DNA OR array.

Consider the following **quantum logic functions** using Quantum PAL.

F1 = AB + BC + AC
F2 = A'C' + AB'.

The given two functions are in sum of products form. The number of product terms present in the given Boolean functions F1 and F2 are three and two, respectively. The truth table of Quantum-DNA PAL for functions F1 and F2 is given in Table 6.2.

Six programmable Quantum AND operations and two fixed DNA OR operations are required for producing those two functions. But, it is required to perform an extra two Quantum OR operations as three product terms for each function are required to do OR operation. The corresponding Quantum PAL is shown in the following figure. Consider the realization of the Boolean expression F1 = AB + BC + AC F2 = A'C' + AB' using Quantum Programmable Array Logic.

For the given problem, there are three inputs (A, B, C) and two outputs (F1 and F2). The complement of three inputs is obtained through Quantum NOT operation. Thus, the realization has six input lines (input with its complement).

The given first expression has three product terms and the second expression has two product terms. But, as the DNA OR operation is fixed, the fuses are placed in the corresponding literal to obtain the product terms. Six quantum AND operations are designed using quantum computing. Quantum computing provides fast computation in logic devices than DNA computing. So, it is required to use a cache memory is used to store the quantum qubits. It is required to transform qubit into DNA molecule. For this, NMR Relaxation is used by which an excited magnetic state returns to its equilibrium distribution. Here, the inputs of DNA OR operations are fixed. All the outputs of Quantum AND operations are applied as inputs to each DNA OR operation. Therefore, the outputs of Quantum-DNA PAL will be in the form of the

Table 6.2 Truth table of quantum-DNA PAL for functions F1 and F2

$\vert A\rangle$	$\vert B\rangle$	$\vert C\rangle$	$\vert F1\rangle$	$\vert F2\rangle$
$\vert 0\rangle$	$\vert 0\rangle$	$\vert 0\rangle$	TGGATC	ACCTAG
$\vert 0\rangle$	$\vert 0\rangle$	$\vert 1\rangle$	TGGATC	TGGATC
$\vert 0\rangle$	$\vert 1\rangle$	$\vert 0\rangle$	TGGATC	ACCTAG
$\vert 0\rangle$	$\vert 1\rangle$	$\vert 1\rangle$	ACCTAG	TGGATC
$\vert 1\rangle$	$\vert 0\rangle$	$\vert 0\rangle$	TGGATC	ACCTAG
$\vert 1\rangle$	$\vert 0\rangle$	$\vert 1\rangle$	ACCTAG	ACCTAG
$\vert 1\rangle$	$\vert 1\rangle$	$\vert 0\rangle$	ACCTAG	TGGATC
$\vert 1\rangle$	$\vert 1\rangle$	$\vert 1\rangle$	ACCTAG	TGGATC

sum of products form. The architecture of Quantum-DNA PAL for functions F1 and F2 is given in Fig. 6.4.

6.3.3 *Working Principle*

According to the truth table (Table 6.2) of Quantum-DNA PAL, it is necessary to do the following operations to get the desired output qubits:

1. For input qubits |A>, |B>, |C> =|0>, |0>, and |0> function F2 produces ACCTAG.
2. For input qubits |A>, |B>, |C> =|0>, |0>, and |1> none of function produces ACCTAG.
3. For input qubits |A>, |B>, |C> =|0>, |1>, and |0> function F2 produces ACCTAG.
4. For input qubits |A>, |B>, |C> =|0>, |1>, and |1> function F1 produces ACCTAG.
5. For input qubits |A>, |B>, |C> =|1>, |0>, and |0> function F2 produces ACCTAG.
6. For input qubits |A>, |B>, |C> =|1>, |0>, and |1> functions F1 and F2 produce ACCTAG.
7. For input qubits |A>, |B>, |C> =|1>, |1>, and |0> function F1 produces ACCTAG.
8. For input qubits |A>, |B>, |C> =|1>, |1>, and |1> function F1 produces ACCTAG.

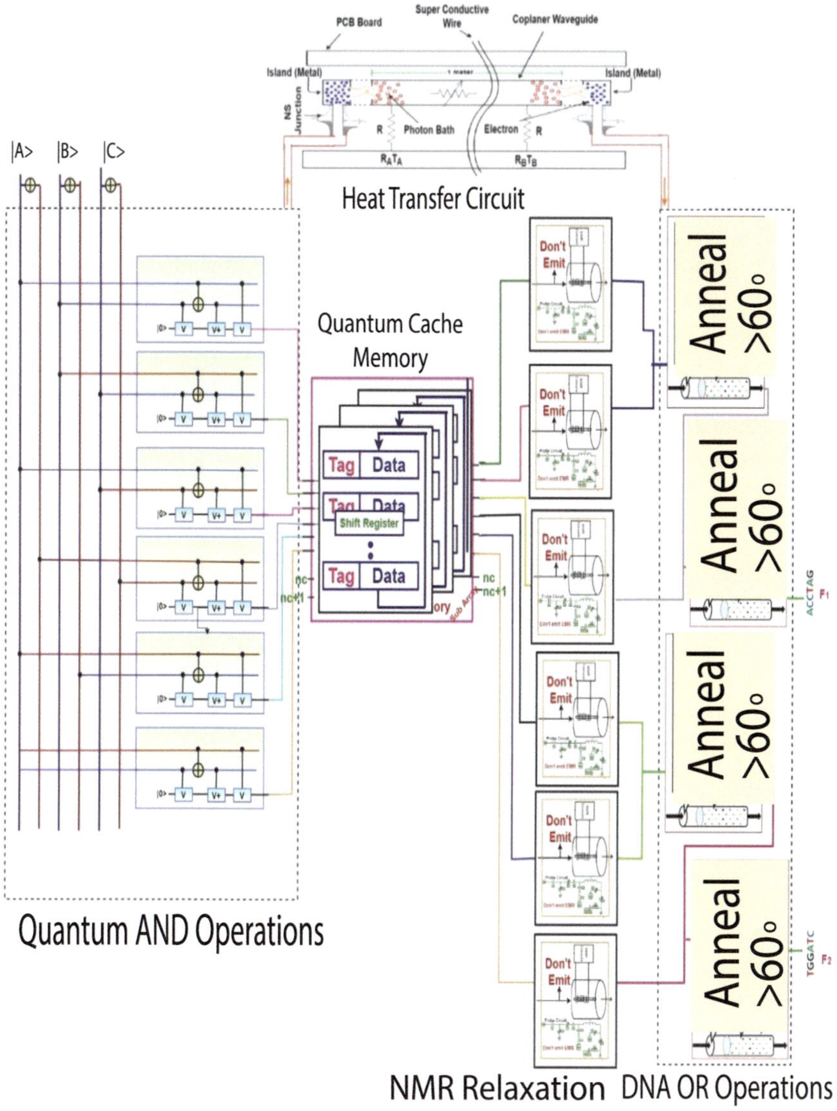

Fig. 6.4 Quantum-DNA PAL for functions F1 and F2

6.4 Quantum-DNA Field Programmable Gate Arrays

Quantum-DNA FPGAs have a hierarchy of reconfigurable interconnects that allow blocks to be linked together to form an array of Quantum-DNA programmable logic blocks. Logic blocks can be configured to execute sophisticated combinatorial operations or to work as simple logic gates such as quantum or DNA AND and XOR. The memory components which might be simple flip-flops or larger memory blocks, are included in most Quantum-DNA FPGA logic blocks. Many Quantum-DNA FPGAs can be reprogrammed to execute various logic tasks, allowing for flexible reconfigurable computing similar to that done in software. In other way, FPGAs are not directly used in quantum biological computing in the way they are used in general-purpose quantum computing. However, they can play a crucial role in supporting and enabling research and development in this field, particularly for simulating and controlling quantum systems. FPGAs offer the flexibility and speed needed for high-performance parallel processing, which is essential for complex quantum algorithms and simulations. In summary, while FPGAs aren't directly part of the "quantum biological computer" itself, they are valuable tools for enabling research, development, and simulation in this field. Their ability to perform high-speed parallel processing and be customized for specific tasks makes them ideal for supporting the complex challenges of quantum biology.

6.4.1 Block Diagram

Quantum computing use Qubits (|0> and |1>) and DNA Computing system use Molecule to represent information. In Quantum-DNA computing FPGA design, the qubit will work as input for the programmable logic devices. Quantum computing provides faster computation in logic devices than DNA computing. So, it is required to use a storage device for making a balance between these two computing systems. DNA computing gives more storage capacity, as a result, the output operation of the logic devices will perform in DNA computing. It is required to transform qubit into DNA molecule. For this, a transformation process needs to be done by which an excited magnetic state returns to its equilibrium distribution. Quantum Computing produces more heat in circuits which can be used in DNA computing to perform DNA logic operations. For this, a heat transfer circuit is needed to use between Quantum and DNA operations. The Quantum-DNA FPGA implementation methods will improve the result of any logic device. A general organization using block diagram is shown in Fig. 6.5.

6.4.2 Circuit Architecture

Quantum-DNA FPGA logic block is designed by connecting Flip-Flops, Look-up Tables (LUT), and Multiplexers. Here a simple Quantum-DNA FPGA logic block is designed by using a D Flip-Flop, a Look-up Table (LUT), and a multiplexer. A two-input LUT consists of one Quantum AND, one Quantum NOT and one Quantum OR operation. The output of the LUT will go through the D Flip-Flop and multiplexer as input. The D Flip-Flop is a sequential circuit that consists of four Quantum NAND

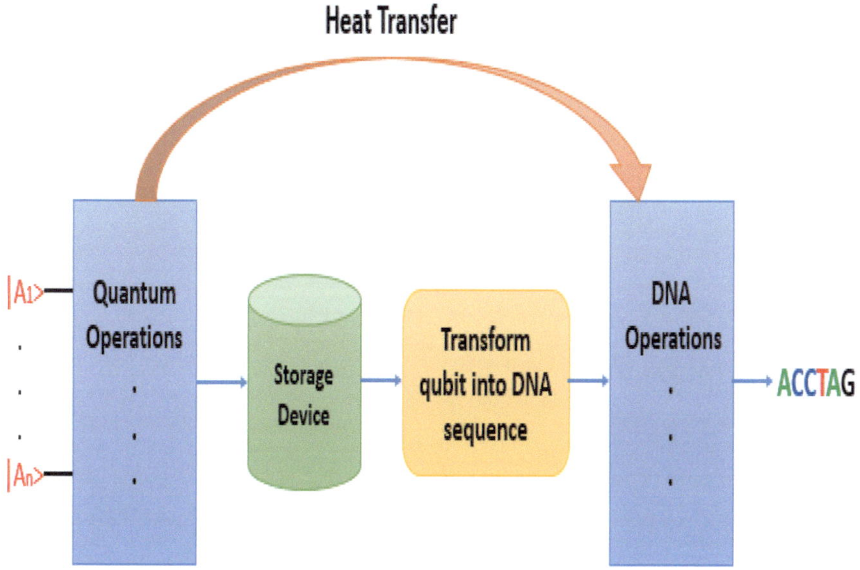

Fig. 6.5 General organization of quantum-DNA circuit

operations and one Quantum NOT operation. The output of the Quantum D Flip-Flop will go through the multiplexer as input. A Quantum 2-to-1 MUX is designed using two Quantum AND operations, one Quantum NOT operation and one DNA OR operation. Outputs of the two Quantum AND operations will be stored in a Quantum cache memory. NMR Relaxation is used to transfer the qubit into the DNA sequence. DNA OR operation performs OR of these two DNA sequences. A heat transfer circuit is used to transfer the extra heat produced by the Quantum circuit to the DNA circuit. The circuit architecture of the Quantum-DNA FPGA logic block is shown in Fig. 6.6.

6.4.3 Working Principle

Two inputs |A0> and |A1> first go through the Look-up Tables (LUT). Then, the LUT uses one Quantum AND operation by applying the Quantum NOT operation to the output of the Quantum NAND operation, one Quantum NOT operation and one Quantum OR operation. In LUT, |A0> will go through Quantum NOT operation. |A0> and |A1> will go through Quantum AND operation. The outputs of the Quantum NOT AND operations will go through the Quantum OR operation and produce the LUT output.

2 input, one is the output of LUT and another is clock input will go through the D flip-flop. The D flip-flop uses one Quantum NOT operation and four Quantum NAND

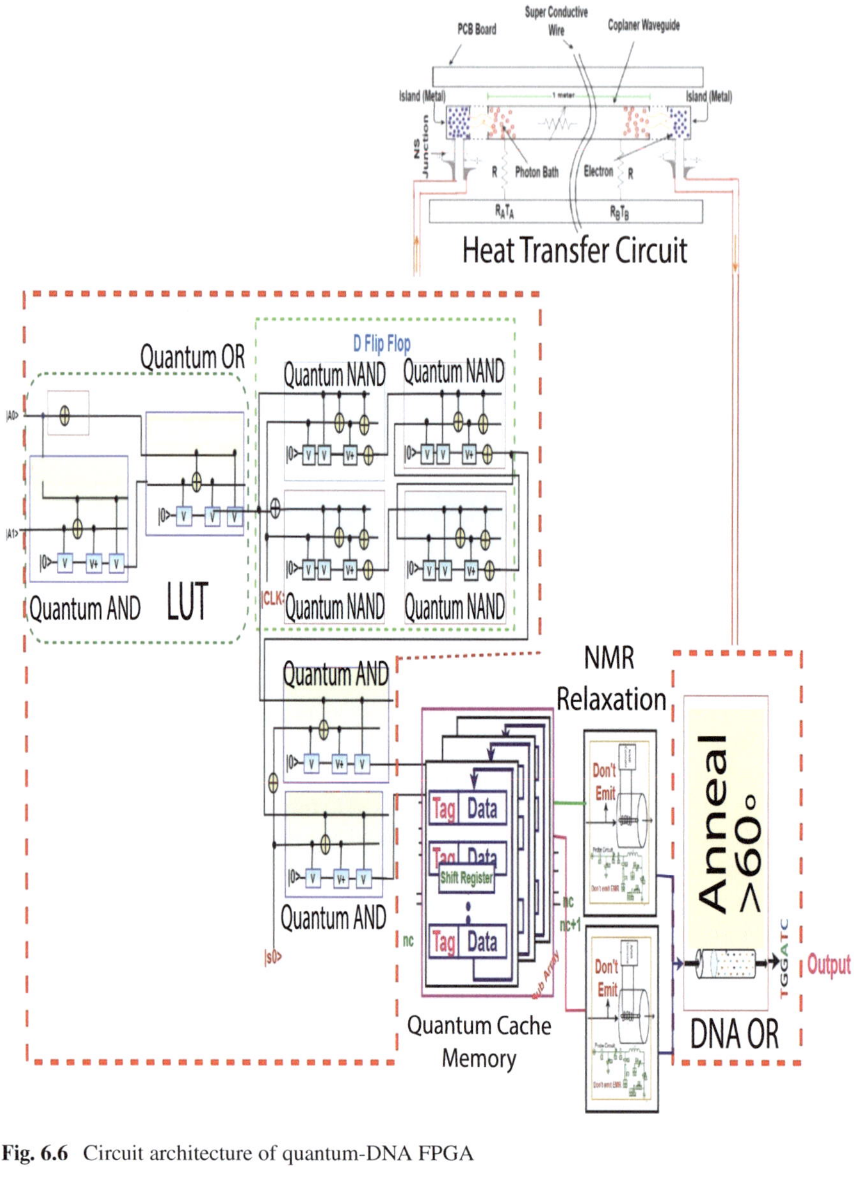

Fig. 6.6 Circuit architecture of quantum-DNA FPGA

operations. The D flip-flop transfers the LUT output, if the CLK input sequence is |1>. If the CLK input is |0>, one of the inputs to each of the last two Quantum NAND operations will be |1>. Thus, the output of the D flip-flop remains unchanged regardless of the values of the LUT output.

Two Quantum AND operations, one Quantum NOT operation, and one DNA OR operation are used in 2-to-1 MUX. S_0' (Complement of S_0) is created by applying the Quantum NOT operation for input S_0. The NOT gate and LUT output will go through as inputs for the first Quantum AND operation. S_0 and D flip-flop (DFF) output will go through as inputs for the second Quantum AND operation. The output of these two Quantum AND operations will store in a Quantum cache memory. Cache memory stores these qubits as quantum computing is faster than DNA computing. NMR Relaxation is used to transfer the qubit into the DNA sequence. DNA OR operation performs OR of these two DNA sequences. Quantum circuits produce more heat than DNA circuits. For this, a heat transfer circuit is used to transfer the extra heat produced by the Quantum circuit to the DNA circuit. DNA OR operation uses this heat to produce the FPGA output.

6.5 Quantum-DNA Complex Programmable Devices

Complex programmable logic device (CPLD), is one kind of integrated circuit that application designers build-up to implement digital hardware like mobile phones. These can handle knowingly higher designs than SPLDs (simple programmable logic devices) but it offers less logic than FPGAs (field-programmable gate arrays). The CPLDs include numerous logic blocks; each of the blocks includes 8 to 16 macrocells. Because every logic block executes a specific function, all of the macrocells in a logic block are fully connected. Depending upon the use, these blocks may or may not be connected to one another.

In terms of complexity, CPLD (complex programmable logic device) lies in between SPLD (simple programmable logic device) and FPGA and thus, it inherits features from both these devices. CPLDs are more complex than SPLDs but they are complex than FPGAs. In another way, a CPLD in quantum biological computing refers to a quantum computer, specifically designed to handle the complexities of biological systems. This device, leveraging the principles of quantum mechanics, aims to solve problems in biological sciences that are computationally intractable for classical computers. It involves manipulating and controlling qubits (quantum bits) to perform computations on a quantum level, enabling potential advancements in areas like drug discovery, protein folding, and understanding complex biological processes.

For this, it is required to use a heat transfer circuit between Quantum and DNA operations. In this way, the Quantum-DNA CPLD implementation methods will improve the result of any logic device.

6.5.1 Circuit Architecture

The CPLD function block is designed by connecting Flip-Flops, Logic function and Multiplexers. Here a simple Quantum CPLD functional block is designed by using a D Flip-Flop, a logic function and a multiplexer. A simple logic function F = AB + BC + CA consists of three Quantum AND, and two Quantum OR operations. The output of the logic function will XOR with zero. The output of the XOR will go through the D Flip-Flop and multiplexer as input. The D Flip-Flop is a sequential circuit that consists of four Quantum NAND operations and one Quantum NOT operation. The output of the Quantum D Flip-Flop will go through the multiplexer as input. A Quantum 2-to-1 MUX is designed using two Quantum AND operations, one Quantum NOT operation and one DNA OR operation. The outputs of the two Quantum AND operations will store in a Quantum cache memory. The NMR Relaxation is used to transfer the qubit into a DNA sequence. The DNA OR operation performs OR of these two DNA sequences. A heat transfer circuit is used to transfer the extra heat produced by the Quantum circuit to use in the DNA circuit. The circuit architecture of Quantum-DNA CPLD is shown in Fig. 6.7.

6.5.2 Working Principle

Three inputs |A>, |B> and |C> first go thought the logic function (F = AB + BC + AC). Then, the logic function is implemented using three Quantum AND operations and two Quantum OR operations. Finally, the outputs of the logic function will go through the Quantum XOR (XOR with |0>) operation and produce the output.

Two inputs, one is the output of XOR and another is clock input that will go through the D flip-flop. The D flip-flop uses one Quantum NOT operation and four Quantum NAND operations. The D flip-flop transfers the XOR output, if the CLK input sequence is |1>. If the CLK input is |0>, one of the inputs to each of the last two Quantum NAND operations will be |1>. Thus, the output of the D flip-flop remains unchanged regardless of the values of the XOR output.

Two Quantum AND operations, one Quantum NOT operation and one DNA OR operation are used in 2-to-1 MUX. The complement of S_0 is created by applying the Quantum NOT operation for input S_0. The NOT gate and XOR outputs will go through as inputs for the first Quantum AND operation. S_0 and D flip-flop (DFF) output will go through as inputs for the second Quantum AND operation. The output of these two Quantum AND operations will store in a Quantum cache memory. Cache memory stores these qubits as quantum computing is faster than DNA computing. NMR Relaxation is used to transfer the qubit into the DNA sequence. DNA OR operation performs OR of these two DNA sequences. Quantum circuits produce more heat than DNA circuits. For this, a heat transfer circuit is used to transfer the extra heat produced by the Quantum circuit to the DNA circuit. DNA OR operation used this heat to produce the CPLD output.

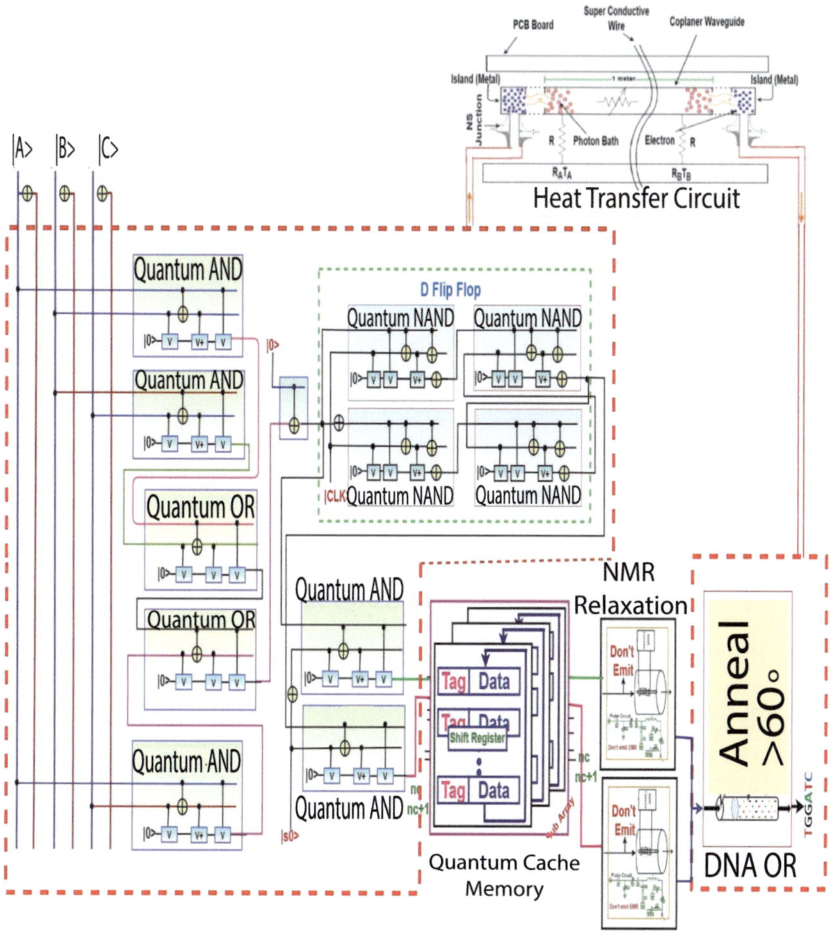

Fig. 6.7 Circuit architecture of quantum-DNA CPLD

6.6 Applications

The applications of Quantum PLD and DNA PLD are described in the previous two chapters in individual section. So, both of their applications can be achieved in Quantum-DNA computing. All applications of Quantum PLD and DNA PLD can be implemented in future.

6.7 Summary

This chapter has discussed about the PLD in Quantum-DNA computing. In each circuit, there were two parts. The first part is Quantum part, where the qubits are used as inputs and the second or last part is the DNA part where DNA sequences are used to express the outputs. A quantum cache memory is used to store quantum bits temporarily. A data conversion circuit is used to convert the qubits into DNA sequences. A heat transfer circuit is also used here to transfer excessive heat from the quantum part to DNA part. The next chapter is the opposite of this chapter which is called DNA-Quantum PLD.

Chapter 7
Programmable Devices in DNA-Quantum Computing

7.1 Introduction

In this chapter, the topic of DNA-Quantum PLD (Programmable Logic Device) will be discussed. It is a logic device that has both Programmable DNA AND array and Programmable Quantum OR array. Hence, it is the most flexible PLD. The design inputs of PLD in DNA operations are programmable. That means each DNA operation has both normal and complemented inputs of variables. So, based on the requirement, any of those inputs can be programmed. So, it is possible to generate only the required product terms by using these Quantum operations. A storage device needs to store DNA sequences between these two computing systems. DNA computing produces less heat than quantum computing. As a result, the extra heat from the outside is needed to supply the DNA circuit. DNA molecules are needed to transform into qubits and for this, a transformation process (Trap-Ion) is used by which an excited magnetic state returns to its equilibrium distribution. Quantum computing produces more heat in circuits which can be used in further DNA computing the heat can be transferred to a cooler circuit to cool down.

Quantum operations are also programmable in this case. Because all the outputs of DNA operations are applied as inputs to each Quantum OR operation, any number of needed product terms may be coded. As a result, DNA-Quantum devices' outputs will be in the form of a sum of product forms. In other way, programmable devices in biological quantum computing or DNA-quantum computing or bio-quantum computing explore using biological systems like DNA/RNA to create programmable computers, similar to how quantum computers use qubits. These biological systems offer potential for revolutionary advancements in fields like pharmaceuticals. While quantum computers can perform complex calculations and simulations, biological computers offer a different approach to computation by leveraging the inherent complexity and information storage capabilities of biological molecules. The field of biological quantum computing is still in its early stages of development, with ongoing research exploring the potential of these systems and creating stable and reliable pro-

H. Md. Hasan Babu, *Quantum Biocomputing in Quantum Biology Volume II*,
https://doi.org/10.1007/978-981-97-5349-9_7

grammable biological devices presents significant challenges, but ongoing research is making progress in this area. If successful, biological quantum computing could offer a paradigm shift in how we approach computation, with potential applications in various scientific and technological fields.

7.2 DNA-Quantum Programmable Logic Array

In this section, the first part of the DNA-Quantum PLA considers DNA operations and the last part considers quantum operations. It is important to note that the both parts (Quantum Part and DNA Part) are taken from the structure of the DNA-Quantum PLA. In other way, PLAs are a type of programmable logic device used to implement combinational logic circuits. They can be used in both classical and quantum computing, and their application in biological quantum computing explores the intersection of these fields. This field explores the use of biological systems (like DNA and RNA) to perform quantum computations. It combines the principles of quantum mechanics with the capabilities of biological molecules. The use of PLAs or similar programmable logic devices in biological quantum computing could lead to the development of novel quantum algorithms, quantum simulators, and even quantum-DNA computers. In addition, this field of multiple-valued computing, which deals with systems that use more than two states (like 0 and 1), is also relevant to both quantum and biological computing. A book titled âŁœMultiple-Valued Computing in Quantum Molecular BiologyâŁž explores this area in detail.

7.2.1 Block Diagram

DNA-Quantum PLA is a programmable logic device that has both Programmable DNA AND array and Programmable Quantum OR array. Hence, it is the most flexible PLD. The block diagram of DNA-Quantum PLA is shown in Fig. 7.1.

Here, the inputs of DNA AND operations are programmable. That means each DNA AND operation has both normal and complemented inputs of variables. A storage device needs to store DNA sequences between these two computing systems. DNA computing produces less heat than quantum computing. As a result, it is required to supply extra heat from outside in the DNA circuit. It is required to transform DNA molecules into qubits. For this, it is necessary to use a storage device and a transformation process (Trap-Ion) by which an excited magnetic state returns to its equilibrium distribution.

Here, Quantum OR operations are also programmable. So, any number can be programmed for required product terms, since all the outputs of DNA AND operations are applied as inputs to each Quantum OR operation. Therefore, the outputs of DNA-quantum PAL will be in the form of the sum of products form.

Table 7.1 Truth Table of DNA-Quantum PLA for Functions F1, F2, and F3

A	B	C	F1	F2	F3
TGGATC	TGGATC	TGGATC	|0>	|1>	|0>
TGGATC	TGGATC	ACCTAG	|1>	|0>	|0>
TGGATC	ACCTAG	TGGATC	|0>	|1>	|1>
TGGATC	ACCTAG	ACCTAG	|1>	|0>	|1>
ACCTAG	TGGATC	TGGATC	|0>	|1>	|1>
ACCTAG	TGGATC	ACCTAG	|0>	|1>	|1>
ACCTAG	ACCTAG	TGGATC	|1>	|0>	|0>
ACCTAG	ACCTAG	ACCTAG	|1>	|0>	|0>

7.2.2 Circuit Architecture

DNA-quantum PLA is a programmable logic device that has both Programmable DNA AND array and Programmable quantum OR array.

The following logic functions are used to construct the DNA-quantum PLA:
F1 = AB + A'C
F2 = A'C' + AB'
F3 = A'B + AB'.

The given three functions are in sum of products form. The number of product terms present in the given Boolean functions F1, F2 and F3 is two. The truth table of DNA-quantum PLA for functions F1, F2, and F3 is given in Table 7.1.

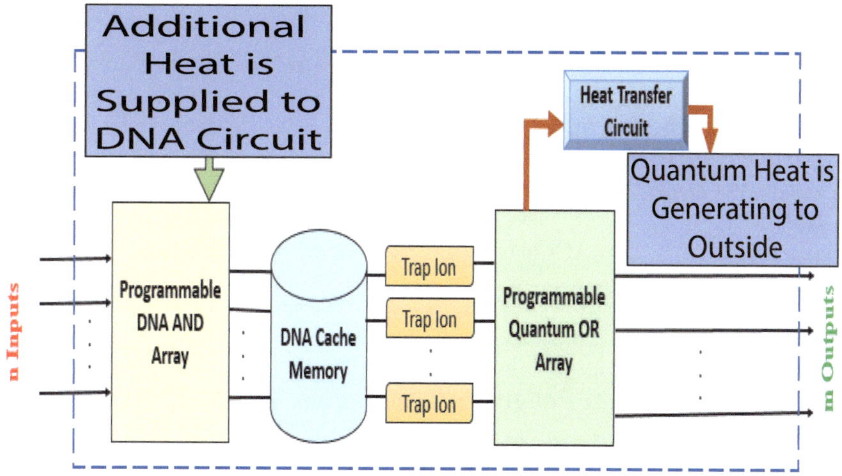

Fig. 7.1 Block diagram of DNA-quantum PLA

A.B's products are used in both functions F2 and F3. So, five programmable Quantum AND gates and three programmable DNA OR gates are required for producing those three functions. For the given problem, there are three inputs (A, B, C) and three outputs (F1, F2, F3). The complement of three inputs is obtained through DNA NOT operation. Thus, the realization has six input lines (input with its complement). The given expression has six product terms and so the fuses are placed in the corresponding literals to obtain the product terms. Five DNA AND operations are designed using DNA computing. It is required to use a DNA cache memory to store the DNA sequences. It is also needed to transform DNA molecules into qubits and for this, Trap-Ion is used by which an excited magnetic state returns to its equilibrium distribution. Here, Quantum OR operations are also programmable. Therefore, the outputs of DNA-Quantum PAL will be in the form of the sum of products form. Circuit architecture of DNA-Quantum PLA is shown in Fig. 7.2.

7.2.3 Working Principle

According to the truth table as shown in Table 7.1 of the DNA-Quantum PLA, it is necessary to do the following operations to obtain the desired output sequence:

1. For inputs A, B, C = TGGATC, TGGATC, TGGATC function |F2> produces |1>.
2. For inputs A, B, C = TGGATC, TGGATC, ACCTAG function |F1> produces |1>.
3. For inputs A, B, C = TGGATC, ACCTAG, TGGATC functions |F2> and |F3> produce |1>.
4. For inputs A, B, C = TGGATC, ACCTAG, ACCTAG functions |F1> and |F3> produce |1>.
5. For inputs A, B, C = ACCTAG, TGGATC, TGGATC functions |F2> and |F3> produce |1>.
6. For inputs A, B, C = ACCTAG, TGGATC, ACCTAG functions |F2> and |F3> produce |1>.
7. For inputs A, B, C = ACCTAG, ACCTAG, TGGATC function |F1> produces |1>.
8. For inputs A, B, C = ACCTAG, ACCTAG, ACCTAG function |F1> produces |1>.

7.3 DNA-Quantum Programmable Array Logic

Programmable Array Logic (PAL) is a logic device, which has a programmable DNA AND array and fixed DNA OR array. It is used to realize a logic function. In this DNA PLD, only DNA AND operations are programmable and hence it is easier to work

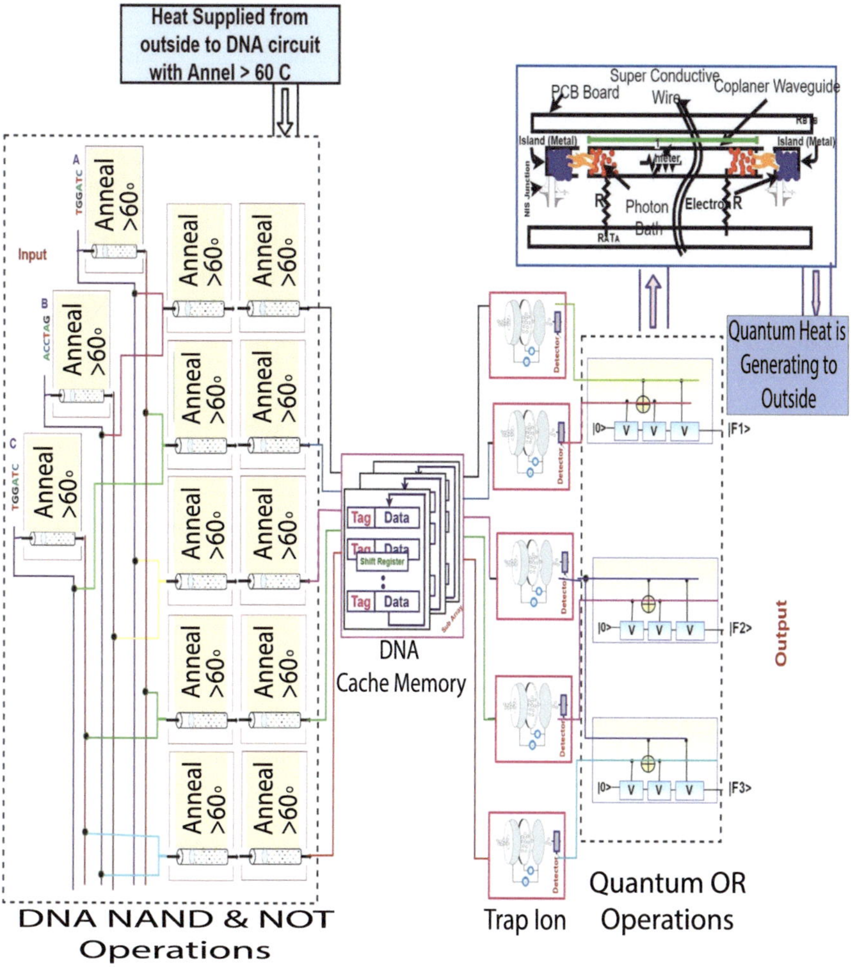

Fig. 7.2 Circuit architecture of DNA-quantum PLA

with DNA PAL. Depending upon the required function, the output line of the DNA AND operation is connected to the corresponding input of the DNA OR operation.

7.3.1 Block Diagram

DNA-Quantum PLA is a programmable logic device that has Programmable DNA AND array and fixed Quantum OR array. Hence, it is the most flexible PLD. The block diagram of the DNA-Quantum PAL is shown in Fig. 7.3.

Here, the inputs of DNA AND operations are programmable. That means each DNA AND operation has both normal and complemented inputs of variables. DNA computing produces less heat than quantum computing. As a result, it is required to supply extra heat from outside in the DNA circuit. It is required to transform DNA molecules into qubits and, for this, a storage device is used and a transformation process (Trap-Ion) by which an excited magnetic state returns to its equilibrium distribution. Quantum computing produces more heat in circuits which can be used in further DNA computing can transfer the heat in a cooler circuit to cool down.

Here, Quantum OR operations are fixed. All the outputs of DNA AND operations are applied as inputs to each Quantum OR operation. Therefore, the outputs of DNA-quantum PAL will be in the form of the sum of products form.

7.3.2 Circuit Architecture

DNA-quantum PAL is a programmable logic device that has both Programmable DNA AND array and fixed quantum OR array.

The following logic functions are used to construct the DNA-quantum PAL:

F1 = AB + BC + AC; and
F2 = A'C' + AB'.

The given two functions are in sum of products form. The number of product terms present in the given Boolean functions F1 and F2 are three and two, respectively. The truth table of DNA-quantum PAL for Functions F1 and F2 is given in Table 7.2.

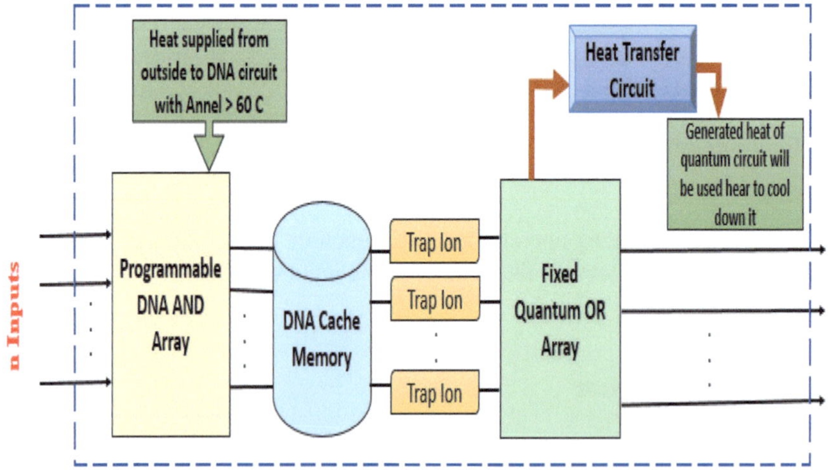

Fig. 7.3 Block diagram of the DNA-Quantum PAL

Six programmable DNA AND operations and two fixed quantum OR operations are required for producing those two functions. But, it is required to perform an extra two Quantum OR operations as three product terms for each function are required to do OR operation.

Consider the realization of the Boolean expression F1 = AB + BC + AC F2 = A'C' + AB' using Quantum Programmable Logic Array.

It is required to use a cache memory is used to store the DNA sequences. It needs to transform DNA molecules into qubits and for this, Trap-Ion is used by which an excited magnetic state returns to its equilibrium distribution. Here, Quantum OR operations are fixed. All the outputs of DNA AND operations are applied as inputs to each Quantum OR operation. Therefore, the outputs of DNA-Quantum PAL will be in the form of the sum of products form. Circuit architecture of DNA-Quantum PAL is shown in Fig. 7.4.

7.3.3 Working Principle

According to the truth table (Table 7.2) of the DNA-Quantum PAL, it is necessary to do the following operations to obtain the desired output sequence.

1. For inpus A, B, C = TGGATC, TGGATC, TGGATC function |F2> produces |1>
2. For inputs A, B, C = TGGATC, TGGATC, ACCTAG none of function produces |1>.
3. For inputs A, B, C = TGGATC, ACCTAG, TGGATC function |F2> produces |1>.
4. For inputs A, B, C = TGGATC, ACCTAG, ACCTAG function |F1> produces |1>.
5. For inputs A, B, C = ACCTAG, TGGATC, TGGATC function |F2> produces |1>.
6. For inputs A, B, C = ACCTAG, TGGATC, ACCTAG functions |F1> and |F2> produce |1>.

Table 7.2 Truth Table of DNA-Quantum PAL for Functions F1 and F2

A	B	C	F1	F2
TGGATC	TGGATC	TGGATC	**|0>**	**|1>**
TGGATC	TGGATC	ACCTAG	**|0>**	**|0>**
TGGATC	ACCTAG	TGGATC	**|0>**	**|1>**
TGGATC	ACCTAG	ACCTAG	**|1>**	**|0>**
ACCTAG	TGGATC	TGGATC	**|0>**	**|1>**
ACCTAG	TGGATC	ACCTAG	**|1>**	**|1>**
ACCTAG	ACCTAG	TGGATC	**|1>**	**|0>**
ACCTAG	ACCTAG	ACCTAG	**|1>**	**|0>**

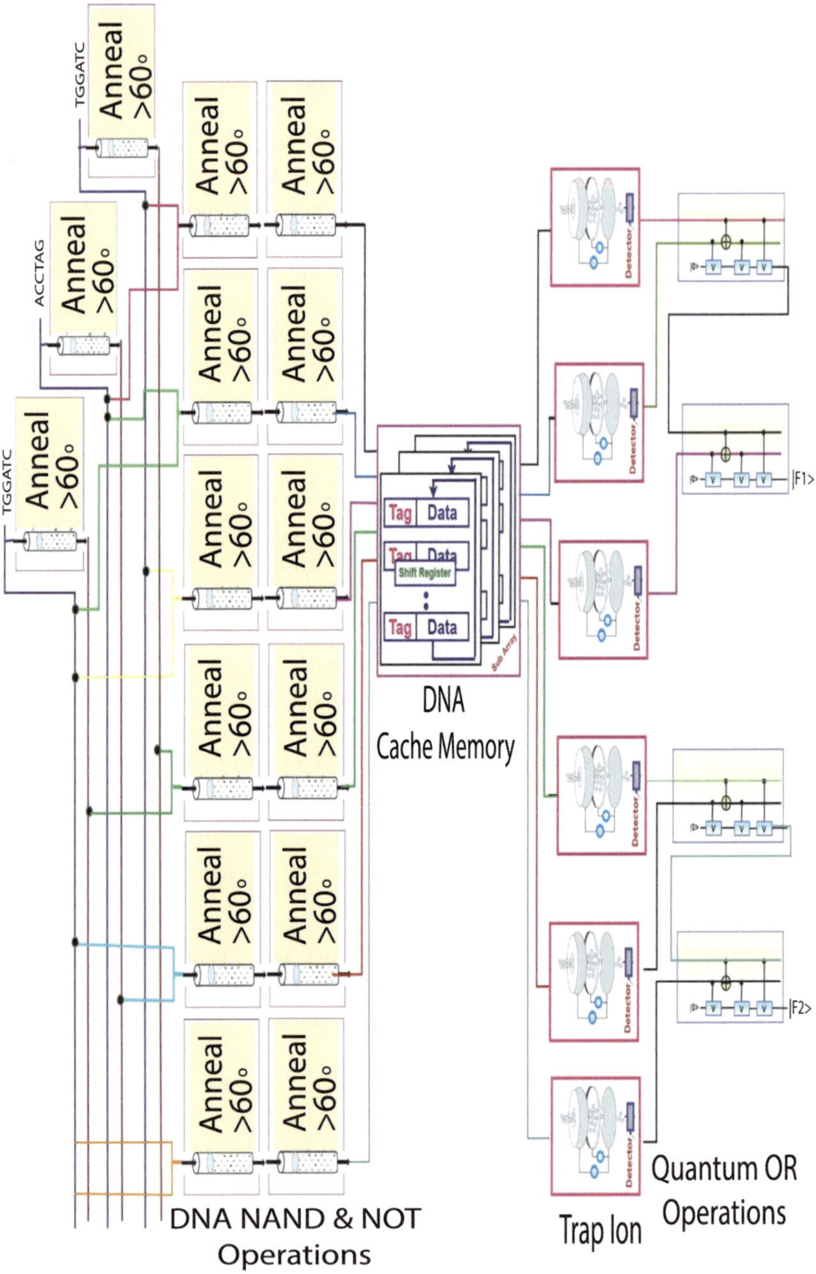

Fig. 7.4 Circuit architecture of DNA-Quantum PAL

7. For inputs A, B, C = ACCTAG, ACCTAG, TGGATC function |F1> produces |1>.
8. For inputs A, B, C = ACCTAG, ACCTAG, ACCTAG function |F1> produces |1>.

7.4 DNA-Quantum Field Programmable Gate Arrays

A DNA-Quantum Programmable Logic Device is a DNA-Quantum Field Programmable Gate Array. It's a programmable digital integrated circuit with adjustable logic blocks and programmable connections between them. The end-users can configure it to implement specific applications. Customers or designers can use programmable linkages to perform certain operations easily. In other way, FPGAs have potential applications in bioquantum computing, primarily for accelerating and improving the efficiency of calculations related to quantum dynamics and simulations. FPGAs' parallel processing capabilities and high speed make them suitable for tasks that are computationally intensive and repetitive. Furthermore, they can offer energy efficiency compared to traditional CPUs and GPUs, which is crucial for supercomputing centers. In summary, FPGAs offer a promising solution for accelerating and optimizing calculations in bioquantum computing, particularly in areas like quantum dynamics simulation and error correction. Their parallel processing, speed, and energy efficiency make them valuable tools for pushing the boundaries of quantum computing research and applications. In this section, DNA-Quantum FPGA is explained in detail.

7.4.1 Block Diagram

Quantum computing uses qubits (|0> and |1>) and DNA computing system uses molecules to represent information. In DNA-Quantum computing FPGA design, DNA sequences will work as input for the programmable logic devices. DNA computing produces less heat than quantum computing. As a result, it needs to supply extra heat from outside in the DNA circuit. It is required to transform DNA molecules into qubits and for this, a storage device is used and a transformation process by which an excited magnetic state returns to its equilibrium distribution. Quantum Computing produces more heat in circuits which can be used in further DNA computing can transfer the heat in a cooler circuit to cool down. The DNA- Quantum FPGA implementation methods will improve the result of any logic device. The block diagram of DNA-Quantum FPGA is given in Fig. 7.5.

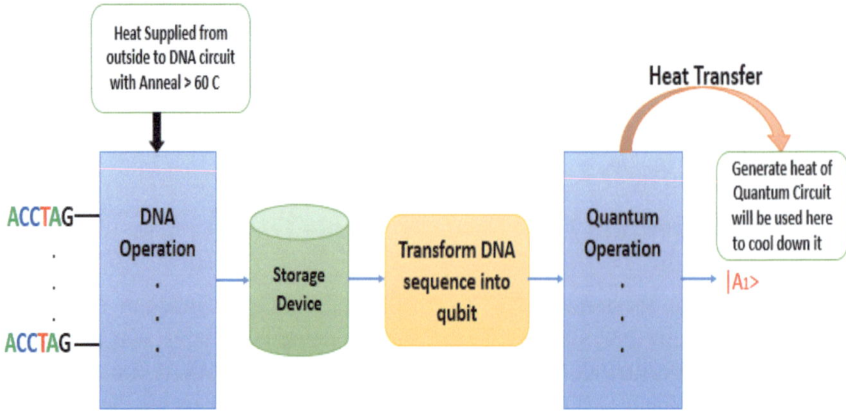

Fig. 7.5 General organization of DNA-quantum circuit

7.4.2 Circuit Architecture

DNA-Quantum FPGA logic block is designed by connecting Flip-Flops, Look-up Tables (LUT), and Multiplexers. Here, a simple DNA-Quantum FPGA logic block is designed by using a DNA D Flip-Flop, a DNA Look-up Table (LUT), and a Quantum multiplexer. The circuit architecture of DNA-Quantum FPGA is shown in Fig. 7.6.

A simple two-input DNA LUT consists of one DNA AND, one DNA NOT, and one DNA OR operation. The output of the LUT will go through the DNA D Flip-Flop and to quantum multiplexer as input. The DNA D Flip-Flop is a sequential circuit that consists of four DNA NAND operations and one DNA NOT operation. The output of the D Flip-Flop will go through the quantum multiplexer as input. Outputs of the LUT and D flip-flop will be stored in a DNA cache memory. Trap Ione is used to transfer the DNA sequence into qubits. A quantum 2-to-1 MUX is designed using two quantum AND operations, one quantum NOT operation, and one quantum OR operation. The multiplexer generates the desired output for the FPGA logic block as qubits. Heat is supplied from outside to the DNA circuit for operation. A heat transfer circuit is used in the quantum multiplexer circuit to transfer the extra heat produced by the quantum circuit to the outside to cool it.

7.4.3 Working Principle

Two inputs A0 and A1 first go through the Look-up Tables (LUT). The LUT uses one AND gate for AND operation by applying the NOT gate to the output of the NAND gate, one NOT gate, and one OR gate. In LUT, if both of the input sequences are "false" (TGGATC), then one will combine with the supplied "true" ACCTAG sequence in DNA NAND gate to produce a double-stranded molecule. DNase will

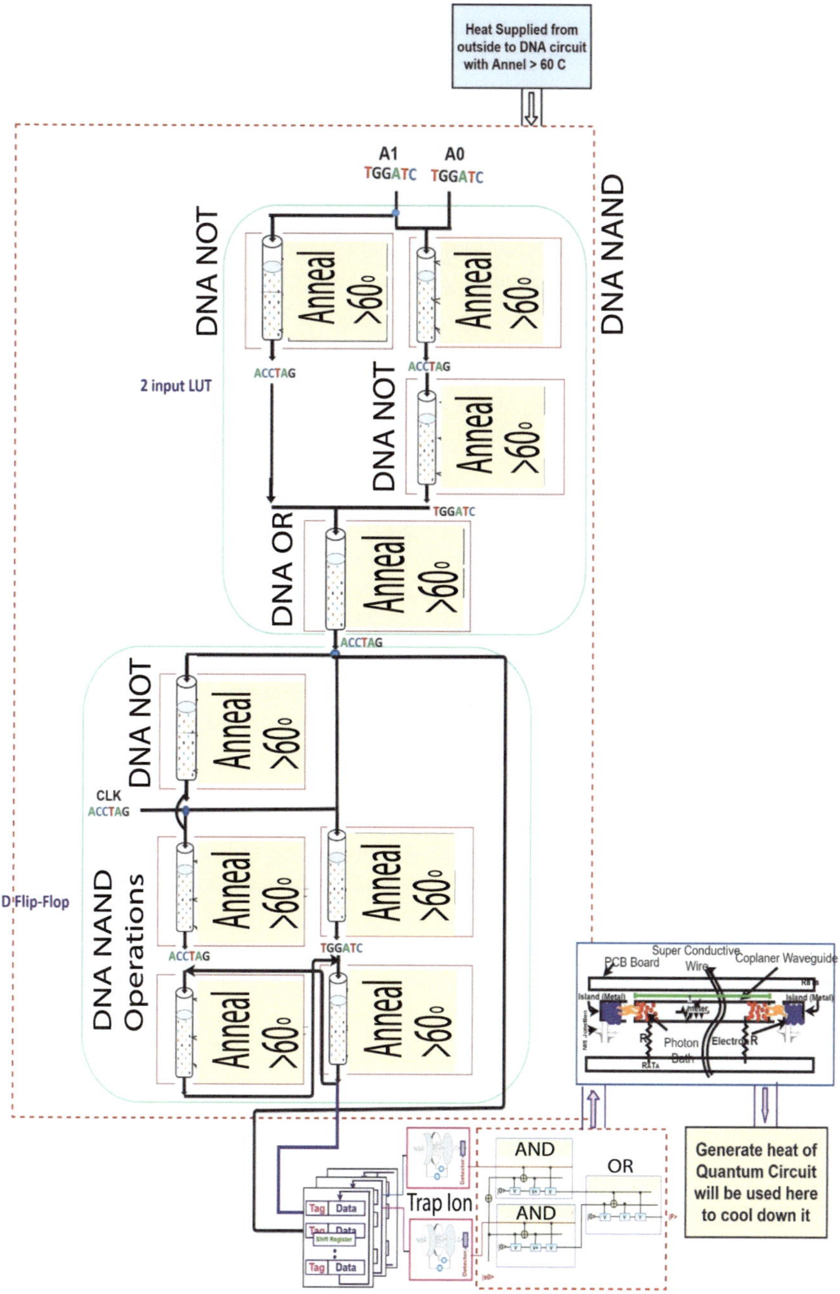

Fig. 7.6 DNA-quantum FPGA

destroy the remaining input sequence and the double-stranded sequence will generate a "true" ACCTAG sequence. This "true" sequence will first go through DNA NOT gate and results in a "false" evaluation. A1 "false" sequence will go through DNA NOT gate and combine with the supplied "true" ACCTAG sequence in DNA NOT gate, then ACCTAG will bind with the provided ACCTAG sequence, representing a "true" evaluation. The output of these two DNA NOT gates (TGGATC, ACCTAG) will go through the OR gate, then the "true" ACCTAG sequence will combine with the "false" TGGATC sequences to produce a double-stranded sequence. DNase will destroy the remaining "false" sequence and the gate will result in a "true" evaluation in the LUT output.

If one input sequence A0 is "false" and the other A1 is "true", then the "false" one will combine with the supplied "true" ACCTAG sequence in DNA NAND gate to produce a double-stranded molecule. DNase will destroy the remaining input sequence and the double-stranded sequence will generate a "true" ACCTAG sequence. This "true" sequence will go through DNA NOT gate and result in a "false" evaluation. A1 "true" sequence will go through DNA NOT gate and combine with the supplied "true" ACCTAG sequence in DNA NOT gate then ACCTAG will not bind with the provided ACCTAG sequence, provided a "false" evaluation. The output of these two DNA NOT gates (TGGATC, TGGATC) will go through the DNA OR gate, then the "false" sequence will not combine with either of the "false" TGGATC sequences to produce a double-stranded sequence. DNase will destroy the "false" sequences and the gate will result in a "false" evaluation in the LUT output.

If one input sequence A0 is "true" and the other A1 is "false", then the "false" one will combine with the supplied "true" ACCTAG sequence in DNA NAND gate to produce a double-stranded molecule. DNase will destroy the remaining input sequence and the double-stranded sequence will generate a "true" ACCTAG sequence. This "true" sequence will go through DNA NOT gate and result in a "false" evaluation. A1 "false" sequence will go through DNA NOT gate and combine with the supplied "true" ACCTAG sequence in DNA NOT gate then ACCTAG will bind with the provided ACCTAG sequence, representing a "true" evaluation. The output of these two DNA NOT gates (TGGATC, ACCTAG) will go through the DNA OR gate, then the "true" ACCTAG sequence will combine with the "false" TGGATC sequences to produce a double-stranded sequence. DNase will destroy the remaining "false" sequence and the gate will result in a "true" evaluation in the LUT output.

Finally, if both of the input sequences are "true" (ACCTAG), then none will combine with the supplied "true" ACCTAG sequence in DNA NAND gate to produce a double-stranded molecule. DNase will destroy all sequences and generate a "false" sequence. This "false" sequence will go through DNA NOT gate and result in a "true" evaluation. A1 "true" sequence will go thought DNA NOT gate and combine with the supplied "true" ACCTAG sequence in DNA NOT gate then ACCTAG will not bind with the provided ACCTAG sequence, provided a "false" evaluation. The output of these two DNA NOT gates (ACCTAG, TGGATC) will go through the DNA OR gate, then the "false" sequence will not combine with either of the "false" TGGATC sequences to produce a double-stranded sequence. DNase will destroy the "false" sequences and the gate will result in a "true" evaluation in the LUT output.

Two inputs, one is the output of LUT and another is clock input will go through the DNA D flip-flop. The DNA D flip-flop uses one DNA NOT gate and four DNA NAND gates. The DNA D flip-flop transfers the LUT output if the CLK input sequence is "true" (ACCTAG). If the CLK input is "false", one of the inputs to each of the last two NAND gates will be "true", thus the output of the D flip-flop remains unchanged regardless of the values of the LUT output.

Outputs of the LUT and D flip-flop will be stored in a DNA cache memory. Trap Ione is used to transfer the DNA sequence into qubits. Two Quantum AND operations, one Quantum NOT operation, and one Quantum OR operation are used in 2-to-1 MUX. S_0' is created by applying the Quantum NOT operation for input S_0. The NOT gate and LUT output will go through as inputs for the first Quantum AND operation. S_0 and D flip-flop (DFF) output will go through as inputs for the second Quantum AND operation. The output of these two Quantum AND operations will go through the Quantum OR operation to generate the FPGA logical block output. Heat is supplied from outside to DNA circuit for operation. A heat transfer circuit is used in the quantum multiplexer circuit to transfer the extra heat produced by the Quantum circuit to the outside to cool it.

7.5 DNA-Quantum Complex Programmable Devices

CPLDs (complex programmable logic devices) are a type of integrated circuit used by application designers to construct digital hardware such as mobile phones. These can handle more design complexity than SPLDs (simple programmable logic devices), but they have less logic than FPGAs (field-programmable gate arrays). CPLDs have a large number of logic blocks, each with 8–16 macrocells. All of the macrocells in a logic block are fully linked because each logic block performs a specific function. These blocks may or may not be linked to one another, depending on the application. In another way, CPLDs with quantum computers are playing an increasingly vital role in bioquantum computing, a field that combines quantum mechanics and biology to solve complex problems in areas like drug discovery and protein folding. These devices, particularly those utilizing qubits (quantum bits), allow for the manipulation and measurement of quantum particles, enabling new computational power for simulating and modeling biological systems. While promising, quantum computing and DNA computing are still in its early stages, and there are challenges in scaling up the number of qubits and DNA sequences, maintaining their coherence (the ability to maintain superposition and other properties), and developing efficient quantum algorithms and DNA algorithms. However, research and development are ongoing, with the potential to revolutionize fields like biology and medicine. In this section, the DNA-Quantum Complex Programmable Logic Device is discussed with the circuit diagram and working procedure.

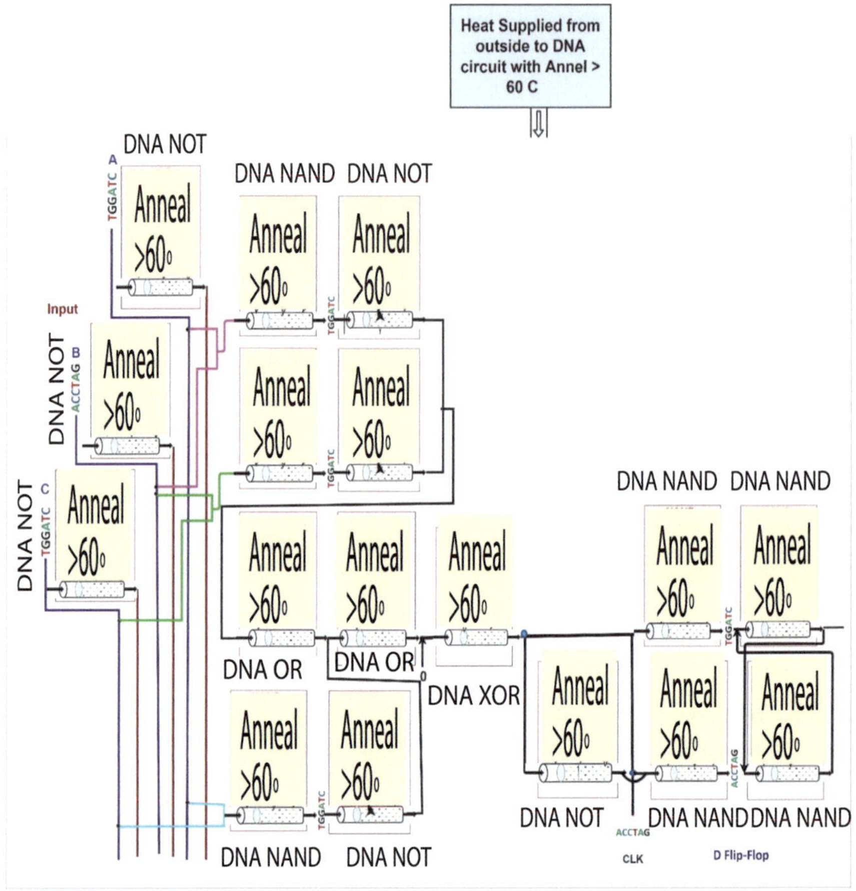

Fig. 7.7 Circuit diagram of DNA part of DNA-quantum CPLD

7.5.1 Circuit Architecture

DNA-Quantum CPLD circuit is divided here into two blocks. One is the DNA part of the CPLD and another is the Quantum part of the CPLD. DNA part consists of DNA operation of logic function and Flip-flop. Here a simple part of the DNA CPLD as shown in Fig. 7.7 is designed by using a DNA D Flip-Flop and a DNA logic function. A simple DNA logic function consists of three DNA NAND, NOT, and two DNA OR operations. The output of the logic function will be XOR with zero. The output of the XOR will go through the DNA D Flip-Flop. The D Flip-Flop is a sequential circuit that consists of four DNA NAND operations and one DNA NOT operation. Heat is supplied from outside to DNA circuit for operation. Diagram of DNA part of DNA to Quantum CPLD is shown in Fig. 7.7.

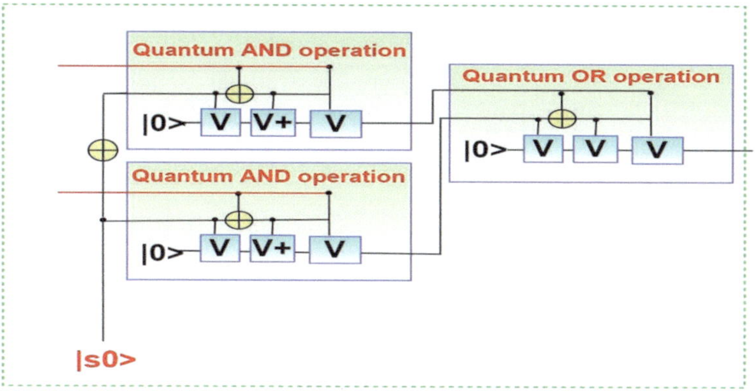

Fig. 7.8 Circuit diagram of quantum part of DNA-quantum CPLD

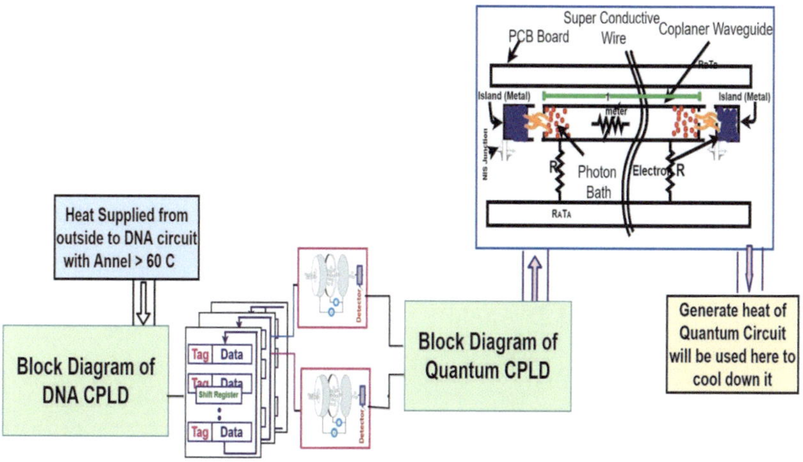

Fig. 7.9 Block diagram of DNA-quantum CPLD

Another part of the DNA-Quantum CPLD circuit is Quantum CPLD part. Quantum part consists of Quantum operation of Multiplexer. Here a simple Quantum CPLD part as shown in Fig. 7.8 is designed by using a Quantum Multiplexer. A simple 2-to-1 Quantum multiplexer consists of two Quantum AND and one Quantum OR operations.

DNA-Quantum CPLD functional block is designed by connecting the DNA circuit and Quantum circuit. A simple DNA to Quantum CPLD functional block as shown in Fig. 7.9 is designed by using DNA block and Quantum block.

The outputs of the DNA block will be stored in a DNA cache memory. Trap Ione is used to transfer the DNA sequence into qubit. Heat is supplied from outside to the DNA circuit for operation. A heat transfer circuit is used in the quantum multiplexer

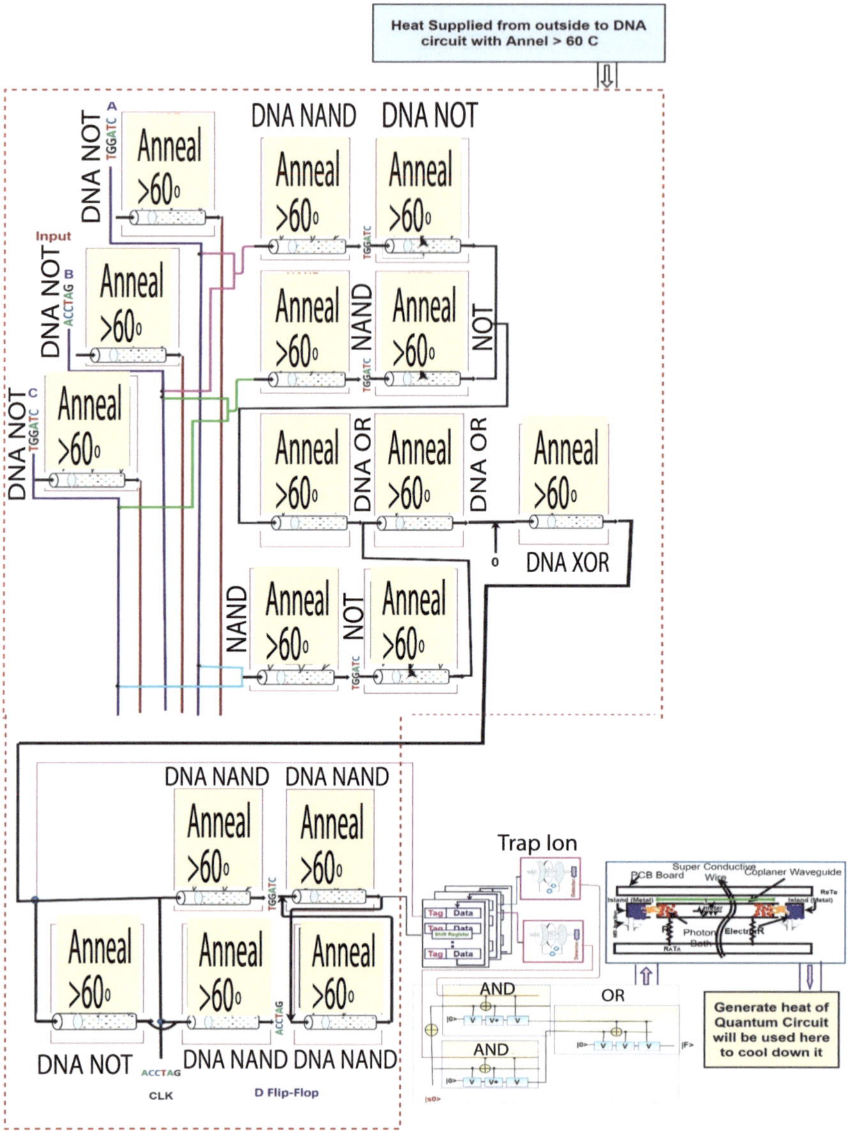

Fig. 7.10 Circuit architecture of DNA-quantum CPLD

circuit to transfer the extra heat produced by the quantum circuit to the outside to cool it. The whole circuit architecture of DNA-Quantum CPLD is depicted in Fig. 7.10.

7.5.2 *Working Principle*

Three inputs A, B, and C first go through the logic block. The logic function is then implemented using three DNA AND operations and two DNA OR operations. Outputs of the logic function will go through the Quantum XOR (XOR with TGGATC) operation and produce the output. If all the three input sequences are "false" (TGGATC), then all the inputs will go through three DNA NAND operations and combine with the supplied "true" ACCTAG sequence in DNA NAND operation to produce a double-stranded molecule. DNase will destroy the remaining input sequence and the double-stranded sequence will generate a "true" ACCTAG sequence. These "true" sequences will go through DNA NOT operation and result in a "false" evaluation. The output of these three DNA NOT gates will go through the DNA OR operation. DNase will destroy the "false" sequence and the operation will result in a "false" evaluation in the logic function output. The output of the logic function will XOR with zero.

Two inputs, one is the output of XOR and another is the clock input will go through the DNA D flip-flop. The DNA D flip-flop uses one DNA NOT operation and four DNA NAND operations. The DNA D flip-flop transfers it to the DNA XOR output, if the CLK input sequence is ACCTAG. If the CLK input is TGGATC, one of the inputs to each of the last two quantum NAND operations will be ACCTAG. Thus, the output of the D flip-flop remains unchanged regardless of the values of the XOR output.

The output of the logic function and DNA D flip-flop will be stored in a DNA cache memory. Trap Ion is used to transfer the DNA sequence into qubits. Two Quantum AND operation, one Quantum NOT operation, and one Quantum OR operation are used in 2-to-1 MUX. S_0' (Complement of S_0) is created by applying the Quantum NOT operation for input S_0. The NOT gate and XOR will go through as inputs for the first Quantum AND operation. S_0 and D flip-flop (DFF) output will go through as inputs for the second Quantum AND operation. The output of these two Quantum AND operations will go through the Quantum OR operation to generate the qubits of the CPLD functional block output.

7.6 Summary

This chapter has presented programmable logic devices in DNA-Quantum computing where the PLD circuits consist of the DNA operations and quantum operations. DNA operations on the first part and the quantum part will be on the last part of the PLD circuit. All necessary circuits and descriptions of the DNA-Quantum PLD have been presented clearly in this chapter. Here, DNA circuits are in the first part, where the extra heat is provided from the external source. On the other hand, the quantum part produces extra heat which is needed to be connected to a cooler to consume. The excessive heat could destroy the whole circuit.

Part III
Nano-Processor in Quantum Biocomputing

Overview

The quantum world is a domain where the smallest atoms and particles may exist. The size is less than 100 nanometers. This is where the foundations of quantum computers are constructed, which are based on two primary principles: superposition and entanglement. And it is because of these factors that quantum computers are so amazing, effective, powerful, and distinct from traditional computers. Biochemistry and molecular elements are employed to achieve computer tasks formerly accomplished by standard silicon-based technology in DNA computing, which is a new sector of study. DNA computing simulates logic gates and Boolean circuits at the molecular level. A DNA computer consists of DNA-based logic gates, a crucial component of a DNA computer. DNA computing nanoprocessor integrates all the components of DNA computing into one machine. An integrated DNA nanoprocessor can be constructed, mimicking the traditional silicon-based computer processor. And this proposed DNA nanoprocessor is also capable of performing computations logically. All types of problems of a DNA computer are encoded using the DNA alphabet A, C, T, and G. The fundamental motto of this nanoprocessor is to enhance the system's performance. Quantum biology (QB) is a combinational field of quantum physics and molecular biology, which has drawn researchers' attention to discover new world as their features are similar. It is now an emerging field to research. A hybrid nanoprocessor that contains both quantum physics and biology properties with similar features of quantum physics and life science. In other way, nanoprocessors play a role in quantum biocomputing by facilitating the creation of efficient quantum light sources and single-photon emitters, which are essential for quantum photonics. They can also be used to build networks of nanobased channels for protein filament traffic, offering a potential alternative to traditional quantum computers. In this part, firstly, a quantum nanoprocessor and a DNA nanoprocessor will be shown, then a complete hybrid quantum-DNA and DNA-quantum nanoprocessor will be shown with NMR-NMR relaxation in zero kelvin temperature with proper explanation. As it is a loamy nanoprocessor, the system needs to convert qubit to DNA form and DNA to qubit form. Here two methods will be needed to complete it. One is NMR relaxation that converts qubit to the DNA sequence, and another is only NMR that converts DNA to the qubit. In addition, a source is used to supply heat that aids DNA sequences in making chemical reactions. Finally, a heat transfer circuit is used to pass heat from the quantum components to the outside. So, this part will describe the details of nanoprocessor in quantum biocomputing, which means quantum nanoprocessor, quantum-DNA nanoprocessor, and DNA-quantum nanoprocessor.

Chapter 8
Quantum Nanoprocessor

8.1 Introduction

Richard Feynman, a physicist, claimed in the 1970s that computers could imitate physics. He was the first to discover nanotechnology, which involves manipulating individual atoms and molecules to create complex atomic structures. Quantum nanotechnology is the term for the technology used to control individual states. However, Eric Drexler delivered the first course on nanotechnology in 1988; he hypothesized the concept of nanoscale, whereas Drexler and Merkle created a molecular machine technique for computing the energy and structure of static systems with no direct knowledge on their dynamics.

Several researchers' groups have been trying to simulate components of nanomachines for the last many years. And, in 1994, Peter Shor discovered an algorithm that is known as Shor's algorithm; this algorithm is capable of factoring a number N in O ((logN) 3) time and O (logN) space. In 2001, a group demonstrated this algorithm at IBM, which factored 15 into 3 and 5. They used a quantum computer with seven qubits to implement this. A seven-qubit register was implemented using Shor's algorithm.

Afterward, Cirac and Zoller proposed a physical system for atomic ions whose electronic states store quantum information. The single-trapped ion can carry the quantum information, which is manipulated. Finally, David Wineland's group demonstrated an ion-trapped quantum computer at the National Institute of Standard Technology. Quantum information or a qubit can be implemented using a two-level system; the first is the electron's spin in a magnetic field, and the other is to use two levels of an atom. The qubit is initialized and manipulated for single-qubit gates, two-qubit gates, qubit state preparation, and readout which have been implemented with the fidelity needed for fault-tolerant Quantum Computing (QC) using high threshold quantum error correction codes. However, despite performing many promises shown by trapped ions, many challenges still need to be addressed to make a functional quantum computer.

H. Md. Hasan Babu, *Quantum Biocomputing in Quantum Biology Volume II*,
https://doi.org/10.1007/978-981-97-5349-9_8

However, in 2000, DiVincenzo showed five critical criteria needed to make a helpful quantum processor. The five criteria are (1) A physical system containing well-characterized qubits. (2) The system's ability must be well defined and determined. (3) Qubit decoherence times are much longer than any gate times. (4) A gaggle of universal quantum gates can be possible to apply to every qubit. (5) The ability to read out the qubit state has to be with accuracy. Only trapped ions fulfill the property mentioned above. This broader term "Nano" encompasses computing at the nanoscale, including quantum nanoprocessors and other nanoscale technologies. Quantum nanoprocessors, also known as quantum processing units (QPUs), are the brains of quantum computers that leverage quantum mechanical principles at the nanoscale to perform calculations differently from traditional processors. These nanoprocessors use the properties of particles like electrons or photons to perform calculations, potentially offering speed advantages for specific tasks. Quantum nanoprocessors could potentially solve certain complex problems much faster than classical computers, especially those involving optimization or simulation. These processors have the potential to impact fields like drug discovery, materials science, and artificial intelligence. The current stage of quantum computing is known as the noisy intermediate-scale quantum (NISQ) era, where processors have a limited number of qubits and are not yet fully fault-tolerant. Researchers are actively working to overcome these challenges and advance the technology towards a fully functional quantum computer. In this chapter, a nanoprocessor is built differently, and in a later chapter, a nanoprocessor will be built using trapped ions that follow the criteria mentioned earlier.

8.2 Basic Definitions

A Central Processing Unit is known as the central processor or main processor. A quantum electronic circuit within a quantum computer can execute the input/output operations. Basic arithmetic and logical units carry out the instructions of the computer program. The CPU controls all kinds of data flow and instructions. However, the CPU consists of five core components such as:

1. Quantum Control unit (CU)
2. Quantum Register
3. Quantum Arithmetic logic unit (ALU)
4. Quantum RAM
5. Quantum Buses.

1. **Quantum Control Unit**: A control unit is a quantum circuit that gives instructions within a computer processor. The control unit consists of many selection circuits like multiplexers and decoders.
2. **Quantum Register**: A nanoprocessor register is the most miniature set of qubit data that can hold the places part of a quantum nanoprocessor. For example, a register can have a qubit sequence, instructions, and a storage address.

3. **Quantum Arithmetic Logic Unit**: At first, Oskin et al. presented ALU as an element in a quantum computer architecture. The arithmetic unit controls the arithmetic operations that are executed by the program. The ALU unit performs mathematical functions such as subtract, divide, and multiply.
4. **Quantum RAM**: RAM stands for "Random Access Memory," called volatile memory. It is one of the CPU components that helps to increase system performance. The main goal of RAM is to store and access data on a short-term basis.
5. **Quantum Buses**: All computers need buses, whether they are quantum computers, supercomputers, or classical computers; they are used for transferring data between processors and other components. That is why the quantum nanoprocessor used in quantum computers will have been described here, where buses will be used like other computers. There are three types of buses such as address bus, control bus, and data bus which are briefly described in the architecture of essential components.

8.3 Block Diagram of a Complete Quantum Nanoprocessor

CPU (Central Processing Unit) is the heart of a computer that manipulates data and executes instructions. It consists of a lot of complete circuits such as IR (instruction register), PC (Program counter), Multiplexer, ALU (arithmetic logic unit), and RAM (Random Access Memory). All circuits are made of quantum logic gates as a quantum nanoprocessor. As quantum circuits produce heat that can create anarchy, it is required to solve it. As a result, a heat transfer circuit is shown in the following figure that will eradicate heat from the nanoprocessor to the outside environment, and a refrigerator will cool down this heat. A complete nanoprocessor block diagram is given in Fig. 8.1. This figure (Fig. 8.1) is the complete two-qubit quantum nanoprocessor where only essential CPU components are shown. CPUs need to have two inputs: instruction and data for performing meaningful work. The task of the instruction register is to tell the CPU what actions need to be performed on the data. Instructions are represented using a qubit. The CPU's inputs are stored in the memory. In Fig. 8.1, it is seen that the RAM data is coming from memory to the instruction register, and the CPU functions are following a cycle of fetching an instruction. After fetching, it is decoded, and finally, it is executed. This process can be called the "fetch decode execute" cycle. The cycle starts when data is transferred from memory to the instruction register. Note that the information sent from memory always uses the data bus for transferring data. The unique qubit patterns are extracted by selecting machine language in the IR and sent to the decoder. The purpose of the decoder is to decode the coded information from one format to another format. The second step of the cycle starts to work because of decoder. The decoder represents which qubit pattern will operate and activate that circuits needed to perform the actual operation. The course will follow the following instructions if the process is thoroughly accomplished.

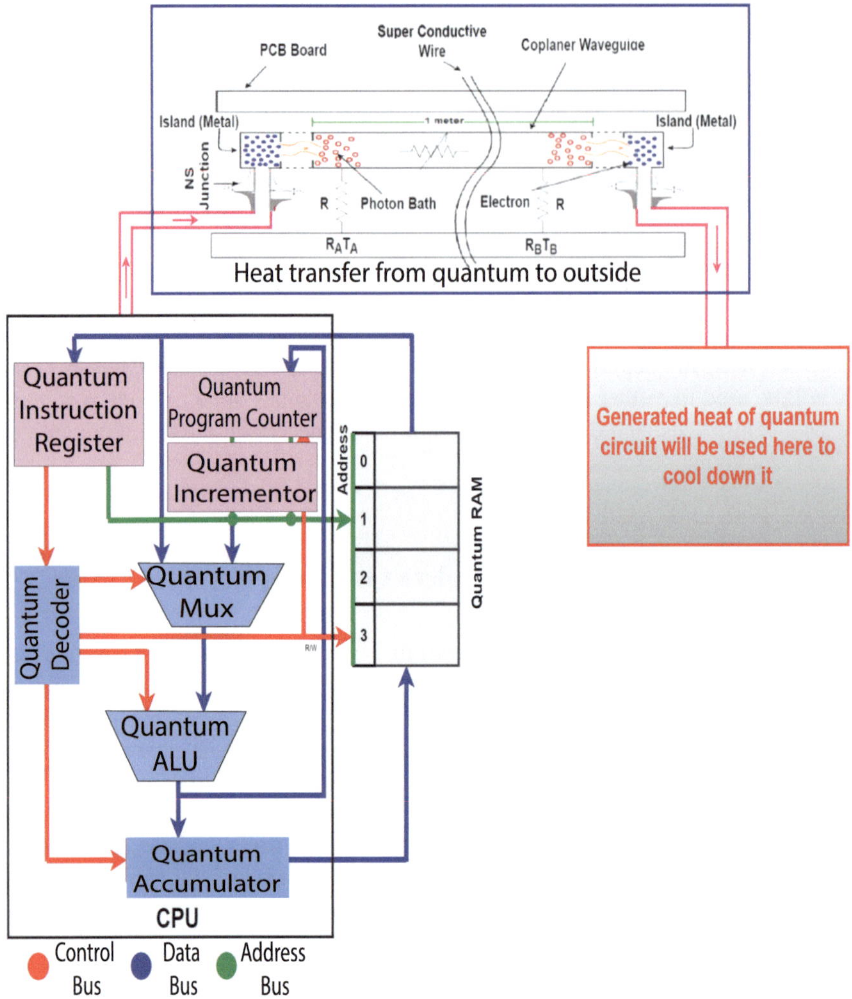

Fig. 8.1 Quantum nanoprocessor

The quantum program counter (PC) is a special-purpose register that holds the address of the next instruction to be executed from memory known to the CPU. If an instruction is completed, the program counter is incremented by one memory location. This theme is the whole working procedure of this nanoprocessor. Notice that a heat transfer circuit has been used as the quantum circuit produces so much heat, creating chaos within the circuit. That is why this heat has been transferred from the circuit to cool it by using a quantum refrigerator.

8.4 Basic Components of Quantum Nanoprocessor

A complete nanoprocessor has been made using the following components shown in Fig. 8.1. The features of this CPU are as follows:

1. Quantum RAM
2. Quantum Instruction Register
3. Quantum Program Counter
4. Quantum Incrementor
5. Quantum decoder
6. Quantum Multiplexer
7. Quantum ALU
8. Quantum Accumulator.

Moreover, the buses are used in the CPU to transfer data from one component to another. There are three types of buses: Data bus, Address bus, and control bus. The data bus is bidirectional, which carries the data back and forth between CPU and RAM. The address bus is unidirectional, taking memory addresses from the nanoprocessor to other components like primary storage and input/output devices. And, the last bus is a control bus used to transfer nanoprocessors to other elements that make sure everything is flowing perfectly from place to place or not. These are also essential components of the CPU that need to be accomplished for meaningful work.

However, all components are explained in detail in this section.

8.4.1 Design Procedure of Quantum RAM

It requires two address lines with one ancilla qubit for simulating 4-to-2-qubit RAM, and each address line needs to be CNOT form as well. These address line combinations will be the input of 2-to-4 decoders which consists of four quantum AND gates, and this decoder has one enable input. So getting four select lines from this decoder, and each select line will go through each RAM cell. Note that word calculation of RAM will be 2^k, where k is the address line and 2^k is the total words of n bit, and decoder combination will be $k \times 2^k$. This two-qubit RAM consists of four separate RAM cells, and each cell has three inputs such as |In0> or |In1>, anyone selects a line and read/write inputs. The obtained output from 4 quantum RAM cells will be the input of a quantum OR gate, which produces the final result. This part is the whole design procedure of 4-to-2 qubit RAM.

8.4.1.1 Working Principle of Quantum RAM

This component is the most crucial element of the CPU that stores the data quickly, a primary and volatile memory. The circuit of 4-to-2 qubit quantum RAM is given in Fig. 8.2.

This Fig. 8.2 represents the implementation of 4-to-2 qubit RAM. This quantum RAM consists of four separate “Words” of memory, and each is 2 qubits wide. The quantum RAM Cell has three inputs and one output. The complete circuit of a quantum RAM cell is described in Fig. 8.3 with proper explanation. A word consists of two quantum RAM cells arranged in such a way that both qubits can be accessed simultaneously. Four words of memory need two address lines. |A0> and |A1> are the two-qubit address lines input that goes through a 2-to-4 decoder that selects one of the four words. The memory-enabled input enables the decoder. If the memory enable is |0>, all output of the decoder will be |0>, and in that case, none of the memory addresses will be selected. But when the memory enable is |1>, one of the four words is preferred. The value specifies the word in the two address lines. The read/write input determines the operation when a word has been selected. During the read operation, the four qubits of the selected word pass to the quantum OR gates to the output |Z0> and |Z1> terminals. However, during the write operation, the data available in the input lines are transferred into the four quantum cells of the selected word. The quantum RAM cells that are not selected become disabled, and their previous qubit never changes. But when the memory-enabled input that passes into the decoder is equal to |0>, none of the words are selected, and then all quantum cells remain unchanged regardless of the value of the read/write input. This is the working procedure of RAM. The quantum RAM cell is given below in Fig. 8.3.
The quantum RAM cell has been designed using an R-S flip-flop. The number of total quantum cells per word will be $m \times n$, where m represents words with n bits. The quantum cell has three inputs such as “Select,” “Read/Write,” and “Input,” and one output line that is labeled by “Output.” The “select” input is used to access either reading or writing. The cell performs the memory operation when the select line is high or |1>. But when the select line of the quantum cell is low or |0>, the cell is not interested in completing a read from or written to.
The following input is “Read/Write,” where a system clock will conduct this input. If the clock value on the read/write line is |0>, this will signify “read,” and when it is |1>, it will perform the “write” phase. Consider the cell that has been selected. In that case, if the clock value is |0>, then the cell contents are to be read, and this time the output value will depend only on the Q value of the flip flop. But if Q is low, the cell output will be |0>, and if Q is high, the cell output will be|1>. It occurs because the quantum AND gate added to the cell’s production has three inputs-negated read/write, select, and Q; and both “negated read/ write” and “select are currently high.

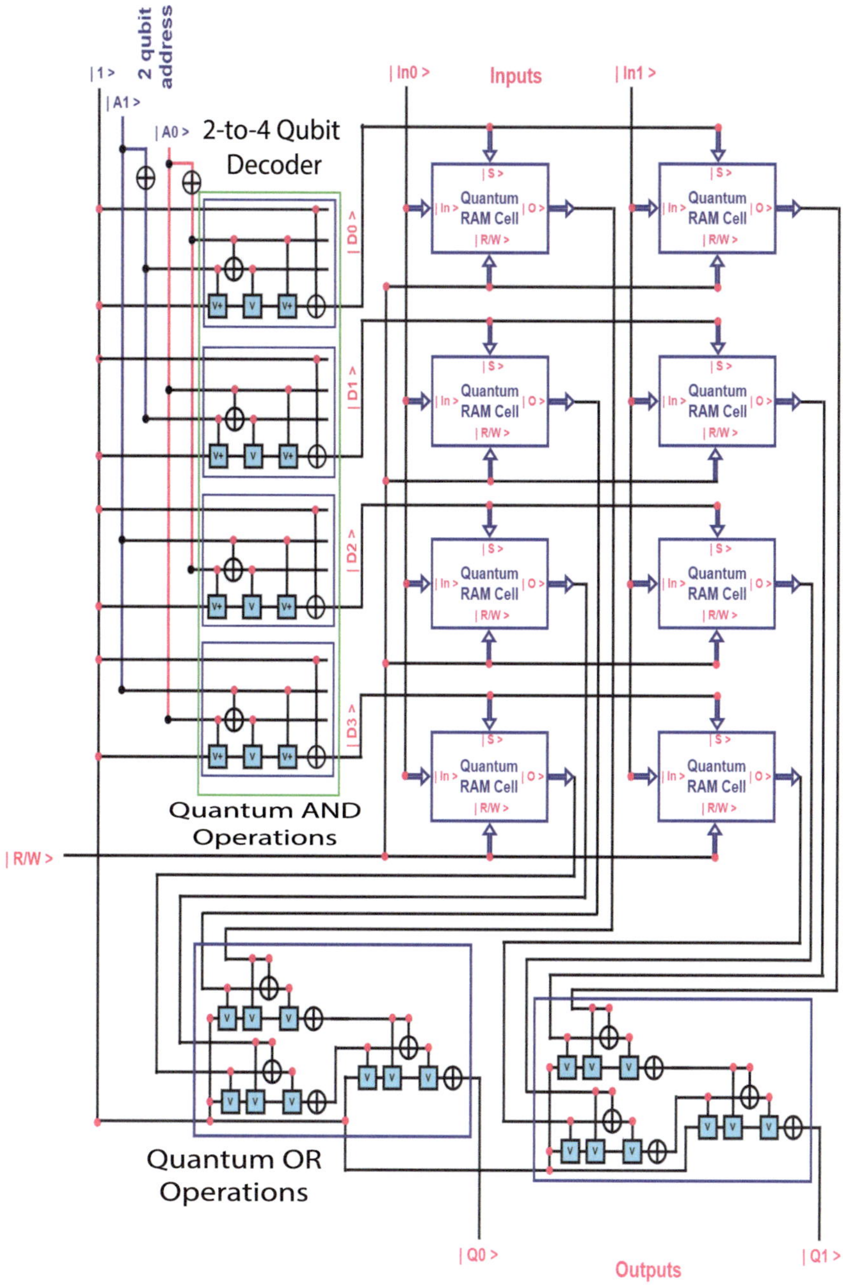

Fig. 8.2 4-to-2 qubits quantum RAM

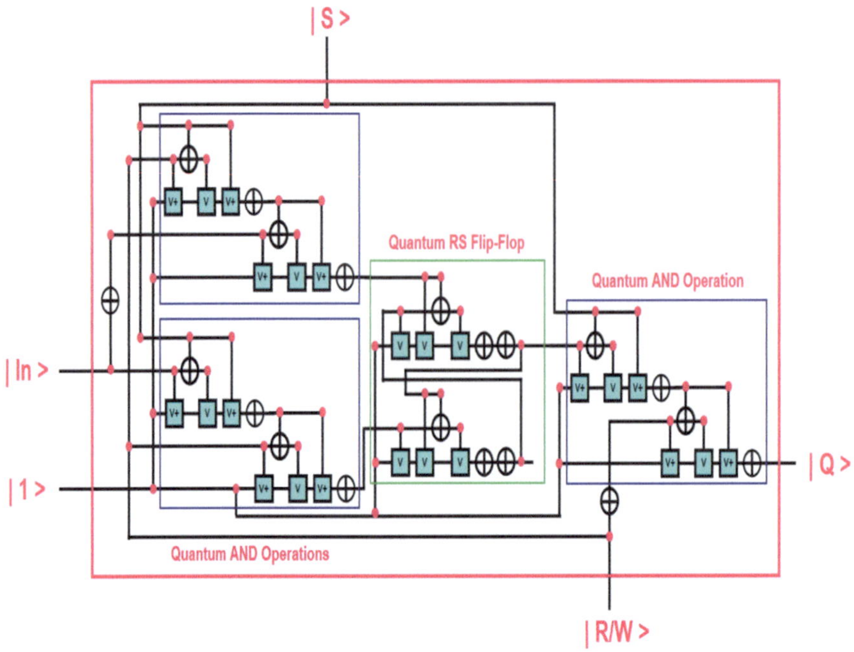

Fig. 8.3 Quantum RAM cell

8.4.2 Design Procedure of Quantum Instruction Register

The quantum instruction register (IR) consists of 16 quantum AND operations shown in Fig. 8.4. As this instruction is a component of a two-qubit CPU, instructions will be $2^2 = 4$. The minimum four instructions can be defined LOAD A, LOAD B, ADD A B, and OUT. An instruction register is a special-purpose register mainly used to store the instructions executed by the quantum computer.

8.4.2.1 Working Principle of Quantum Instruction Register

An instruction register holds that instruction is currently being executed. The quantum IR stores the instruction word. When the CPU fetches any instruction from memory, it is temporarily stored in the instruction register. The instruction can be a qubit word or code that defines a specific operation to be performed. After that, the CPU decodes the instruction and then executes it.

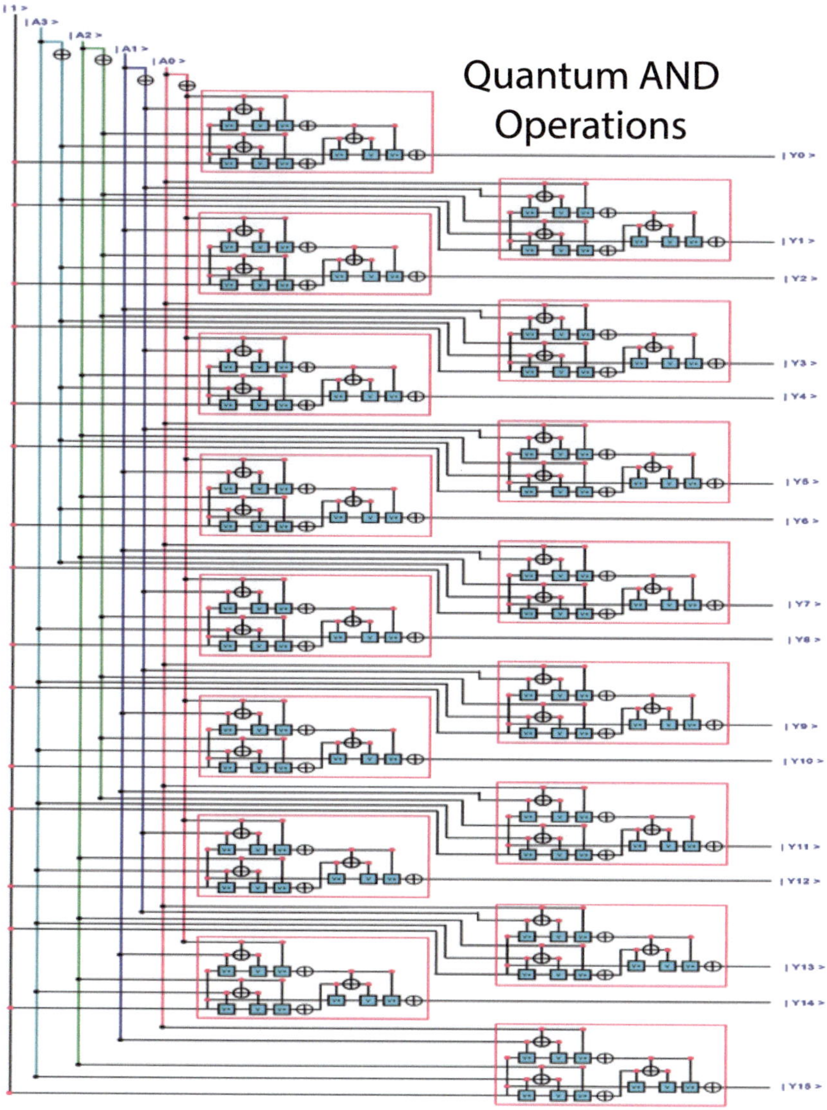

Fig. 8.4 4-qubit instruction register

8.4.3 *Design Procedure of Quantum Program Counter*

The quantum program counter (PC) consists of two D flip-flops where the first D flip-flop is designed using four NAND gates shown in Fig. 8.5 using green color boxes. This D flip-flop generates two outputs where one output has been skipped as there is no need to show it, according to Fig. 8.5. Another D flip flop consists of

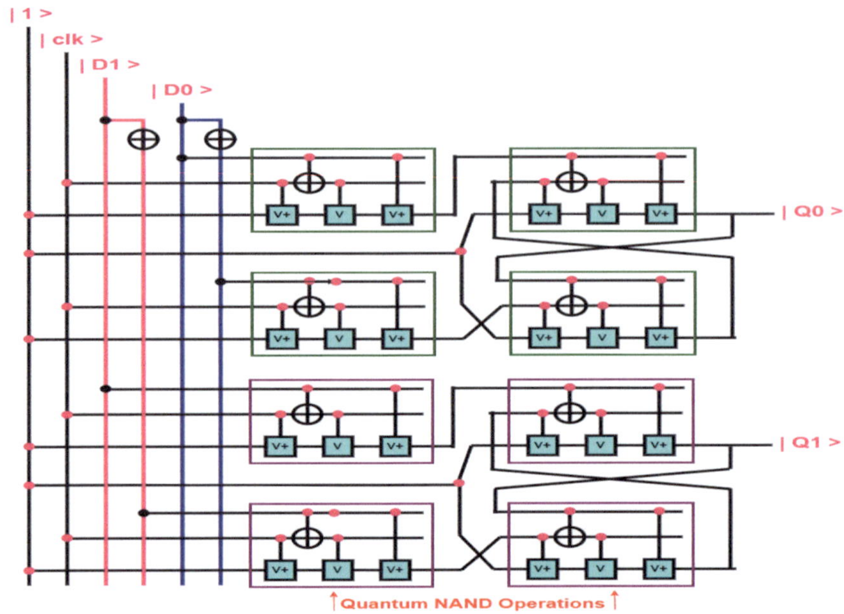

Fig. 8.5 2-qubit program counter register

four NAND gates confined to a violet color box. This D flip-flop also generates two outputs. Figure 8.5 shows that only two outputs that act as inputs are needed. So, the rest of the two outputs have been skipped. This two-qubit program counter is applicable for two-qubit nanoprocessors. The circuit is shown in Fig. 8.5.

8.4.3.1 Working Principle of Quantum Program Counter

Using this program counter is to store the next instruction executed next. The program counter is incremented by one when the current instruction is completed. All instructions and data have a specific address in memory. For example, if a program begins with an instruction stored in memory location 2, the program counter will first be loaded with address 2. When this instruction is executed, the PC is incremented by one to the following address, i.e., 3. The instructions in a program always follow the sequence memory location for storing themselves. Here |D0> and |D1> present inputs that go through the D flip-flop and give the desired outputs |Q0> and |Q1>.

8.4.4 Design Procedure of Quantum Incrementer Circuit

The quantum incrementer circuit is shown in Fig. 8.6.

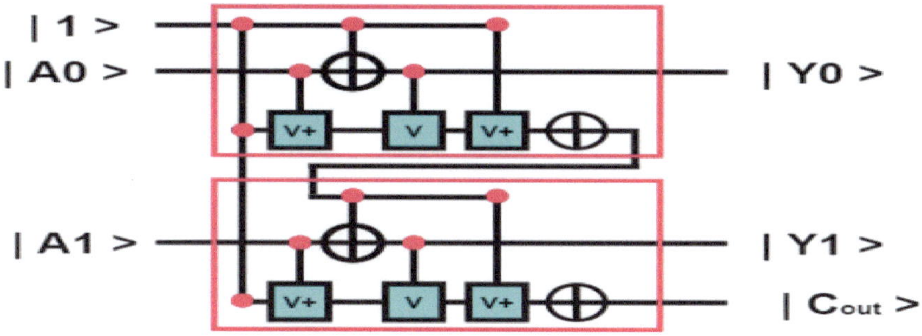

Fig. 8.6 2-qubit quantum incrementer

This circuit has been designed using two quantum half-adders. Each half-adder consists of one quantum XOR gate and one quantum AND gate. The quantum XOR gate gives the sum result, and the quantum AND gate generates the carry-out.

8.4.4.1 Working Principle of Quantum Incrementer Circuit

The program counter holds that the address will be executed next. When an instruction is completed, it needs to be incremented by one. That is why a two-qubit Incrementer circuit is used here, where one qubit value is added to the value of quantum variables stored in a register. The least significant qubit of the half adder is given |1> as a constant direct input and |A0> as the second input. So, |A0> and |A1> are the total qubits of half-adder where |1> is added, and produce the incremented outputs |Y0> and |Y1>.

8.4.5 Design Procedure of Quantum Decoder

The 2-to-4 decoder is used in a two-qubit CPU. The 2-to-4 qubit quantum decoder consists of four quantum AND operations with one enabled input. |A0> and |A1> are inputs with CNOT form and without CNOT form; |D0>, |D1>, |D2>, and |D3> are outputs. Figure 8.7 shows the circuit architecture of the quantum 2-to-4 decoder.

8.4.5.1 Working Principle of Quantum Decoder

The decoder is a combinational quantum circuit that has n input lines and can produce 2^n output lines. Each output has one product, and to achieve this product, quantum AND operations are performed here. For example, the minterm of two-input qubits |A0> and |A1> when enable input is |1>. But if enable input is |0>, all the decoder

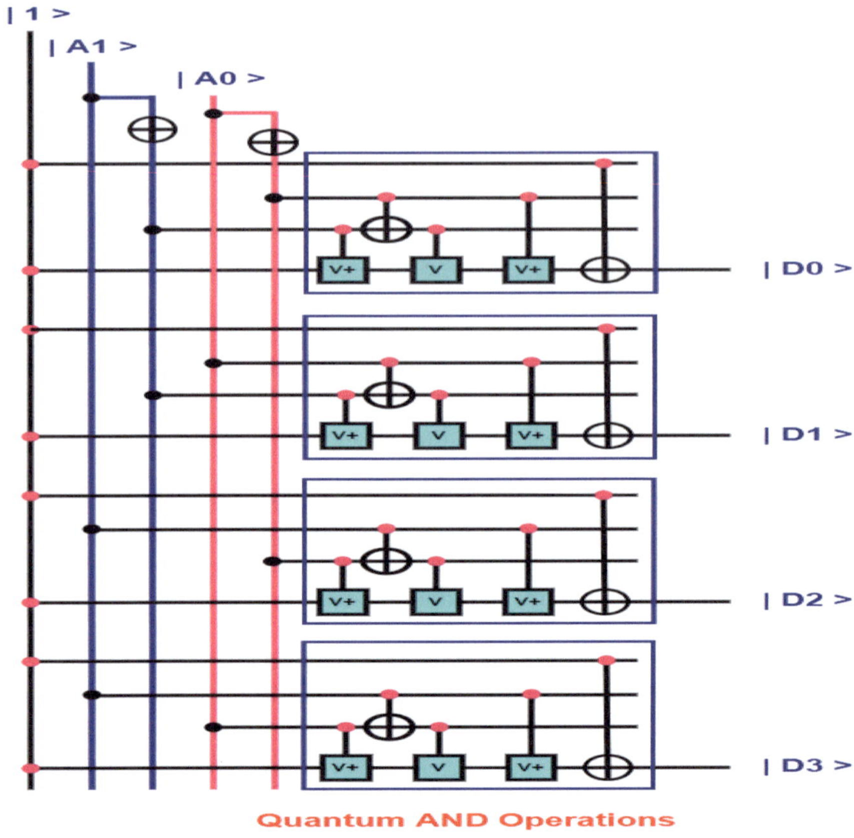

Fig. 8.7 Quantum 2-to-4 decoder circuit

outputs will be equal to zero and when enable is |1>, one of these four outputs will be active, i.e., |1>.

8.4.6 *Design Procedure of Quantum Multiplexer*

4-to-1 quantum MUX is shown in Fig. 8.8. A 4-to-1 MUX (multiplexer) consists of four data inputs lines as |D0> to |D3>, two select lines-|S0> and |S1> and a single output line. There are many ways to draw 4-to-1 MUX, but here designing the 4-to-1 MUX using three 2-to-1 MUX.

The calculation of mux will be 2^n-to-1 MUX requires $(2^n - 1)$ 2-to-1 MUX. The select lines |S0> and |S1> select any four input lines to connect the output lines.

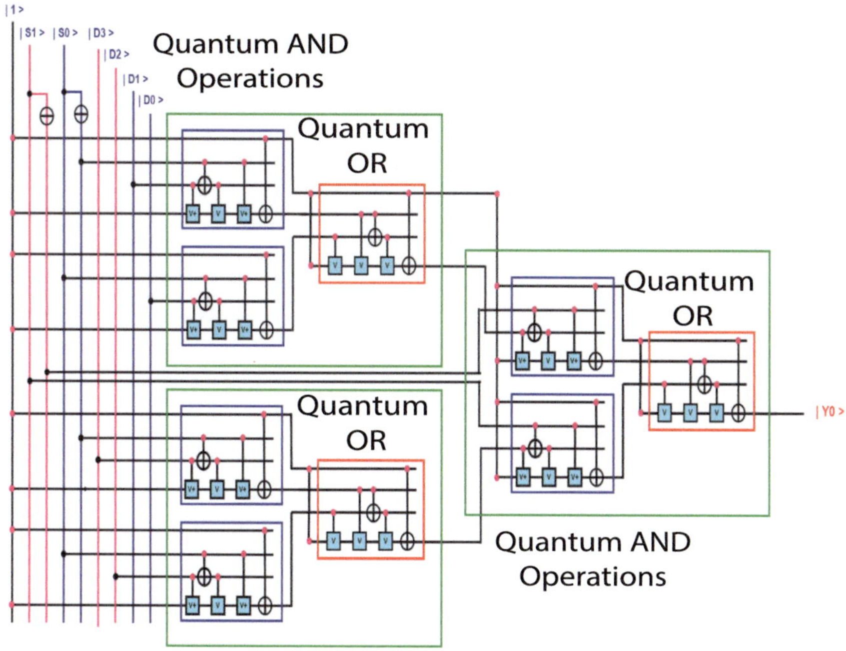

Fig. 8.8 4-to-1 quantum MUX

8.4.6.1 Working Principle of Quantum Multiplexer

The multiplexer has multiple inputs and a single output. Each 2-to-1 MUX consists of two quantum AND operations. The obtained production from two quantum AND gates will be the input of a quantum OR gate. Yield two outputs with the least significant qubits from 2-to-1 MUX, and these two outputs are propagated. In such a way, the outputs are achieved.

8.4.7 Design Procedure of Quantum ALU

This is the two-qubit ALU which consists of 16 quantum AND operations selected by 2-to-4 quantum decoders. According to the 2-to-4 quantum decoder, only one logical operation is performed: addition, multiplier, subtractor, or divider. Two-qubit quantum ALU operation circuit is shown in Fig. 8.9.
Every logical operation like addition and subtraction is executed with the help of a decoder. Each logical operation is described as follows.

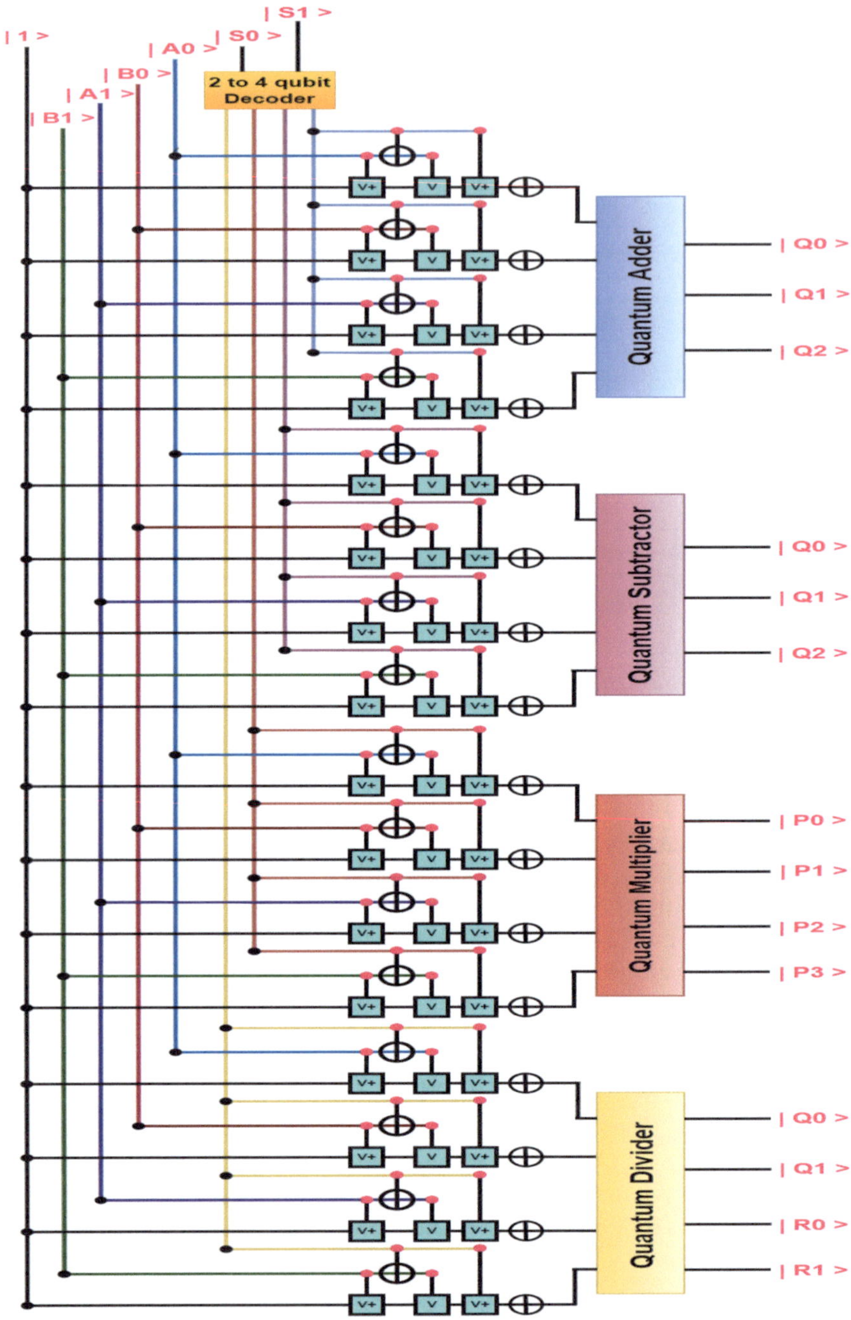

Fig. 8.9 2-qubit quantum ALU operation

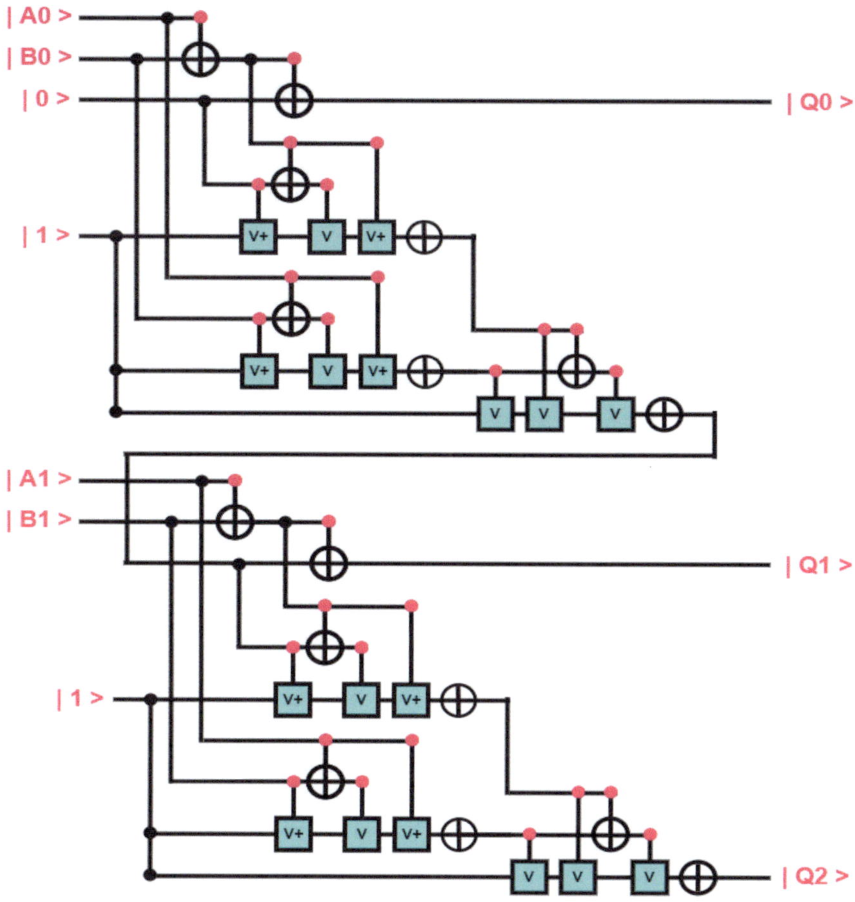

Fig. 8.10 Quantum adder operation

- **Quantum Adder**

This adder consists of quantum X-OR, AND, and OR operations used to perform addition. Figure 8.10 represents the diagram of the quantum adder.
According to Fig. 8.10, the first adder produces a sum, and Cout will be the Cin of the next adder. And this full adder finally has a second output and a Cout.

- **Quantum Subtractor**

This subtractor consists of a quantum full subtractor used to perform subtraction. Figure 8.11 shows the quantum subtractor circuit.
A full subtractor is such a combinational quantum circuit that can perform subtraction of two qubits where |A> and |B> are control bits, |Bin> is the borrow input and output is the difference D, and Bout is the borrow out. Here |A>, |B> indicates

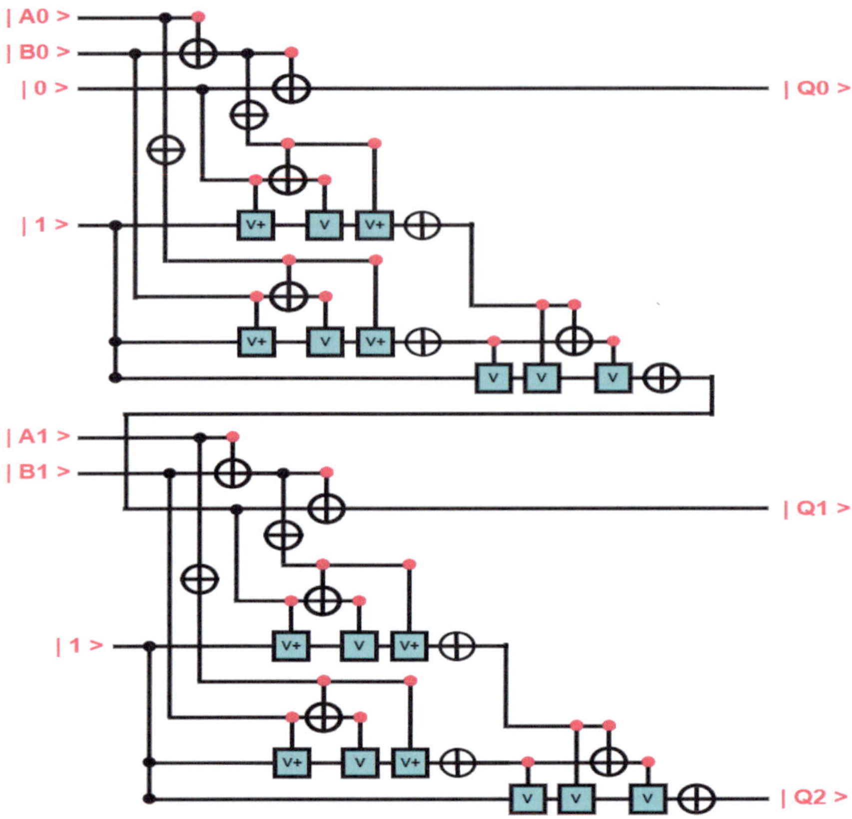

Fig. 8.11 Quantum subtractor operation

respectively minuend and subtrahend and |0> is considered a target qubit which is constant here. According to Fig. 8.11, the first full subtractor produces a difference, and Bout will be the Bin of the next full subtractor. And this full subtractor finally has a second output and a Bout.

- **Quantum Multiplier**

For a two-qubit quantum multiplier, two quantum digits are required. In a two-bit multiplier, four quantum AND gates and two half-adders are required to design the main circuit. Here, the AND gates will perform the multiplication and half-adders will add the partial product terms or carry. The circuit of the quantum multiplier is given in Fig. 8.12.

At first, it is required to perform quantum AND operations. Let $|A_1>$, $|A_0>$ and $|B_1>$, $|B_0>$ be the quantum digits where $|A_1>$, $|A_0>$ is the multiplicand and $|B_1>$, $|B_0>$ is the multiplier. As it is a quantum circuit, $|1>$ is assumed as a constant qubit which acts as a target qubit. In this circuit, the quantum AND gate will perform

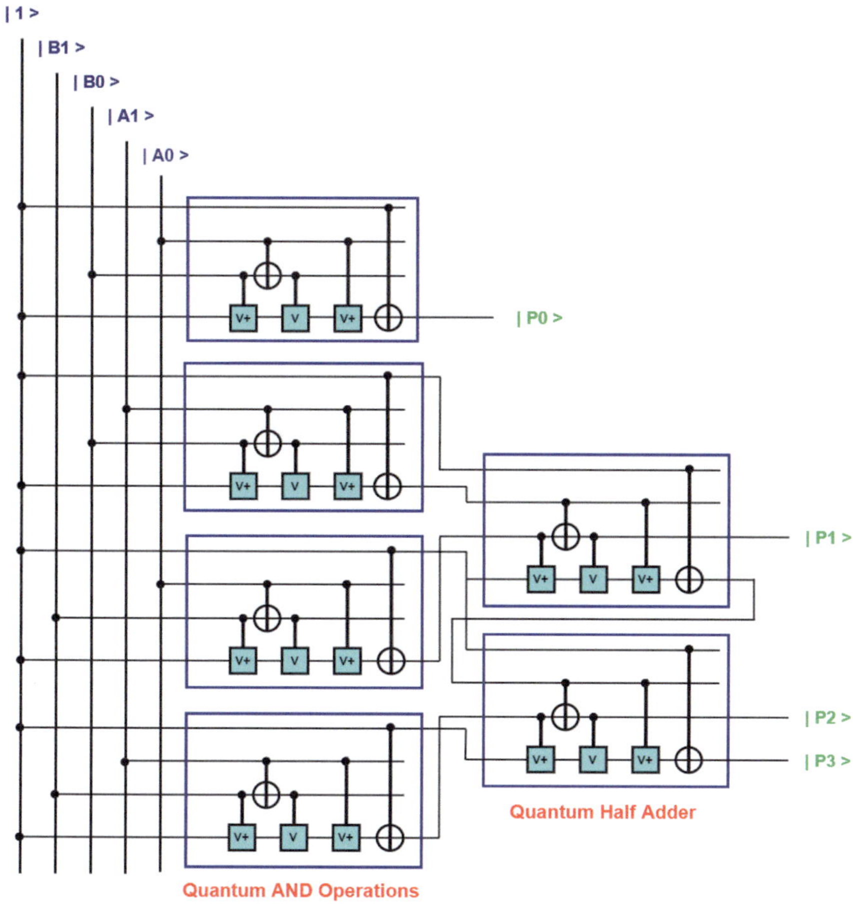

Fig. 8.12 Quantum multiplier operation

the quantum multiplication and the quantum half-adders will add the partial product terms. By doing these logical operations, it is easy to get a four-qubit output.

- **Quantum Divider**

For a two-qubit quantum divider, firstly there needs a quantum AND logical operation. In this AND operation, three inputs have been used. To avoid dilution, the third input of at first AND gate again has been connected with the second AND gate operations which helps us to get the output of AND operations using three inputs. It is the two-qubit divider as shown in Fig. 8.13.

First of all, in this circuit, four quantum AND operations are used where three inputs have been used. To avoid dilution of the circuit, the third input of the first AND gate again has been connected with the second AND gate which helps to get the output of AND operations using three inputs. Then the output of the quantum AND gate has to

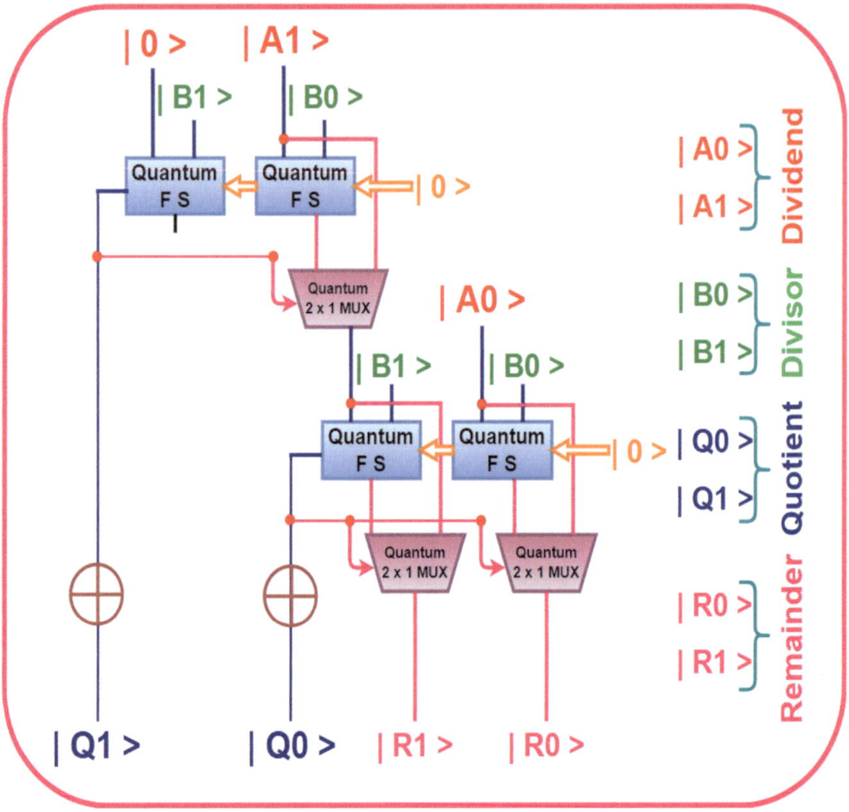

Fig. 8.13 Quantum divider operation

connect with OR gate as input for getting the exact output of $|Y_1>$. The constant $|1>$ is used to get the exact output of AND gate which works as a target qubit and it is the basic properties of the quantum logic circuit. This circuit has been completed using a two-qubit divisor and two-qubit divisible which produce maximum two-qubit output. Let $|A_0>$ and $|A_1>$ be respectively, $|1>$ and $|0>$ and, $|B_0>$ and $|B_1>$ be respectively $|1>$ and $|1>$ where $|A_0>|A_1>$ is divisor and $|B_0>|B_1>$ is dividend. Now, if $|1>|1>$ is dividend by divisor $|1>|0>$, the Quotient, respectively, 0 at $|Y_0>$ and 1 at $|Y_1>$ which works according to the following circuit. This is the basic working principle of the two-qubit quantum divider.

8.4.7.1 Working Principle of Quantum ALU

An arithmetic logic unit performs arithmetic and logic operations such as addition, subtraction, multiplication, and division. All information in a computer is stored and converted either $|0>$ or $|1>$. Open-switch and closed-switch concepts are used here.

An available transistor is a device in which no current passes through, represented by |0>. On the other hand, a closed transistor is a device in which current passes through, represented by |1>. Multiple transistors are connected to accomplish operations. For example, one transistor can be connected to the second one and turn the transistor's switch on or off based on the state of the second transistor. This is called a quantum gate operation that can allow the flow of current.

8.4.8 Design Procedure of Quantum Accumulator

The two-qubit accumulator consists of one quantum AND gate and two D flip-flops, where each D flip-flop is made of four NAND gates. | LOAD> and |Clk> are the inputs of the quantum AND gate, and the obtained output from AND gate will be the input of each D flip-flop. After logical execution, the first and second outputs from the first and second D flip-flops are obtained. The two-qubit quantum accumulator is shown in Fig. 8.14.

8.4.8.1 Working Principle of Quantum Accumulator

An accumulator is a register that acts as a temporary storage location and holds an intermediate value in mathematical and logical operations. For example, in the process "2+3+4", the accumulator will have the first value 2, then the value 5, and finally the value is 9. The data will be stored in the accumulator when the quantum AND gate's output is |1>. And if the quantum AND's output is |0> then no data will be stored in the accumulator.

8.5 Applications

It has already been shown that quantum computing solves several problems using many algorithms. The most well-known quantum algorithms are Shor's integer factoring algorithms and Grover's database search algorithm. It is also applied to error correction, Fourier transforms, and new methods such as quantum key exchange and quantum teleportation can transfer the state from one location to another. In 1994, Peter Shor discovered an algorithm to factor a large integer in polynomial time. He showed that it is possible to break the RSA cryptosystem using this algorithm. This algorithm can give the correct answer with high probability, and the other one can reduce the chance of failure repeating the algorithm. Another the most critical algorithm is Grover's algorithm. This algorithm has been designed for searching unsorted databases. For searching an unsorted array of size N, it needs O ($\sqrt{N}$) operations. In classical algorithms, it requires $O(N)$. This algorithm may be more accurate when it is an inverting function. It can also be used for estimating the median and mean of a set of numbers, and it is also possible to solve the collision problem. It can determine the connectivity of an n-vertex graph using O (3/2). One of the most significant advantages is that it can perform parallel computations, improving system performance. The 2^n inputs can be easily stored in n qubits using the Superposition state. The n qubits can be used as the input for a quantum circuit that performs arbitrary computations. Moreover, it can be used in artificial intelligence and machine learning, computational chemistry, drug design and development, cybersecurity and cryptography, and weather forecasting.

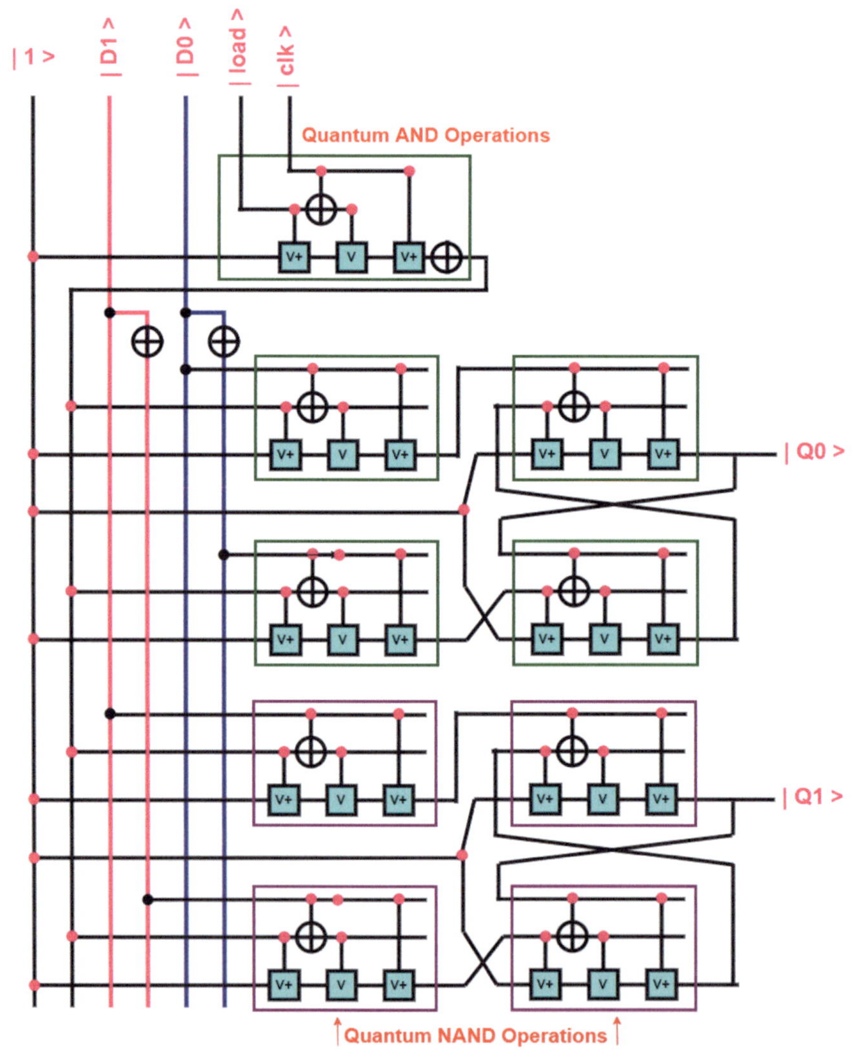

Fig. 8.14 2-qubit quantum accumulator

8.6 Summary

This chapter has tried to give possible architectural approaches for making the quantum nanoprocessor. This chapter explores specific architectural designs for quantum nanoprocessors that can achieve high performance. Then, individual quantum circuits for two-qubit CPUs with their design procedures and working principles are demonstrated. The design of all components of a quantum ALU is also discussed. All these ideas have been implemented theoretically. Finally, applications of this quantum nanoprocessor are also shown here.

Chapter 9
Quantum-DNA Nanoprocessor

9.1 Introduction

Quantum theory is a fundamental theory that describes the nature and behavior of matter and energy on the atomic and subatomic levels. Neils Bohr and Max Planck are the fathers of quantum theory, and both received Nobel prizes in Physics for their work on quanta. Einstein is the third founder of quantum theory who described light as quanta in his theory of the photoelectric effect and also won the Nobel prize in 1921. Biology, on the other hand, is a discipline of science concerned with the study of living creatures and their vital processes. This area is concerned with all elements of life that are physicochemical. All living systems are made up of molecules, and quantum mechanics can describe all molecules. Therefore, it is clear that quantum theory must play an essential role in biology because of having organisms. Over the last decades, scientists have been trying to merge quantum and biology. Recent research has shown us that phenomena such as photosynthesis, respiration, and how people think are all influenced by quantum mechanics. Quantum biology is a field where quantum mechanics and theoretical chemistry are described to biological objects and problems. Quantum biology may be used for computations, and it is concerned with the influence of non-trivial quantum phenomena. Quantum biology was firstly discussed by Erwin Schrodinger. In 1994, he debated applications of quantum mechanics in biology in his book named "What is Life?" However, in 1963 proton tunneling was published by Per-Olov Lowdin as another mechanism for DNA mutation. And, in his paper, he discovered "quantum biology" as a new field. Recently, many researchers have shown that DNA molecules have the property of coherence. They are also capable of having superposition, entanglement, and tunneling. As quantum physics and life science characteristics are identical, a hybrid nanoprocessor is proposed using DNA molecules sequence and quantum bit. This chapter emphasizes how a qubit is converted into a DNA molecule sequence. To convert a qubit into a DNA sequence, NMR relaxation is used. NMR relaxation is when an exciting magnetic state returns to its equilibrium distribution. When a molecule

H. Md. Hasan Babu, *Quantum Biocomputing in Quantum Biology Volume II*,
https://doi.org/10.1007/978-981-97-5349-9_9

drops into the NMR probe as a sample, it goes to an excited state with the help of a magnetic field. When the electromagnetic resonance doesn't emit, the magnetic field becomes weak. In that case, the superposition state molecule loses its energy and comes into the ground state. This whole process is known as NMR relaxation. This process is executed at room temperature. In other way, quantum biological computing and nanoprocessors explore the intersection of quantum mechanics, biology, and nanoscale technology to develop novel computing paradigms. It combines the power of quantum entanglement and superposition with biological systems and nanotechnology to create advanced computational tools. It use the theme of quantum biology that investigates how quantum phenomena influence biological processes, such as enzyme catalysis and sensory perception. It aims to leverage quantum mechanical principles in biological systems for computation. So, this chapter will describe the quantum-DNA nanoprocessor.

9.2 Basic Definitions

A quantum-DNA nanoprocessor is a hybrid circuit that can execute all logical instructions on the quantum qubit as well as DNA molecule size programming. It can also handle any kind of logical problems and logical units that carry out the computer program's instruction. The quantum-DNA nanoprocessor controls all types of data flow and instructions. Inputs are fed into a qubit form in each component, and output is achieved in DNA molecule form. However, The CPU consists of five core components are given as follows:

1. Control Unit (CU)
2. Register
3. Arithmetic Logic Unit (ALU)
4. RAM
5. Buses.

1. **Control unit**: The control unit is the main component of a nanoprocessor in a computer that directs the operation of the nanoprocessor. The primary function is to fetch and execute instructions from the main memory.
2. **Register**: Registers are one kind of memory to accept, store, and transfer data and instructions which are used immediately by the nanoprocessor. The main function is to hold an instruction, a storage address, or any data. The register mainly containing the memory location will be used to calculate the address of the next instruction when the current instruction is completed.
3. **Arithmetic logic unit**: In computing, an arithmetic logic unit is a combinational DNA circuit that handles all the calculations the nanoprocessor may need. It performs all types of arithmetic operations like addition, subtraction, and division.
4. **RAM**: RAM stands for "Random Access Memory," called volatile memory. It is one of the components that contain information as a DNA sequence or qubit, and

the nanoprocessor can read or write to those sequences or qubits as information depending on whether the READ or WRITE line is signaled. The main goal of RAM is to store and access data on a short-term basis.

5. **Buses**: Buses are high-speed internal connections used to send control signals and data between the nanoprocessor and other components of the quantum-DNA computers. All computers need buses, whether they are DNA computers, quantum computers, supercomputers, or classical computers; and these buses are used for transferring data between processors and other components. That is why The DNA nanoprocessor used in DNA computers will have been described here, where buses will be used like other computers. There are three types of buses—address bus, control bus, and data bus which are briefly described in the architecture of basic components.

9.3 Block Diagram of Quantum-DNA Nanoprocessor

The CPU (Central processing unit) is the computer's brain, which manipulates data and performs all the basic arithmetic and logical operations. It contains many quantum combinational circuits such as IR (instruction register), PC (Program counter), Multiplexer, ALU (arithmetic logic unit), and RAM (Random access Memory). A source is applied here to provide heat that helps DNA molecules perform a chemical reaction. In addition, a heat transfer circuit is shown here to transfer quantum heat to the DNA CPU. In this section, a quantum-DNA nanoprocessor is presented that consists of the components mentioned above.

Figure 9.1 shows the complete Quantum-DNA nanoprocessor. There are two inputs which are in CPUs to accomplish meaningful work instruction and data. The task of the instruction register is to tell the CPU what actions need to be performed on the data. Instructions are represented using qubit. The CPU's inputs are stored in the memory, i.e., RAM. The CPU functions a cycle of fetching instructions. After fetching an instruction, it is decoded. After decoding, the CPUs execute the instruction. At this time, the control unit helps to move the data from memory to registers in the arithmetic logic unit and then ALU performs the instructions. Finally, the control unit stores the result of this operation in memory or a register. Now, let's describe the above working principle in this combinational quantum-DNA nanoprocessor.

Notice that Fig. 9.1 shows the data as a qubit passed from quantum RAM to the DNA instruction register. It cannot be possible to pass qubit to the DNA instruction register directly. Before passing the DNA instruction register, the qubit must be converted to a DNA sequence. NMR relaxation is used here to convert a qubit to a DNA sequence. It is known to us that quantum computing is faster than DNA computing. Quantum cache memory has been used here to store qubits. Quantum RAM sends data quickly, but DNA cannot accept the information soon. Quantum qubits generate so much heat during computations, and on the other hand, heat needs to provide DNA for chemical reactions. The quantum heat obtained from the quantum logical operations is used in this nanoprocessor to perform chemical reactions of DNA combinational circuits. That is why a heat transfer circuit has been used here.

When an instruction is fetched into the DNA decoder, it is executed. The fetching and executed parts are completed at the DNA nanoprocessor. Biocomputing or DNA computing or biological computing utilizes biological materials and processes for computational tasks, aiming to leverage the inherent complexity and efficiency of biological systems. Nanoprocessors in this context are nanoscale devices that can be programmed to perform specific computations by manipulating the interactions between biomolecules. Nanoprocessors in biocomputing refer to nanoscale computational devices that utilize biological components like DNA, RNA, or proteins to perform computations. These devices are designed to process information based on biological interactions and stimuli, offering potential for highly parallel and energy-efficient computation. When an instruction is completed, data as a DNA sequence is sent to the quantum RAM. In that case, DNA sequences cannot pass directly to quantum RAM. Therefore, it is required to store the DNA sequence to DNA cache memory, and from this cache memory, the DNA sequence is fed into the NMR circuit to convert the DNA sequence to the qubit. It is required to follow these steps in this nanoprocessor:

1. Pass data as qubit to quantum cache memory.
2. Do NMR relaxation to convert qubit to the DNA sequence.
3. After executing an instruction, pass DNA sequence to DNA cache memory.
4. Use only NMR to convert the DNA sequence to the qubit.

The whole nanoprocessor will work in this way. The working procedure of quantum cache memory, NMR relaxation, NMR, and DNA cache memory will be described in detail later.

9.4 Basic Components of Quantum-DNA Nanoprocessor

A quantum-DNA nanoprocessor has been designed using the following components shown in Fig. 9.1. The components of this nanoprocessor are as follows:

1. Quantum RAM
2. DNA Instruction Register
3. DNA Program Counter
4. DNA Incrementor
5. DNA decoder
6. DNA Multiplexer
7. DNA ALU
8. DNA Accumulator.
9. Quantum cache memory
10. DNA cache memory
11. NMR relaxation

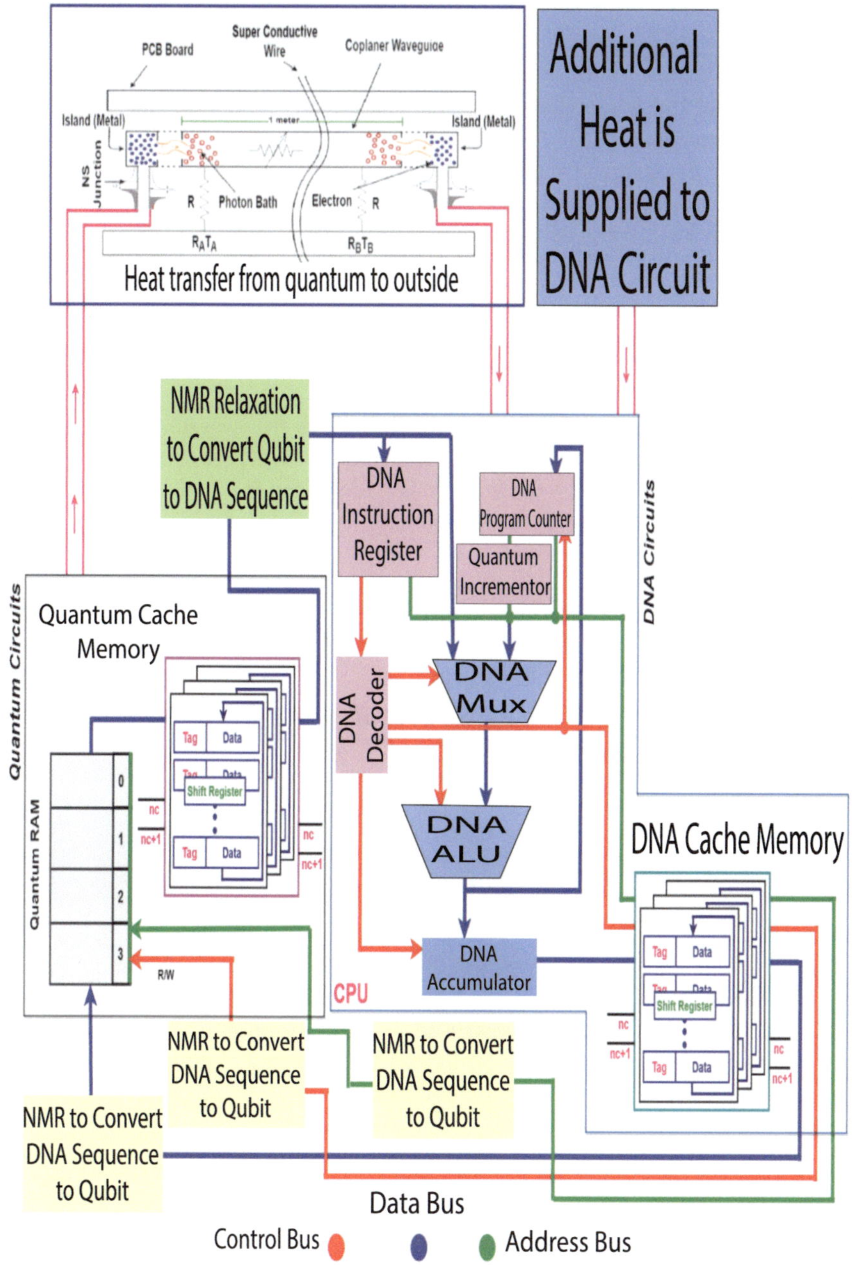

Fig. 9.1 Quantum-DNA Nanoprocessor

12. NMR and
13. Heat transfer circuit.

Except for the above components, three buses are used in this nanoprocessor—data bus, address bus, and control bus. These buses are used to transfer data from one part to another. These are also essential components of the nanoprocessor that help to accomplish meaningful work.

However, now all components are explained in this section.

9.4.1 Design and Working Principles of Quantum RAM

RAM is the most crucial CPU component that stores the data quickly, a primary and volatile memory. It is required to design two-qubit RAM for this quantum-DNA nanoprocessor. This two-qubit RAM is already illustrated in Sects. 8.4.1 and 8.4.1.1 in Chap. 8.

9.4.2 Design and Working Principles of DNA CPU

According to Fig. 9.1, it is observed that all nanoprocessor elements except RAM are in DNA structures. A DNA nanoprocessor consists of a DNA Instruction register, program counter, incrementor, multiplexer, decoder, accumulator, and ALU.

9.4.3 DNA Instruction Register

The DNA instruction register consists of 16 DNA AND operations shown in Fig. 9.2. The instruction bit is double the CPU bit. So, as this instruction is a component of a two-bit DNA CPU, Instructions will be $2^2 = 4$. Therefore, the minimum four instructions can be defined-LOAD A, LOAD B, ADD A B, and OUT. Instruction register is a particular register mainly used to store the instructions currently being executed by the DNA computer.

9.4.4 DNA Program Counter

The DNA program counter consists of two D flip-flops where the first and second D flip-flops are designed using four DNA NAND gates. Each input also has to be in DNA, NOT form. The top four DNA NAND circuits represent the first D flip-flop. This D flip-flop generates two outputs where one output has been skipped as there

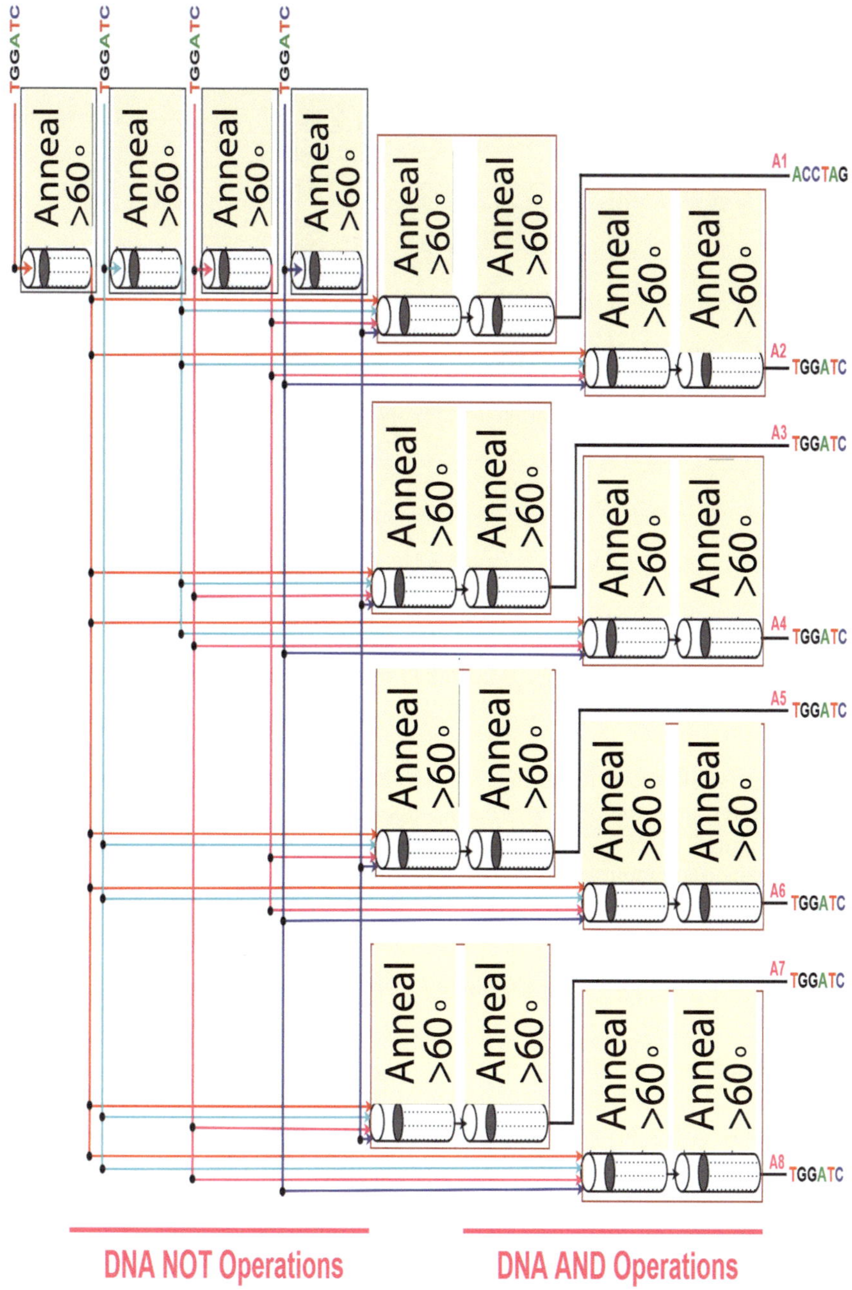

Fig. 9.2 DNA instruction register

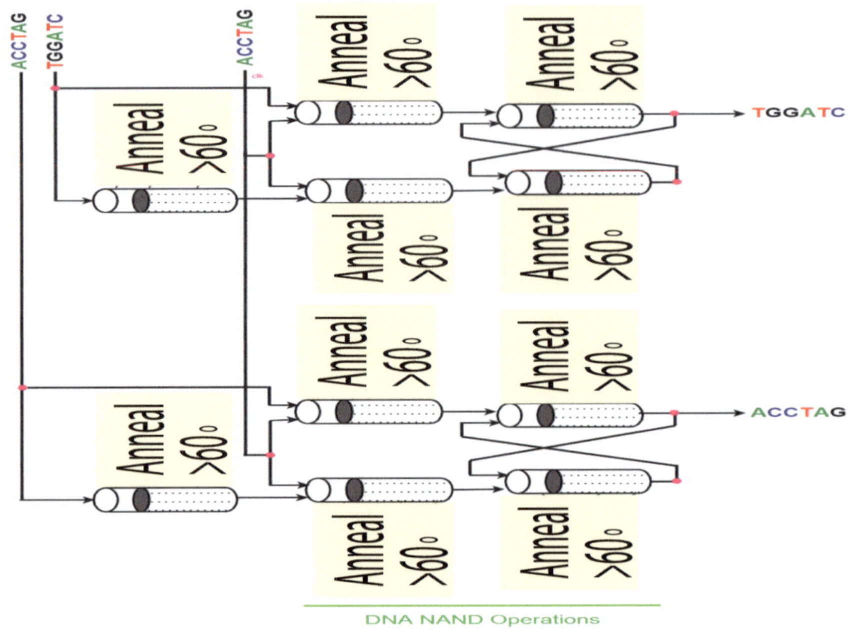

Fig. 9.3 2-Molecular DNA program counter

is no need to show according to Fig. 9.3. Another D flip-flop consists of four NAND gates delivered after the first DNA D flip-flop. This D flip-flop also generates two outputs. Figure 9.3 shows that only two results that act as inputs are required. So, the rest of the two outputs have been skipped. The circuit is as follows.

9.4.5 DNA Incrementer Circuit

The DNA incrementer circuit is designed only using two DNA half-adder circuits shown in Fig. 9.4, where each half-adder consists of an individual DNA XOR circuit and individual DNA AND circuit. The DNA XOR gate is used to compute the sum result, and DNA AND gate produces the carry-out result.

9.4.6 DNA Decoder

To implement a two-molecular DNA CPU, a 2-to-4 DNA decoder is used where this decoder consists of four DNA AND operations gates with one enable input. Two inputs are fed into this decoder, and each input must be a CNOT form. And

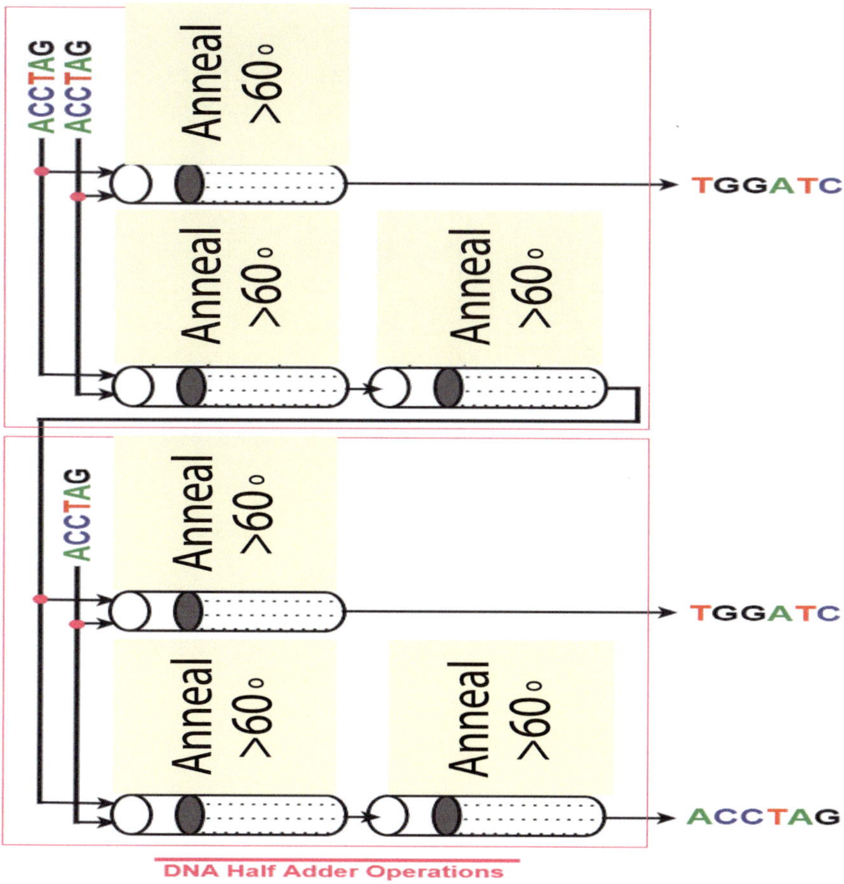

Fig. 9.4 2-Molecular DNA incrementer

this decoder produces four outputs based on given input sequences. The following circuit is a two-molecular DNA decoder as shown in Fig. 9.5.

9.4.7 DNA Multiplexer

A 4-to-1 MUX (multiplexer) consists of four data input lines defined as ACCTAG, TGGATC, ACCTAG, and TGGATC; two select lines are expressed here TGGATC and TGGATC and a single output line. There are many ways to draw 4-to-1 MUX, but here designing the 4-to-1 MUX by using three 2-to-1 MUX. The calculation of MUX will be 2^n-to-1 MUX which requires $(2^n - 1)$2-to-1 MUX. S_0 and S_1 select any

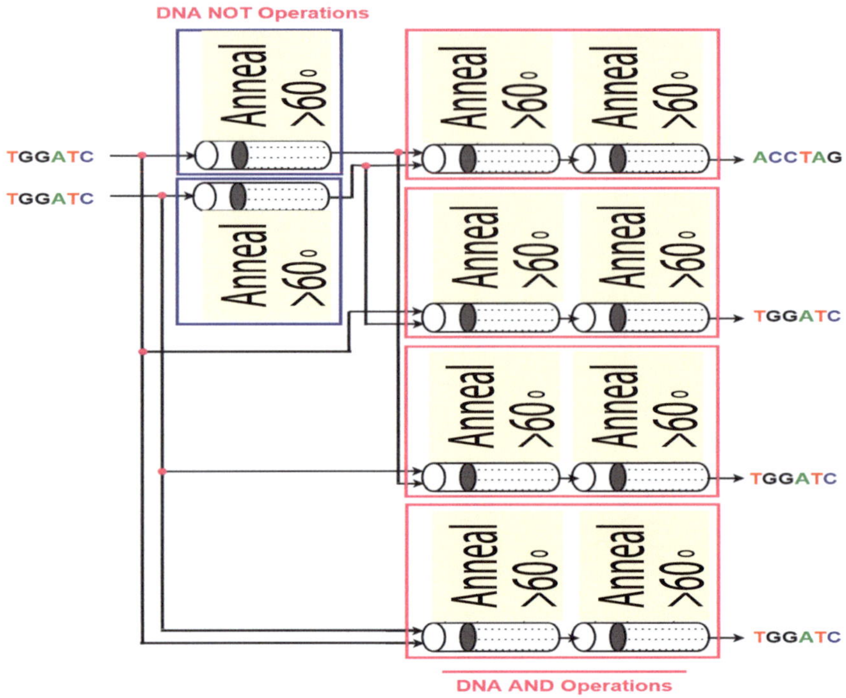

Fig. 9.5 2-Molecular DNA decoder

four input lines to connect the output lines. The circuit of 4-to-1 DNA Multiplexer is shown in Fig. 9.6.

9.4.8 DNA ALU

An arithmetic logic unit is the core component of a nanoprocessor. It is a combinational circuit that performs arithmetic and logic operations. The control unit indicates to ALU what process it needs to achieve on the data and stores the result in an output register. It is required to construct a two-molecular DNA ALU for a two-molecular DNA nanoprocessor. This two-molecular DNA ALU consists of sixteen DNA AND operations selected by DNA 2-to-4 decoder. According to the 2-to-4 decoder, only one logical operation is performed, such as addition, multiplier, subtractor, or divider. The following circuit is DNA two-molecular ALU as shown in Fig. 9.7.
Every logical operation like addition and subtraction is executed with the help of a decoder. Each logical operation is described below.

- **DNA Adder**

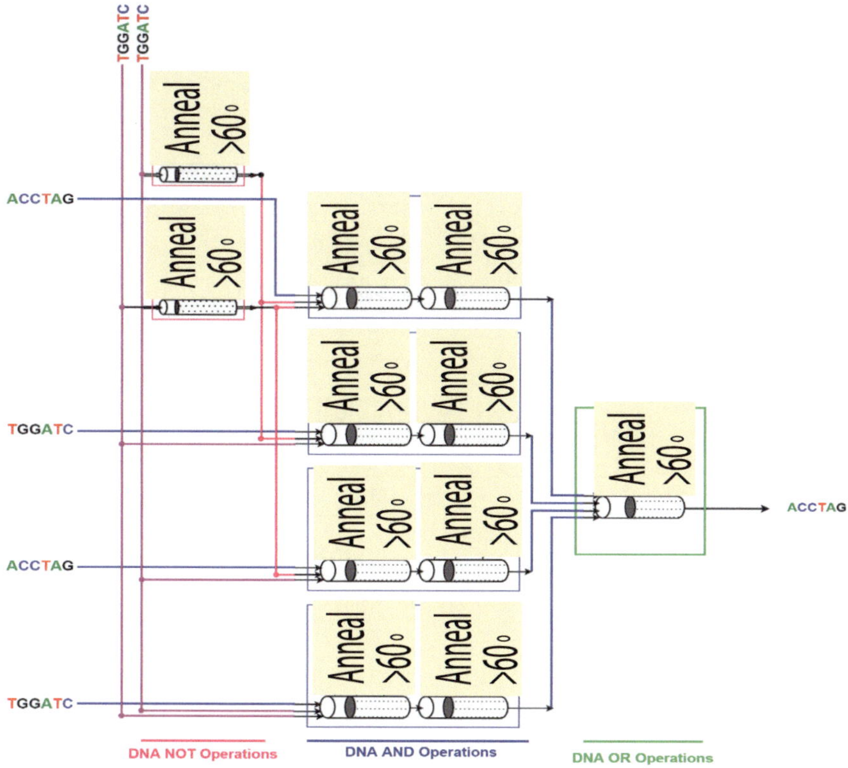

Fig. 9.6 4-to-1 DNA multiplexer

This DNA Adder consists of two DNA full adders used to perform addition. The circuit architecture of the DNA adder is shown in Fig. 9.8.
According to the diagram in Fig. 9.8, the first full adder produces the sum and the C_{out} that will be the C_{in} of the next full adder. And this full adder finally produces a second output and a C_{out}. The design procedure and working principle of a DNA full adder is discussed in Fig. 9.8.

- **DNA Subtractor**

Figure 9.9 shows the DNA full subtractor. A full subtractor is such a combinational DNA circuit that can perform subtraction of two molecules, where A and B are control bits, B_{in} is the borrow input, output is the difference D, and B_{out} is the borrow out. Here A and B indicate, respectively, minuend and subtrahend, and 0 is considered a target bit which is constant here. According to Fig. 9.9, the first full subtractor produces a difference, and B_{out} will be the B_{in} of the next full subtractor.

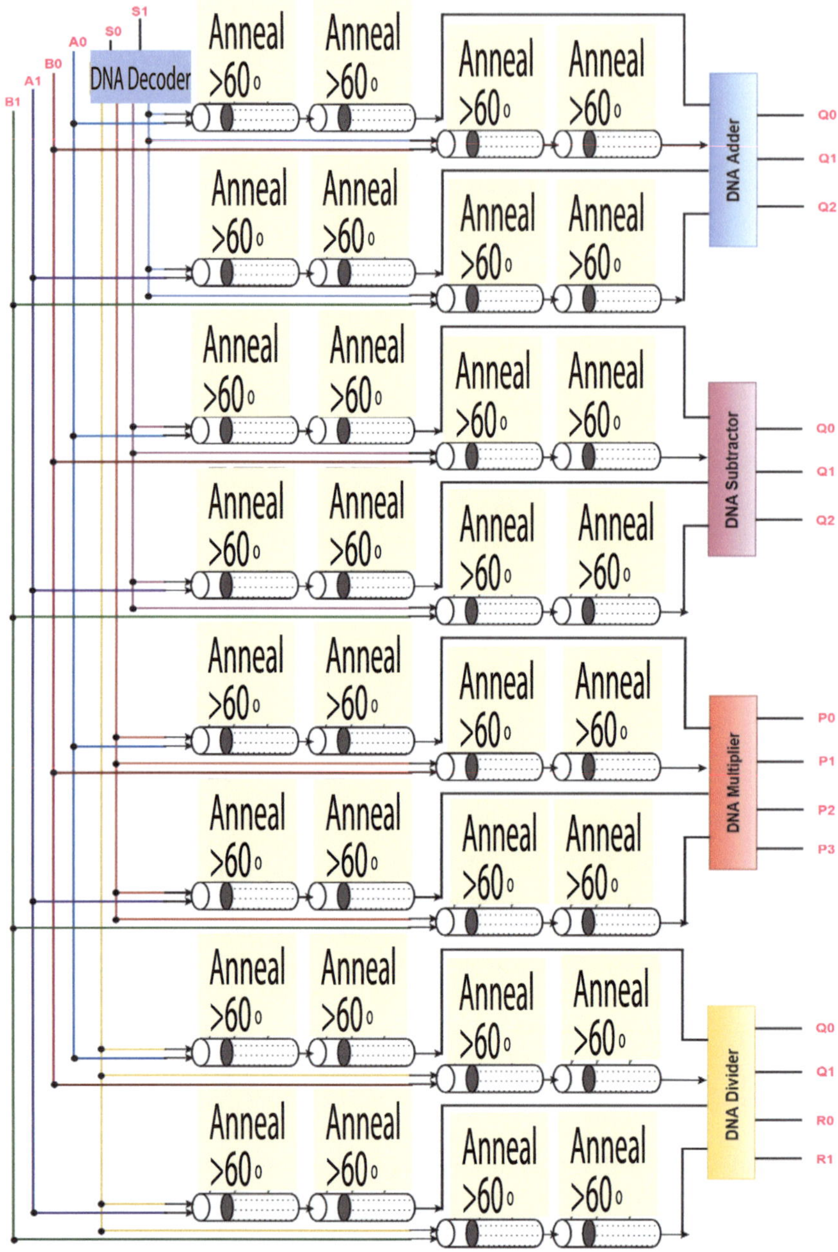

Fig. 9.7 DNA 2-Molecular ALU

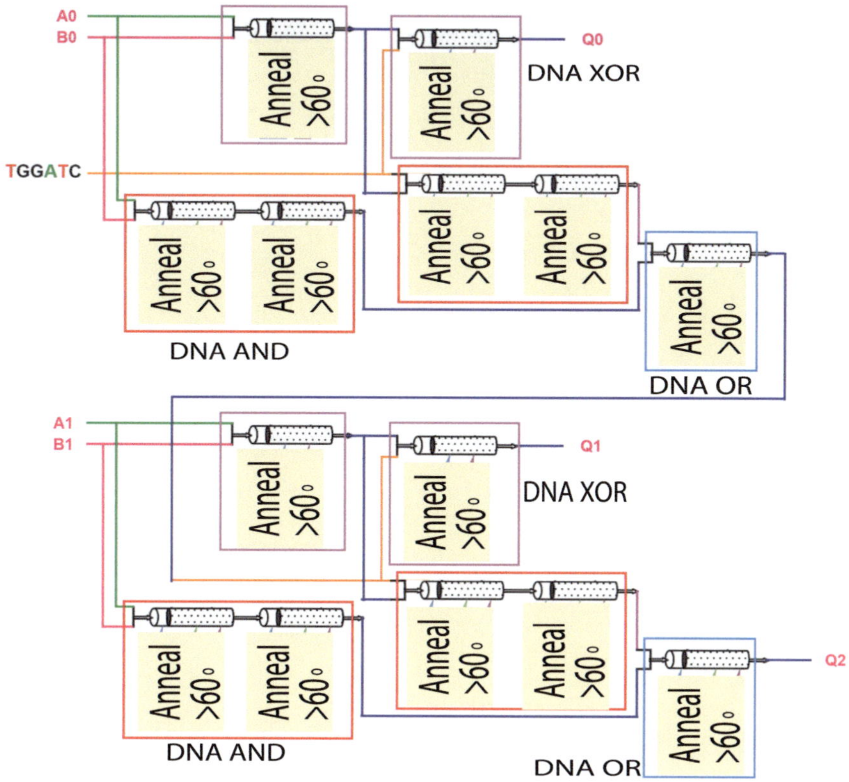

Fig. 9.8 DNA Adder

- **DNA Multiplier**

For performing a two-molecular DNA multiplier, it is required two DNA sequences ACCTAG and TGGATC that represent, respectively, 1 and 0. In this circuit, four AND gate operations have been implemented with two half-adders. Here, the AND gates will perform the multiplication and half-adders will add the partial product terms or carry. Hence, the circuit obtained is given in Fig. 9.10.
At first, it is required to perform four DNA AND gate operations for performing multiplication and half-adders will add the partial product. In this circuit, an input portion the first two DNA sequences of the left sides are ACCTAG, TGGATC which are multipliers and the second two DNA sequences of the right sides are also ACCTAG, TGGATC which are multiplicand. ACCTAG and TGGATC represent, respectively, 1 and 0. There is a constant DNA base sequence in AND gate tubes which makes either TRUE or FALSE bond with the upcoming input sequence. Their TRUE bond will be double-strand but if they do not make any bond, it will make a single bond which is destroyed by the DNase enzyme. They will make a true (1) bond when ACCTAG and TGGATC coincide. But when two ACCTAG sequences or two TGGATC sequences

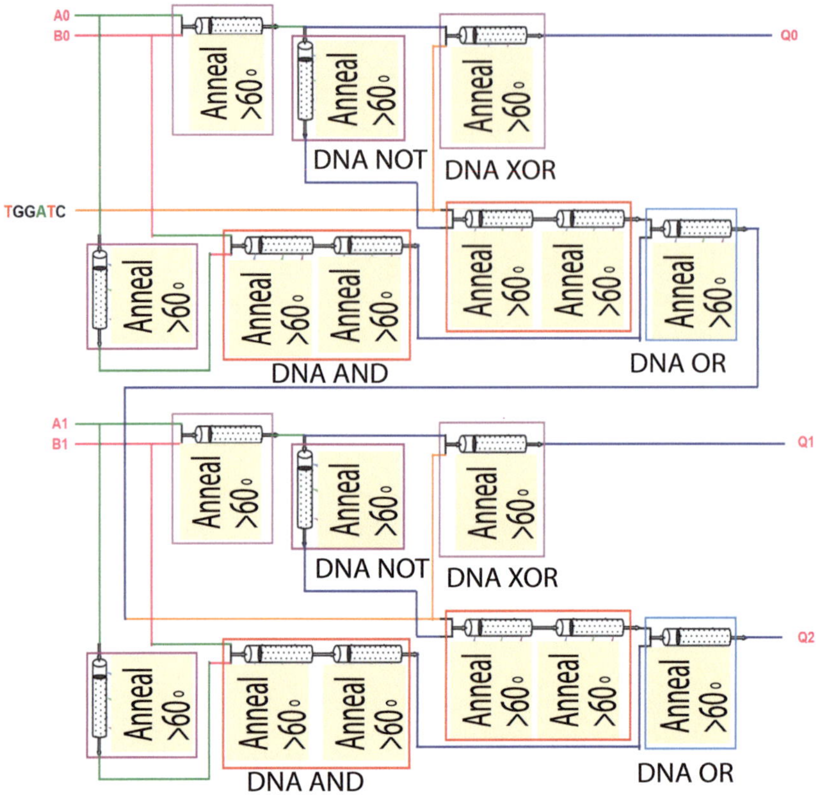

Fig. 9.9 DNA subtractor

try to make a bond, then their bonding will be false because the same sequence will never be able to make a double-strand bond. So, in that case, the output will be 0. The TRUE and FALSE bonds, respectively, represent 1 and 0. According to the given circuit, the first DNA sequence of input is B_1 which has been represented by ACCTAG, the second DNA sequence of input is B_0 (TGGATC), the third DNA sequence of input represents A_1 (ACCTAG), and the last input represents A_0 (TGGATC).

From the circuit, the first output will be the multiplication of TGGATC (A_0) and TGGATC (B_0). As TGGATC represents 0, the multiplication of these two bits will also be 0 (TGGATC) which is the first output.

The first partial product of LSB is the first output/product of LSB. Therefore, there is no need to add any partial product. So, output is obtained directly.

The inputs of the second AND gate are ACCTAG(A_1) and TGGATC(B_0) which will give an output TGGATC (0) and the inputs of the third AND gate are ACCTAG(B_1) and TGGATC(A_0) which gives output TGGATC (0). The obtained two outputs from the two AND gates will be the input of the DNA XOR gate which perform addition

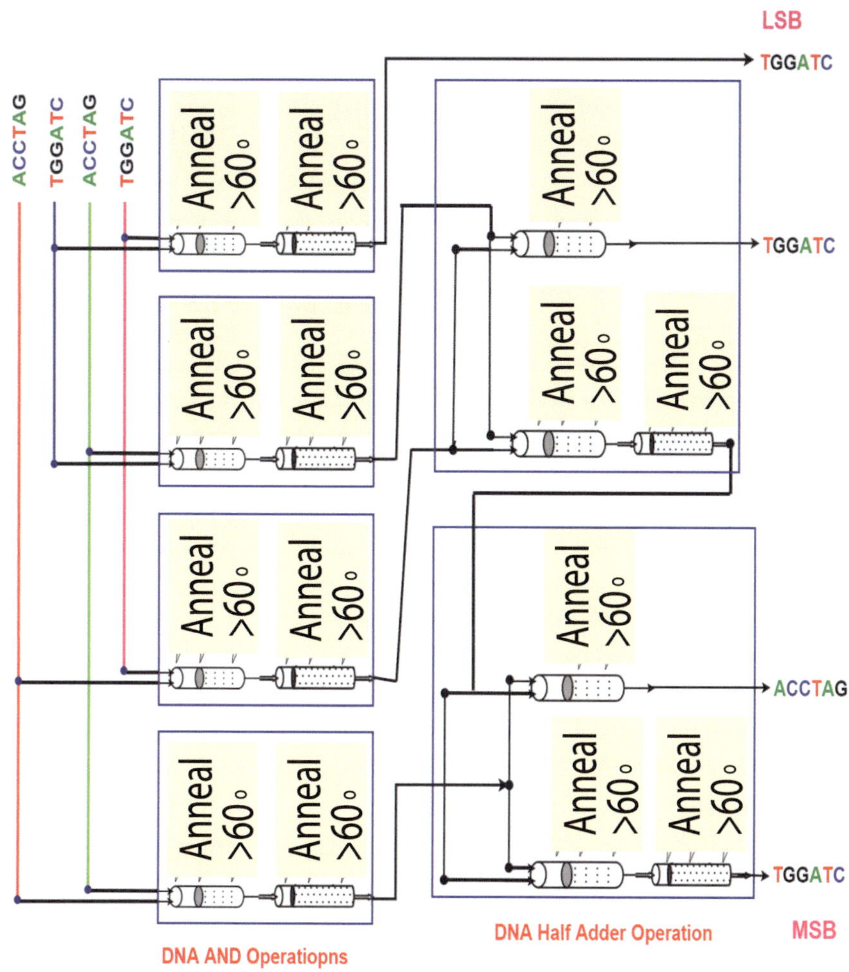

Fig. 9.10 DNA multiplier

and generate output TGGATC (0). Then AND gate is performed the operation which generates a carry bit. This is the half-adder operation. And finally, the last DNA AND gate produces the output ACCTAG (1) which is the multiplication of ACCTAG (B_1) and ACCTAG (A_1). This output and the previous carry bit from the first half-adder will be the input of XOR produces output ACCTAG (1) and the output of the DNA AND gate of the half-adder will act as the carry bit. As there is no carry bit, it simply represents TGGATC (0). Now, the movement from the MSB (Most significant bit) to LSB (least significant bit) then the output will be respectively, TGGATC (0), AGGTAC (1), TGGATC (0), and TGGATC (0), respectively that is the production of ACCTAG (1) and TGGATC (0) multiplicand and ACCTAG (1) and TGGATC (0) multiplier. According to the binary logic, the product of 10 molecules and 10 molecules is 0100 which has been implemented here. In this circuit, the input can be anything. So, it is the whole working procedure of a 2-molecular DNA multiplier.

A_1 (ACCTAG) A_0 (TGGATC)

B_1 (ACCTAG) B_0 (TGGATC)

A_1 B_0 (FALSE) A_0 B_0 (FALSE)$_{(Partial\ product)}$

B_1 A_1 (TRUE) B_1 A_0 (FALSE) × (Left shift)

C_2 (FALSE) B_1 A_1 (TRUE) A_1 B_0 (FALSE) A_0 B_0 (FALSE)

\+

B_1 A_0 (FALSE)

P_3 (FALSE) P_2 (TRUE) P_1 (FALSE) P_0 (FALSE) $_{(Total\ output)}$

If the multiplication of any two DNA sequences is 0, it will be false which is expressed by F and if production is 1 of any two DNA sequences, it will be true which is presented by 1 in the above multiplication portion.

- **DNA Divider**

The logical operations of the circuit depend on the number of DNA sequences. This circuit has been completed using a 2-molecular divisor and 2-molecular divisible which produces a maximum 4-bit output. Here, bit means sequence ACCTAG (1) and TGGATC (0). Let, the divisor be ACCTAG (1) and TGGATC (0) and the dividend be also ACCTAG (1) and ACCTAG (1). Now, if ACCTAG (1) and ACCTAG (1) are divided by divisor ACCTAG (1) and TGGATC (0), then Quotient, respectively, 0 at the first output and 1 at the second output which works according to the following circuit. Figure 9.11 shows the circuit architecture of a DNA divider operation.

9.4.9 Accumulator

One DNA AND gate and two D flip-flops are fed into the accumulator as input; four DNA NAND operations are performed to implement each D flip-flop. The DNA AND gate has two inputs named LOAD and CLOCK. The output of the AND gate will be the input of each D flip-flop. And finally, after performing all operations, each D flip-flop evaluates the first and the second outputs. Figure 9.12 shows the DNA accumulator circuit.

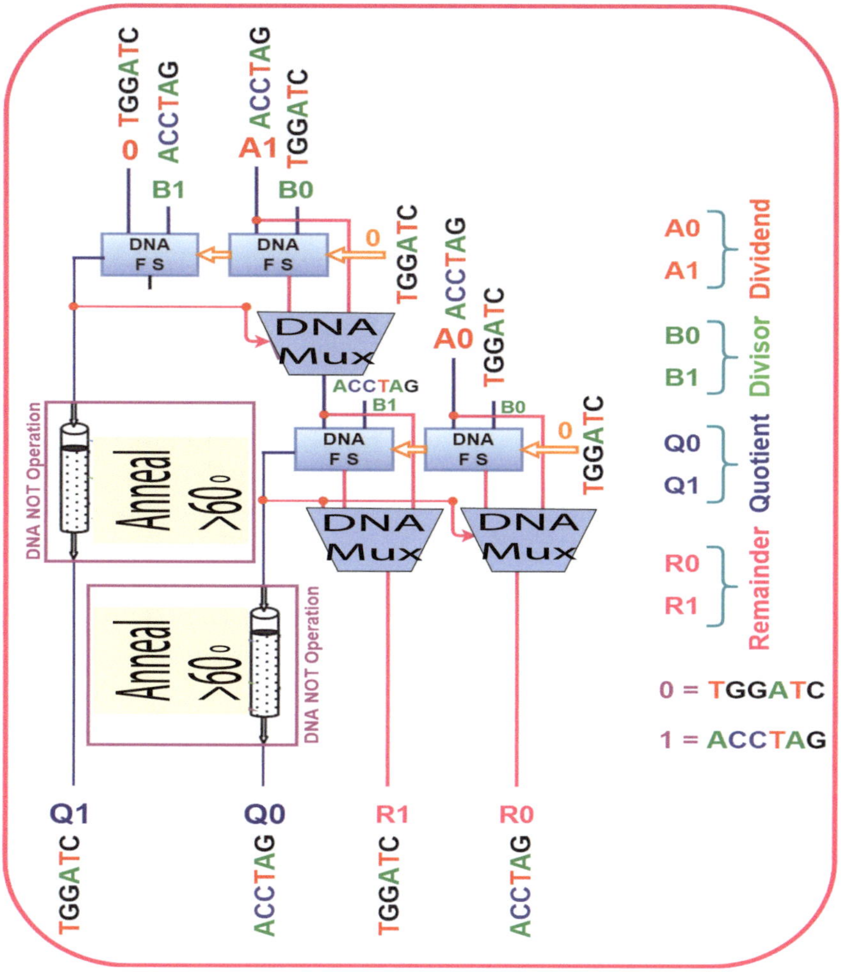

Fig. 9.11 DNA divider

9.4.10 Quantum Cache Memory

Cache memory can be used in the quantum system due to its speed and reliability. It can pass and get data very frequently. The mapping and swapping techniques in the cache memory are much more optimized. But to use a cache memory in the quantum system, it needs to be designed. The Figures and description of Quantum cache memory are discussed in Part I, Chap. 2, Sect. 2.5.

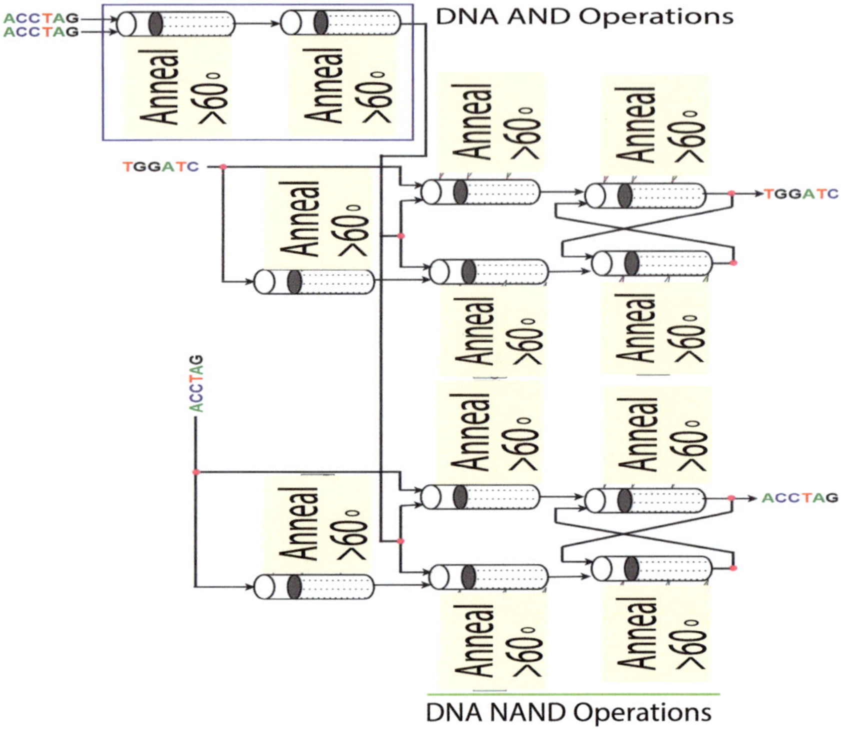

Fig. 9.12 DNA accumulator

9.4.11 DNA Cache Memory

DNA cache memory, which will store DNA sequences after coming from the DNA circuit. The detailed figures and description are discussed in Part II, Chap. 4 and Sect. 4.5.

9.4.12 NMR at 0-K for Converting DNA Sequence to Qubit in DNA-Quantum Nanoprocessor

NMR is a physical phenomenon wherein the nucleus in a magnetic field absorbs and emits electromagnetic radiation. NMR creates a strong magnetic field and makes molecules' nuclei excited, and then molecules exist in superposition. This energy is at a particular resonance frequency that relies upon the strength of the magnetic field and the magnetic properties of the isotope of the atoms. NMR permits the observation of specific quantum mechanical and, magnetic properties of the atomic nucleus. The details of NMR are described in Part IV, Chap. 14, Sect. 14.3.

9.4.13 NMR Relaxation at 0-K for Converting Qubit to the DNA Sequence in Quantum-DNA Nanoprocessor

NMR relaxation is when an exciting magnetic state returns to its equilibrium distribution. When a molecule drops into the NMR probe as a sample, it goes to an excited state with the help of a magnetic field. NMR relaxation is used to convert qubit to a DNA sequence at 0-K. The details of this are described in Part IV, Chap. 14, Sect. 14.3.

9.4.14 Heat Transfer Circuit

Quantum circuits produce heat, and DNA requires heat to perform a DNA operation in quantum computation and DNA computation. In the quantum process, qubits generate heat when they become isolated and estimated. So, this heat can be transferred to DNA circuits. It transmits heat from quantum RAM to DNA CPU. This is the main reason for using the heat transfer circuit here. The details of the heat transfer circuit will be discussed in Part IV, Chap. 13, and Sect. 13.2.

9.5 Applications

Quantum computing is incredibly faster than DNA computing. Furthermore, DNA computing has more and more storage capacity than quantum computing. It is known that quantum qubits generate much heat, and DNA molecules need to provide temperatures for chemical reactions. Therefore, the two are connected so that the temperature can be adjusted. The proposed nanoprocessor can work better in the medical field. They will detect any disease as this nanoprocessor inherits DNA features. Calculations with quantum computing are particularly promising and can solve complex problems with vast amounts of data.

On the other hand, DNA computing can also solve NP combinatorial problems promisingly. So, it can ensure that this nanoprocessor can solve complex mathematical problems and equations faster than the expectations. Moreover, this nanoprocessor will store more and more information as it will bear the DNA's features.

9.6 Summary

This chapter shows a loamy nanoprocessor using quantum computing and DNA computing. Quantum computing and DNA computing have potential advantages in computing technology. Although there are so many advantages, limitations are not more minor. The prime motto of implementing this nanoprocessor is to reduce demerits and to increase merits. The algorithm has been shown here, and speed has been measured to check the power of its performances. DNA can inherit the computing property, and it can be an incredible combination that can solve more unimaginable problems than the expectations. DNA resources are available in nature. So, the cost of this nanoprocessor will be less. Data search will be easy, and this will ensure high privacy. But during implementation many problems will arise. This implementation is complex, and it will be challenging to secure privacy on the internet. Another cause can be heating problems because sometimes it needs to provide more heat in DNA logical operations from outside, which will be more challenging. But the uses of this nanoprocessor will take computer technology a big step forward. But for this, more research in this field is needed.

Chapter 10
DNA-Quantum Nano Processor

10.1 Introduction

DNA computing is one interdisciplinary research field that is growing fast as it can simulate the biomolecular structure of DNA and compute using this DNA sequence. At first, Adleman showed that DNA could solve a problem. He cracked a complex problem using the Hamiltonian path problem, and this approach has been extended by Lipton, who can solve another NP-complete problem. Many researchers are working to produce a computer based on DNA molecules to replace or beneficially complement a silicon-based computer. The primary function of DNA is to transmit data, and they can process this data simultaneously. Data is encoded in a DNA strand form in this DNA computing. It has several merits such as high storage capacity, faster speed because of parallelism, and low power consumption. A mix of 1018 DNA strands could operate 104 times faster than today's advanced computer, which is incredible.

On the contrary, there is another computing system known as quantum computing. It also has several advantages such as data security, high storage capacity, and less power. And the most important thing is that these two systems can inherit their properties from each other. That is why to implement such a nanoprocessor that will be loamy, i.e., it will have the property of two computing systems at a time. It will be a beautiful combination that will add a new dimension to computer technology progress. This nanoprocessor is quite the same as the quantum DNA nanoprocessor. RAM will be kept in DNA form in this nanoprocessor implementation, and other components will be held in quantum form. As RAM will be in a DNA state, it must have a vast and massive storage capacity. Therefore, it will store more and more data in a small area. This is the most significant advantage of this nanoprocessor. However, it is required to convert DNA sequence to qubits and qubits to DNA sequence like the previous nanoprocessor that were already discussed in Chap. 9. A quadrupole ion trap is used to convert the DNA sequence to the qubit. This quadrupole ion trap will accept DNA sequence and convert it into qubit. A quadrupole ion trap is such an ion trap that uses dynamic electric fields to trap any charged particles.

H. Md. Hasan Babu, *Quantum Biocomputing in Quantum Biology Volume II*,
https://doi.org/10.1007/978-981-97-5349-9_10

On the other hand, it is possible to use trap ions to convert qubits to the DNA sequence. An ion trap is an "electric-field-test-tube" that contains gaseous ions. These gaseous ions can be either positively charged or negatively charged. Nanoprocessors, which are devices at the nanoscale, play a crucial role in building quantum computers. They are used to fabricate qubits, which are the fundamental units of information in quantum computing. On the other hand, nanoprocessors can also be used to build bio-inspired computing systems, which mimic the way biological systems process information. For example, networks of nanoscale channels can be created to regulate the flow of molecules, potentially leading to a new type of computing. Bioquantum computing or DNA-quantum computing or biological quantum computing combines the principles of quantum mechanics with biological systems. It explores how quantum phenomena can be used to perform computations in biological contexts, such as simulating complex biological systems or developing new drugs. This field aims to leverage the power of quantum computers to tackle challenges in biology and medicine, such as drug discovery, disease modeling, and personalized medicine. Nanoprocessors in bioquantum computing refer to the use of nanoscale devices and quantum mechanical principles to perform computations, particularly in biological contexts. This field leverages the unique properties of quantum systems, like superposition and entanglement, to tackle problems in biology and medicine that are computationally intractable for classical computers. Nanoprocessors, being extremely small, can be used to create the physical components of quantum computers, such as qubits, and can also be used to build bio-inspired computing systems. This chapter will describe the DNA-quantum nanoprocessor.

10.2 Basic Definitions

A DNA-Quantum nanoprocessor is a combinational nanoprocessor that has been designed using DNA sequence and quantum qubits. Biological quantum computing or DNA-quantum computing or bio-quantum computing explores the potential for using biological systems for quantum computation, either by leveraging naturally occurring quantum phenomena within living organisms or by building quantum computers using biological materials or structures. Nanoprocessors, specifically, could play a role in both of these approaches, providing nanoscale tools for manipulation and control of biological components or for creating small-scale quantum computing devices. Some researchers hypothesize that certain biological processes, like photosynthesis or brain function, might utilize quantum phenomena for enhanced efficiency or information processing. This could involve using quantum entanglement or superposition within living cells. In essence, biological quantum computing and nanotechnology are intertwined. Nanotechnology provides the tools to build and control the complex structures needed for quantum computation, while biological systems offer a potential source of inspiration and materials for quantum computing. A nanoprocessor will execute all types of logical instructions on behalf of programs. A nanoprocessor consists of five core components: control unit, register, arithmetic logic unit, RAM, and buses. All these components help nanoprocessors to accomplish work properly. The basic definitions of these components are already described in Sect. 8.2 in Chap. 8, Sect. 8.3 in Chap. 8, and Sect. 9.2 in Chap. 9.

10.3 Block Diagram of DNA-Quantum Nano Processor

The following nanoprocessor consists of a quantum instruction register, program counter, multiplexer, arithmetic logic unit, and DNA RAM. In addition, there is a heat transfer circuit shown here to eliminate heat from the quantum combinational circuit and an additional power source to provide heat into the DNA combinational circuit to perform the chemical reaction with each other. Figure 10.1 represents the DNA-quantum nanoprocessor.

The above nanoprocessor is called DNA-quantum nanoprocessor, and it is a two-molecular DNA sequence qubit. A DNA sequence passed from DNA RAM to the quantum instruction register in the above nanoprocessor. This is because DNA cache memory is used and a quadrupole ion trap between DNA RAM and quantum IR where cache memory is used to store data, and a quadrupole ion trap has been used here to convert the DNA sequence to the qubit. An ion trap is an "electric-field-test-tube" that contains gaseous ions. These gaseous ions can be either positively charged or negatively charged. On the contrary, a quadrupole ion trap is such an ion trap that uses dynamic electric fields to trap any charged particles.

When an instruction is fetched into the quantum decoder, it is executed. After completing an instruction, data as a qubit is sent to the DNA RAM. But the problem is that one cannot send qubit to DNA RAM directly. It is required to convert it. Before passing and performing this operation, a standard trap ion has been used to convert qubit to DNA RAM address. But in that case, it is required to store the qubit in quantum cache memory. This nanoprocessor will follow the given steps:

1. Pass data as DNA sequence to DNA cache memory.
2. Do quadrupole for converting DNA sequence to the qubit.
3. After executing an instruction, pass the qubit to quantum memory.
4. Use only trap ions for converting qubits to a DNA sequence.

10.4 Basic Components of DNA-Quantum Nanoprocessor

A DNA-quantum nanoprocessor consists of the RAM, IR, ALU, MUX, incrementor, PC, accumulator, and decoder, where RAM is the only component implemented using DNA sequence, and other components of this nanoprocessor have been designed using qubit. DNA RAM is demonstrated below, and the rest of the components have been appropriately illustrated with their circuits in the subsection of 8.4 Section in Chap. 8.

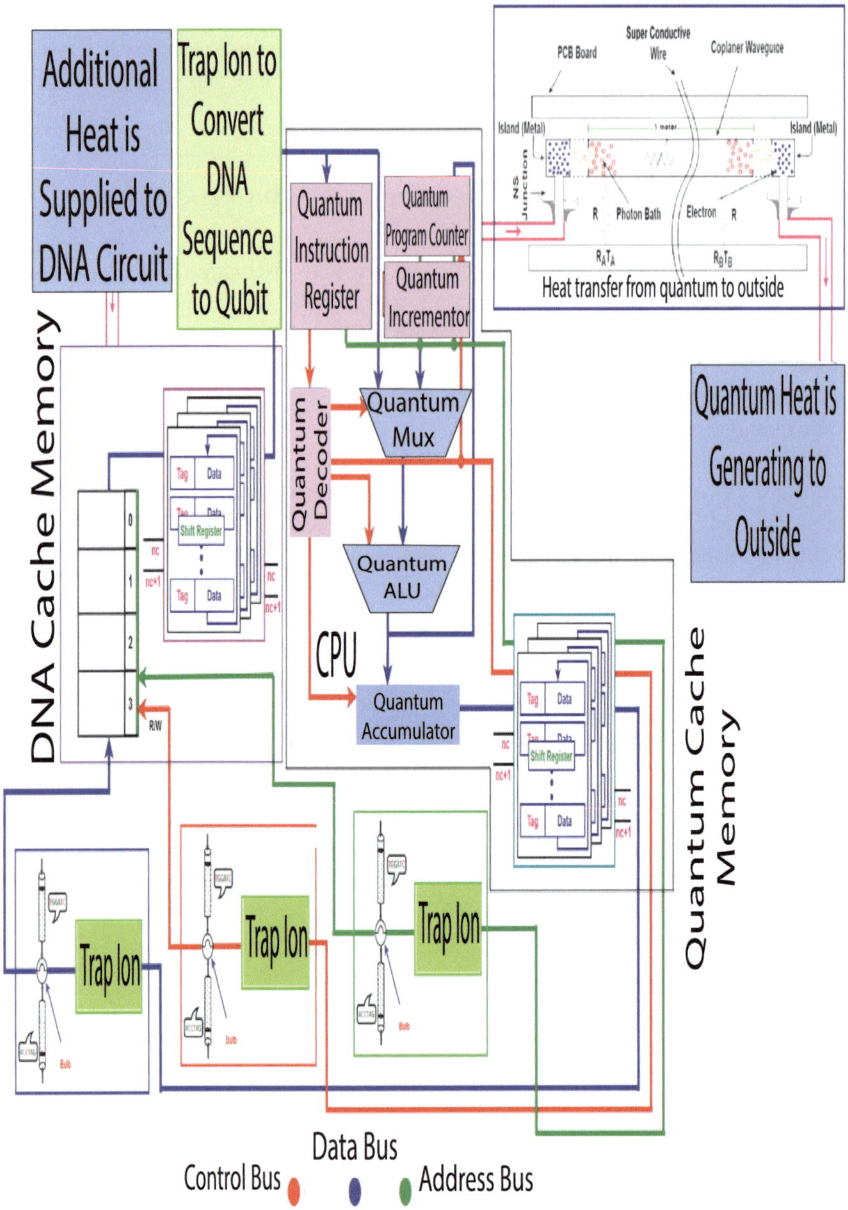

Fig. 10.1 DNA-quantum nanoprocessor

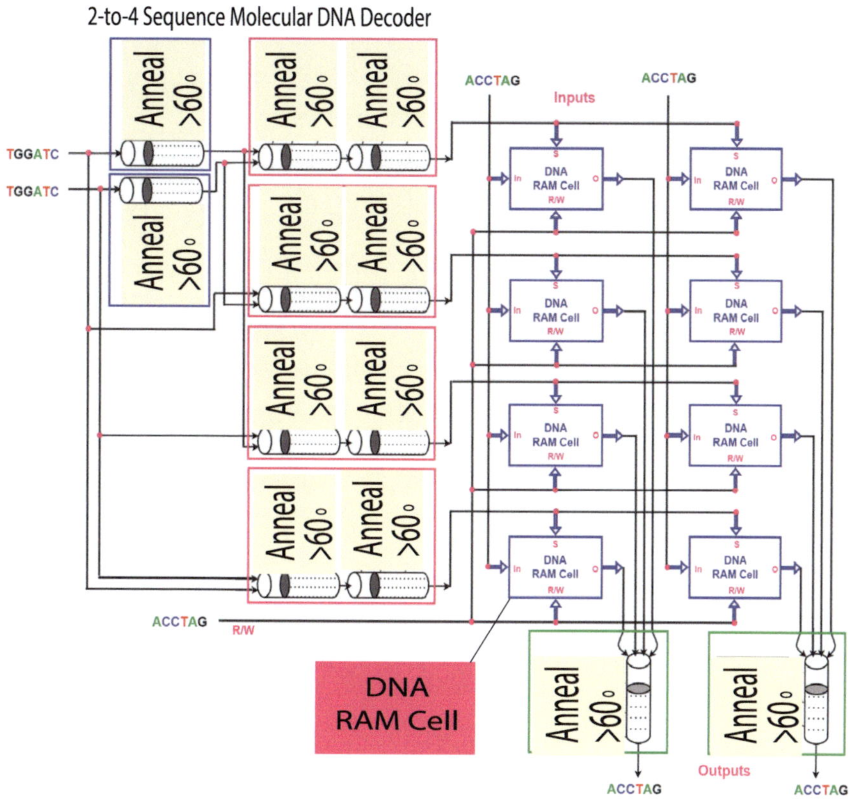

Fig. 10.2 2-molecular DNA RAM

10.4.1 DNA RAM

It requires two DNA sequences as address lines, and each address line has to be connected with a DNA NOT operation for simulating 4-to-2-molecular DNA RAM. These address line combinations will be the input of 2-to-4 DNA decoders which consists of four DNA AND functions, and the DNA decoder must enable input. Four select lines are achieved from the decoder and each select line is attached to each RAM cell. The word calculation of RAM will be 2^k, where k is the address line, and 2^k is the total words of n-bit, and the decoder combination will be $k \times 2^k$. This two-molecular RAM consists of four separate DNA RAM cells, and each cell has three inputs such as D_0 or D_1 or others. Anyone selects lines and can read/write inputs. The obtained output from four DNA RAM cells will be the input of a DNA OR operation, which produces the final output. This is the whole design procedure of 4-to-2-molecular RAM. The circuit of DNA RAM is given in Fig. 10.2.

Figure 10.3 presents the implementation of a 4-to-2-molecular sequence DNA RAM. The DNA RAM cell is given in Fig. 10.3.

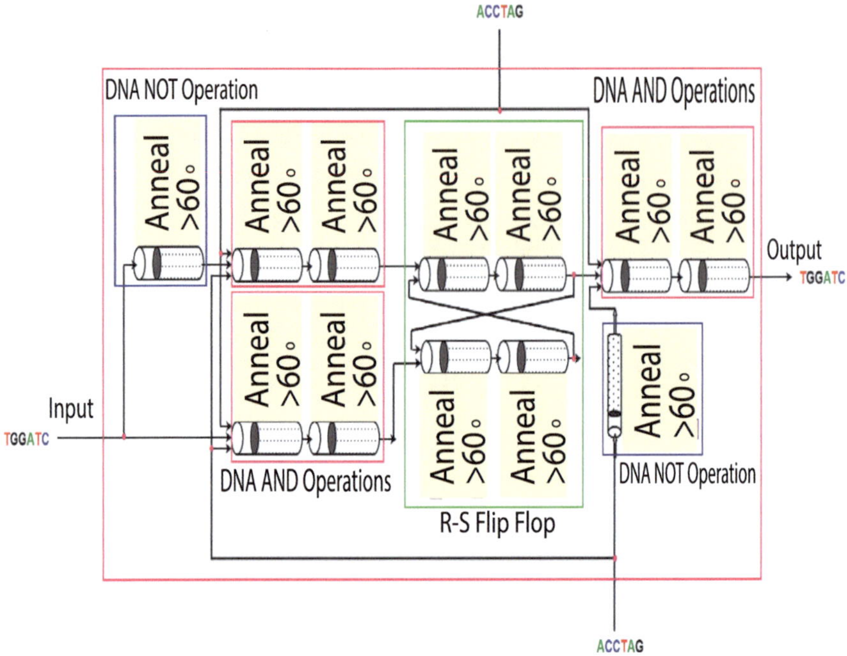

Fig. 10.3 DNA RAM cell

10.5 Quadrupole Ion Trap

A quadrupole ion trap is such an ion trap that uses dynamic electric fields to trap any charged particles. They are also called Paul traps or radio-frequency traps (RF). This trap is used as a component of a mass spectrometer. It is used to convert DNA sequences to qubits. The detailed description is given in Part 2, Chap. 9, and Sect. 9.3.6.

10.6 Paul Trap Ion

An ion trap is an "electric field test tube" that has gaseous ions. These gaseous ions can be either positively charged or negatively charged. The two most common ion traps are the Penning trap, which is used to comb through static and oscillating electric fields. Here the Paul trap is used. The details of this is discussed in Chap. 9, Sect. 9.2.3.

10.7 Design Procedure and Working Principle of DNA Cache Memory and Quantum Cache Memory

Cache memory is a high-speed random-access memory that is built into the nanoprocessor. Data is transferred into cache memory more quickly than RAM. The cache memory holds the temporary data in the CPU which may require for manipulation. It is important to note that the design and working principle of DNA cache memory and quantum cache memory have the same manner. The design procedure and working principle of these cache memories have already been demonstrated in Chap. 4 and Chap. 2, Sects. 4.5 and 2.5.

10.8 Applications

The main role of DNA molecules is that they can store information for long-term. They use DNA strands to store information and use its properties of DNA to perform logical operations. A small test tube of DNA strands can operate billions of operations simultaneously, and thus this DNA computing achieves a high speed. DNA's molecule performance is assumed to be faster than today's supercomputers. On the contrary, the power of quantum computing is impressive. This is because these computers are so effective for data simulations.

Furthermore, these computers ensure high privacy because they are good at cryptography. From this discussion, it is said that these two systems can open a new door to the incredible invention together. Based on figuring out these, a DNA-quantum nanoprocessor is made. Any complex and mathematical operations are calculated using these nanoprocessors within a second. This will ensure high security and can play a significant role in online security depending on the difficulty of large numbers into primes. Furthermore, by combining this nanoprocessor, one can solve any problem with the help of both nanoprocessors, such as optimization problems, weather forecasting, and chemical solutions.

10.9 Summary

This chapter has given a detailed explanation of the nanoprocessor and introduced a new cutting-edge technology-based nanoprocessor. This new nanoprocessor should be fast, error-resistant, and robust. This chapter first shows a block diagram of a two-molecular DNA nanoprocessor and then demonstrates the algorithm to construct this latest technology-based processor. Moreover, heat is measured and the speed is checked. The main motto of this nanoprocessor is to achieve more advantages. This design of DNA-quantum nanoprocessor follows the previous nanoprocessor. It is known that quantum computing and DNA computing already have many potential benefits. Despite having many advantages, if these two logical computations are combined, the users will gain more advantages that will progress the computer technology. This new computing method will add a new dimension to DNA computing and the quantum computing arena. But the problem is that it is not easy to implement practically. Many obstacles have been faced during implementation because the working environment of these two systems is different. For example, it is required to provide heat from outside the territory for performing chemical reactions, which can be challenging to handle. Another thing is that the entanglement of many qubits and DNA sequences is currently as tricky as maintaining the necessary state, and there is a probability of increasing high error rates in this nanoprocessor.

Part IV
Heat, Speed, and Data Related Issues in Quantum Biocomputing

Overview

The main objective of this part is to provide information about Heat Measurement, Heat Transfer, Speed Calculation, Data Conversion and Data Management in quantum computing and DNA computing. The importance of heat calculation cannot be described in limited words. Depending on the produced heat from DNA operational circuit, quantum operational circuit, DNA-quantum operational circuit, and quantum-DNA operational circuit, one can identify the useful one for daily life or research. In addition, produced heat or required heat for quantum computing and DNA computing is an arresting matter for researchers. Chapter 11 describes the procedure for calculating heat from the mentioned operational circuits. In the case of calculating heat from quantum operational circuits, thermodynamics law and its theory has been used for to capture produced heat of each quantum qubit. The produced heat from a circuit is proportional to the number of quantum qubits used in the circuit. In the DNA operational circuit where a certain amount of heat is needed to perform any computing operation. It performs some specific steps as synthesizing, mixing, annealing, melting, amplifying, separating, extracting, cutting, ligating, subtracting, marking DNA sequence, destroying, detecting, and reading DNA sequence. According to the research of Zheng and Yang, each specific step of DNA computation needs different amounts of heat and it never exceeds 98 degrees Celsius. The melting step requires heat which depends on its sequence where the number of adenine, guanine, cytosine, and thymine molecules is considered. Besides the accuracy of a system, speed is a metric that can be used to measure the performance of a system. The main objectives of Chap. 12 are to calculate the operational speed of DNA operational circuit, quantum operational circuit, DNA-quantum operational circuit, and quantum-DNA operational circuit. The heat transfer is another issue here to be discussed. The heat produced by quantum computing can be transferred to DNA computing where heat is needed to perform computation. Excessive heat is bad for quantum circuits, so the surplus heat is transferred to the DNA circuit to keep it balanced. This heat transfer has happened only in quantum-DNA and DNA-quantum circuits and is discussed in Chap. 13. Data conversion is very important in whole working with quantum-DNA circuits and DNA-quantum circuits. Chapter 14 describes the data conversion circuits like trap ion, quadrupole trap ion, NMR, and NMR relaxation circuits. The conversion process of quantum data (qubit) to DNA data (DNA sequences) and DNA data to quantum data is shown here. The last chapter (Chap. 15) of this part describes the data management system during computation. The cache memory to control data is an important matter to explain. DNA cache memory and quantum cache memory are needed for the computation of quantum-DNA circuits and DNA-quantum circuits (quantum biocomputing circuits).

Chapter 11
Heat Calculation

11.1 Introduction

Quantum computing is a field of study that focuses on the creation of computer-based technologies based on quantum-theoretical principles. On the quantum (atomic and subatomic) level, quantum theory describes the nature and behavior of energy and matter. To execute certain computational tasks, quantum computing employs a combination of qubits. All of this are done at a far higher rate than their traditional computing equipment. Quantum computers represent a significant advancement in computing capability, with enormous performance benefits for specific use cases. The ability of bits to be in several states at the same time gives the quantum computer a lot of computing capability. They can accomplish jobs with a mix of $|1\rangle$, $|0\rangle$, and both $|1\rangle$ and $|0\rangle$ at the same time. So, quantum computing can be defined as an area of computing that is focused on the development of computer technology based on the principles of quantum theory. In addition, quantum computing is much faster than classical bit-wise computing.

On the other hand, instead of using typical silicon chips, DNA computing uses biological molecules to do computations. The four-character genetic alphabet (A-adenine, G-guanine, C-cytosine, and T-thymine) is used in DNA computing instead of the binary alphabet (1 and 0) utilized by standard computers. This is possible due to the ability to create small DNA molecules with any arbitrary sequence. The input of any DNA operation can be represented by DNA molecules with specific sequences. The instructions are carried out by laboratory operations on the molecules, and the result is defined as some property of the final set of molecules. DNA computing promises significant and meaningful linkages between computers and life systems, as well as massively parallel computations. DNA computing can actually carry out millions of operations at the same time.

The produced heat from a circuit is always an important topic for each type of computing circuit. Nowadays, produced heat or required heat for quantum computing and DNA computing is an arresting matter for researchers. This chapter is going

H. Md. Hasan Babu, *Quantum Biocomputing in Quantum Biology Volume II*,
https://doi.org/10.1007/978-981-97-5349-9_11

to express some ways to find out the amount of heat produced from a quantum computing circuit and calculate the amount of required heat for DNA computation.

11.2 Basic Definitions for Heat Calculation in Quantum Circuits

This section discusses basic information and theory to calculate the produced heat from quantum circuits and the heat required to perform a DNA operation in quantum and DNA computations.

In quantum operation, qubits generate heat when they become isolated and start to compute. In quantum physics, thermodynamics exists and the thermodynamics rule is quite the same for qubit also. When a quantum system is a single qubit then the Hamiltonian matrix can be written as follows:

$$H = \frac{-1}{2}€\sigma \tag{11.1}$$

This may correspond to an electric spin in a vertical magnetic field where is the energy difference between the states $|\uparrow >$ = |0> and $|\downarrow >$ = |1>. The same Hamilton matrix may also refer to a two-level atom where the ground and excited states are denoted as |0> and |1>, respectively. The Gibbs state of the qubit takes the form:

$$\rho_\beta = \frac{1}{2cosh\left(\beta\frac{€}{2}\right)} e^{\beta€\frac{\sigma}{2}} = \frac{1}{1 + e^{-\beta€}} \left(|0 > 0| + e^{-\beta€}\,|1 > 1|\right) \tag{11.2}$$

Researchers have introduced the inverse temperature $\beta = 1/K\ _B$T. Then the occupation number,

$$n_\beta = tr\left(\hat{n}\hat{\rho}_\beta\right) = \frac{1}{1 + \exp(\beta\epsilon)}$$

So, the average energy of thermal qubit,

$$E = \frac{1}{1 + exp\,(\beta€)}, 0 < E < \frac{1}{2}€ \tag{11.3}$$

This is identified as the thermodynamic energy of the thermal qubit. Now the von Neumann entropy of the Gibbs state can be calculated according to, and with one eye on thermodynamics, the right-hand side can be expressed in terms of the energy E.

$$S(E) = \frac{-€ - E}{€} log \frac{€ - E}{€} - \frac{E}{€} log \frac{E}{€}, 0 < E < \frac{1}{2}€$$

Then S_{th} Energy will, $S_{th}(E) = (k_B ln2) S(E)$.

So, the entropy of a single thermal qubit is,

$$\frac{dS_{th}(E)}{dE} = \frac{1}{T}$$

And, the entropy for n thermal qubit is,

$$\frac{dS_{th}(E)}{dE} = \frac{n}{T} \tag{11.4}$$

It is known that,

$$\beta = \frac{1}{K_B T}$$

Here is the inverse temperature kb is the Boltzmann constant and T is the room temperature initially.

$$\beta = \frac{1}{K_B T}$$

$$= \frac{1}{8.617 \times 10^{-5} \times 300}$$
$$= 39\ ev^{-1}$$

So, the average energy of thermal qubit, $E = \frac{1}{1 \pm e^{39 \times \epsilon}}$

$$= \frac{1}{1 \pm e^{39 \times 0.9}}$$
$$= 5.7051 \times 10^{-16}$$

Here, ϵ is emissivity and this value will be 0 to 1 concerning the molecule. Consider $\epsilon = 0.9$ for ideal purposes.

Now, according to qubit thermodynamics, the S_{th} energy is
S_{th} (E) = (K_b ln 2) S (E)

$$= -k_B \frac{\epsilon - E}{\epsilon} ln \frac{\epsilon - E}{\epsilon} - k_B \frac{E}{\epsilon} ln \frac{E}{\epsilon}$$
$= 1.9118 \times 10^{-18}$ and S_{th} (E) is quantum mechanics qubit entropy.

This is identified as the thermodynamic energy of the thermal qubit. Now the von Neumann entropy of the Gibbs state can be calculated using Eq. (11.5).

S_{th} (E, N) = N (K_b ln 2) S (E/N)

$$= -k_B \frac{\epsilon - E/N}{\epsilon} ln \frac{\epsilon - E/N}{\epsilon} - k_B \frac{E/N}{\epsilon} ln \frac{E/N}{\epsilon} \quad (11.5)$$

Here € $= 0.9$
$E = 5.7051 \times 10^{-16}$
N = Number of qubits in the operation
K_B = Boltzmann constant, 8.617×10^{-5} evT.$^{-1}$

Now, by using Eqs. (11.3), (11.4), and (11.5), produced heat can be calculated from any quantum circuit. Some basic quantum operational circuits are described below for calculating produced heat, where N is the number of the quantum qubit. Produced heat from CNOT, quantum AND, OR, XOR, NAND, NOR, and XOR operational quantum gate operation is calculated below using the total number of quantum qubits.

11.2.1 Quantum NOT Operation

Quantum NOT operation is a one-qubit gate. So, to measure produced heat from the quantum NOT operational gate, it is needed to follow the following steps where Fig. 11.1 shows the quantum NOT gate.

Here, $\frac{dS_{th}(E)}{dE} = \frac{N}{T}$, where N = 1;

So, T $= \frac{dE \ x \ N}{dS_{th}(E)}$

$$= \frac{5.7051 \times 10^{-16} \times 1}{1.9118 \times 10^{-18}}$$
$$= 298.148 \text{ K}$$

So, the produced heat from the quantum NOT gate is 298.148 K.

11.2.2 Quantum CNOT Operation

Quantum controlled-NOT gate (CNOT) is a double qubit gate. So, to measure the roduced heat from the quantum CNOT operational gate, it is needed to follow the following steps where Fig. 11.2 shows the quantum CNOT gate.

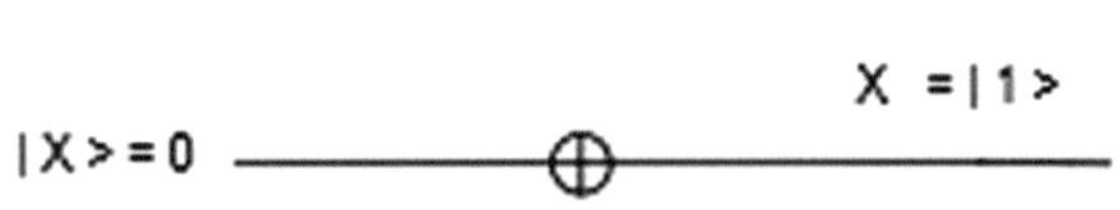

Fig. 11.1 Quantum NOT gate

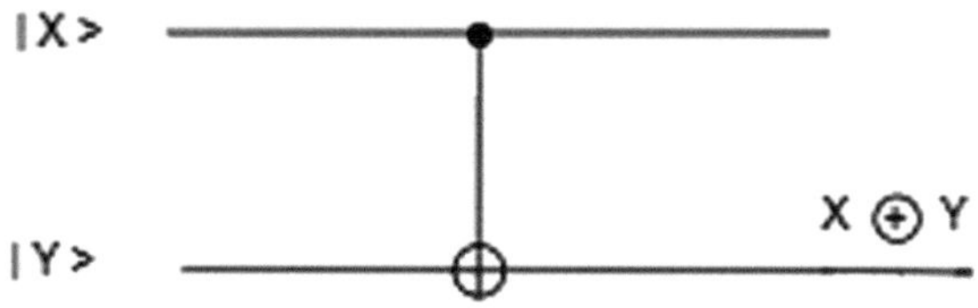

Fig. 11.2 Quantum CNOT gate

It is known that, for N qubit gate, S_{th} (E, N) = N (Kb ln 2) S (E/N)

$$= -k_B \frac{\epsilon - E/N}{\epsilon} ln \frac{\epsilon - E/N}{\epsilon} - k_B \frac{E/N}{\epsilon} ln \frac{E/N}{\epsilon}$$

Quantum CNOT operational gate is a double qubit gate. So, N = 2.

Thus, $S_{th}\ (E, N) = 1.94 \times 10^{-18}$

$T = \frac{dE\ x\ N}{dS_{th}(E)}$

$$= \frac{5.7051 \times 10^{-16} \times 2}{1.948 \times 10^{-18}}$$
$$= 585.32\ \text{K}$$

So, the heat produced from the Quantum CNOT gate is 585.32 K.

11.2.3 Quantum AND Operation

Quantum AND operational gate is a triple qubit gate. So, to measure produced heat from Quantum AND operational gate, the following steps should be followed where Fig. 11.3 shows quantum AND operational gate.

It is known that, for N qubit gate, S_{th} (E, N) = N (Kb ln 2) S (E/N)

$$= -k_B \frac{\epsilon - E/N}{\epsilon} ln \frac{\epsilon - E/N}{\epsilon} - k_B \frac{E/N}{\epsilon} ln \frac{E/N}{\epsilon}$$

Quantum AND operational gate is a triple qubit gate. So, N = 3.

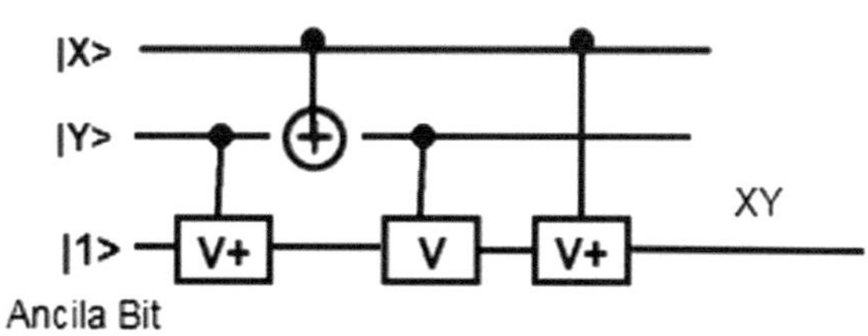

Fig. 11.3 Quantum AND operational gate

Thus, $S_{th}(E, N) = 1.97144x10^{-18}$
$T = \frac{dE \times N}{dS_{th}(E)}$

$$= \frac{5.7051 \times 10^{-16} \times 3}{1.97144 \times 10^{-18}}$$
$$= 868.162 \text{ K}$$

Thus, the produced heat from Quantum AND operational gate is 868.162 K.

11.2.4 Quantum OR Operation

Quantum OR operational gate is a triple qubit gate. So, to measure produced heat from Quantum OR operational gate, the following steps should be followed where Fig. 11.4 shows quantum OR operational gate.

It is known that, for N qubit gate, S_{th} (E, N) = N (Kb ln 2) S (E/N)

$$= -k_B \frac{\epsilon - E/N}{\epsilon} ln \frac{\epsilon - E/N}{\epsilon} - k_B \frac{E/N}{\epsilon} ln \frac{E/N}{\epsilon}$$

Quantum OR operational gate is a triple qubit gate. So, N = 3.

Thus, $S_{th}(E, N) = 1.97144 \times 10^{-18}$

$T = \frac{dE\ x\ N}{dS_{th}(E)}$

$$= \frac{5.7051 \times 10^{-16} \times 3}{1.97144 \times 10^{-18}}$$
$$= 868.162 \text{ K}$$

Thus, the produced heat from Quantum OR operational gate is 868.162 K.

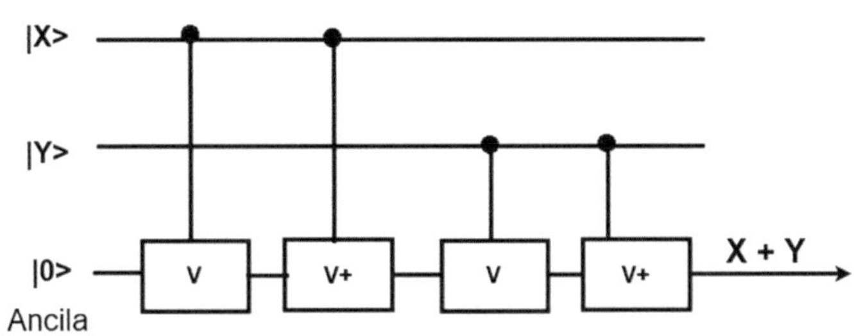

Fig. 11.4 Quantum OR operational gate

11.3 Heat Calculation for Quantum Operational Circuits

This section will describe the way to calculate the total produced heat of a quantum operational circuit. The basic information for calculating produced heat from a particular quantum operational gate is provided in the previous section.

11.3.1 Quantum Full Subtractor

A Full subtractor is a combinational circuit that is developed to overcome the drawback of the half subtractor circuit. It can take three inputs and after subtracting them creates two outputs. Here, a Full subtractor is implemented using the Quantum circuit. To create a Full Subtractor, two quantum NOT, two quantum XOR, two quantum AND, and a quantum OR gate are required. Figure 11.5 shows the quantum circuits of a full subtractor. A full subtractor receives three inputs and produces two outputs containing “|D>” and “|Bout>”.

Full subtractor is a 4 qubit quantum operation (considering 1 ancilla qubit and 3 input qubits) and for this quantum operation heat measurement will be calculated using the following formula.

$$\frac{dS_{th}(E)}{dE} = \frac{n}{T} \tag{11.6}$$

It is known that, for N qubit gate, S_{th} (E, N) = N (Kb ln 2) S (E/N)

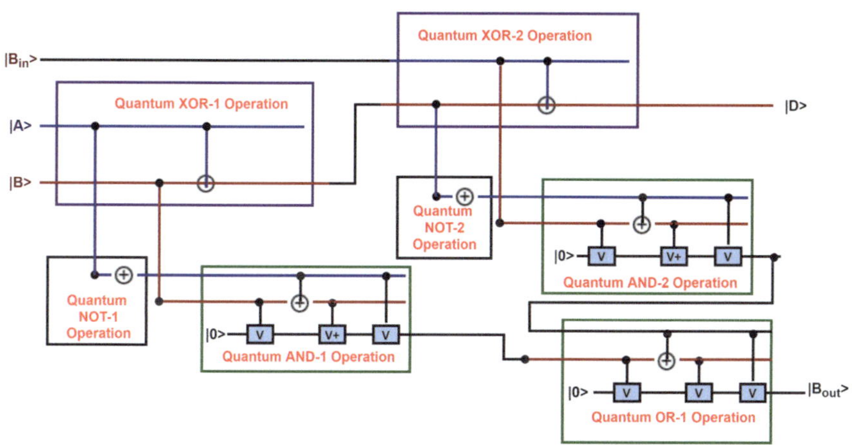

Fig. 11.5 Quantum full subtractor circuit

$$= -k_B \frac{\epsilon - E/N}{\epsilon} ln \frac{\epsilon - E/N}{\epsilon} - k_B \frac{E/N}{\epsilon} ln \frac{E/N}{\epsilon}$$

Quantum Full Subtractor operation has 4 qubits in the figure. So, N = 4.

Thus, $S_{th}\ (E, N) = 1.9791 \times 10^{-18}$

T = $\frac{dE\ x\ N}{dS_{th}(E)}$

$$= \frac{5.7051 \times 10^{-16} \times 4}{1.9791 \times 10^{-18}}$$
$$= 1153.05 \text{ K}$$

Thus, the produced heat from Quantum Full Subtractor operation is 1153.05 K.

11.3.2 Quantum 3-Qubit Even Parity Qubit Checker

A circuit that checks the parity in the receiver is called a Parity Checker. A combined circuit or device of parity generators and parity checkers are commonly used in digital systems to detect single-bit errors in the transmitted data. To create a 3-qubit even parity qubit checker, three XOR gates are required. Figure 11.6 shows the digital and quantum circuits of a 3-qubit even parity-qubit checker. A 3-qubit even parity qubit checker receives four inputs and produces one output containing "E".

3-qubit even parity qubit checker is a 4 qubits quantum operation and for this quantum operation heat measurement will be calculated using the following formula:

$$\frac{dS_{th}\ (E)}{dE} = \frac{n}{T}$$

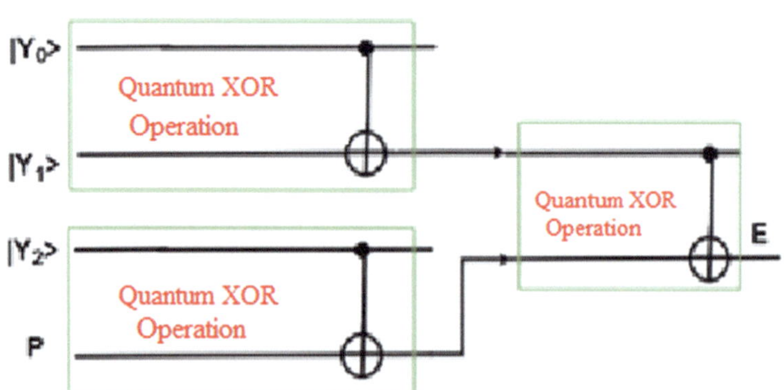

Fig. 11.6 Quantum 3-qubit even parity qubit checker

It is known that, for N qubit gate, S_{th} (E, N) = N (Kb ln 2) S (E/N)

$$= -k_B \frac{\epsilon - E/N}{\epsilon} ln \frac{\epsilon - E/N}{\epsilon} - k_B \frac{E/N}{\epsilon} ln \frac{E/N}{\epsilon}$$

Quantum parity bit checker operation has 4 qubits in the figure. So, N = 4.

Thus, $S_{th}\ (E, N) = 1.9791 \times 10^{-18}$

$T = \frac{dE\ x\ N}{dS_{th}(E)}$

$$= \frac{5.7051 \times 10^{-16} \times 4}{1.9791 \times 10^{-18}}$$
$$= 1153.05\ \text{K}$$

Thus, the heat produced from the quantum parity qubit checker is 1153.05 K.

11.3.3 *Quantum 3-to-1 Multiplexer*

A multiplexer (MUX) is a device that can receive multiple input signals and synthesize a single output signal in a recoverable manner for each input signal. It is also an integrated system that usually contains a certain number of data inputs and a single output. To create a multiplexer, one quantum NOT, two quantum AND, and one quantum OR gate is required. Figure 11.7 shows the quantum circuit of the

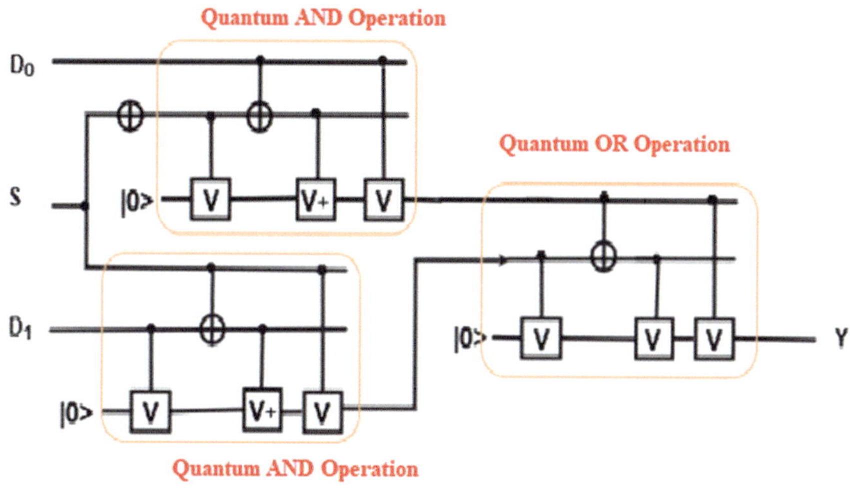

Fig. 11.7 Quantum multiplexer circuit

Multiplexer. A Multiplexer receives three inputs and produces one output containing "Y".

2-to-1 Multiplexer is a 4-qubit quantum operation (considering 1 ancilla qubit and 3 input qubits) and for this quantum operation heat measurement will be calculated using the following formula:

$$\frac{dS_{th}(E)}{dE} = \frac{n}{T}$$

It is known that, for N qubit gate, S_{th} (E, N) = N (Kb ln 2) S (E/N)

$$= -k_B \frac{\epsilon - E/N}{\epsilon} ln \frac{\epsilon - E/N}{\epsilon} - k_B \frac{E/N}{\epsilon} ln \frac{E/N}{\epsilon}$$

Quantum multiplexer operation has 4 qubits in the figure. So, N = 4.
Thus, S_{th} $(E, N) = 1.9791 \times 10^{-18}$

$T = \frac{dE \times N}{dS_{th}(E)}$

$$= \frac{5.7051 \times 10^{-16} \times 4}{1.9791 \times 10^{-18}}$$
$$= 1153.05 \text{ K}$$

Thus, the produced heat from the quantum multiplexer operation is 1153.05 K.

11.4 Basic Definitions for Heat Calculation in DNA Circuits

DNA has the characteristics of enabling classical logical operation using DNA sequence. DNA prefers to be in double-stranded form, while single-stranded DNA naturally migrates towards complementary sequences to form double-stranded complexes. Complementary sequences pair the bases adenine (A) with thymine (T) and cytosine (C) with guanine (G). DNA sequences pair in an antiparallel manner, with the 5' end of one sequence pairing with the 3' end of the complementary sequence.

Each DNA gate input will be the single standard sequence, if one is true, the complementary DNA sequence will be false. Suppose if ACTCGT is the input sequence then the complementary sequence will be TGAGCA. In DNA computing, when designing the logic gate, a predetermined single-strand sequence can be supplied to induce an appropriate chemical reaction. This sequence also helps to evaluate the output value whether it is true or false.

When the mixing step appears, it is needed to mix the two sequences to achieve a union of DNA sequences. In the mixing step, it is needed to give some heat to mix these. Then annealing appears and in annealing, it is needed to cool this little and make a double sequence bond. After annealing the step appears which is melting. In melting, need to heat the double-strand DNA sequence to make them a single strand complementary sequence and this sequence will be used in the DNA logic gate after some steps. So, the DNA melting temperature should be known:

1. **Nearest Neighbors**

   Depending on the nature of the sequence, one of two methods should be used to calculate melting temperature, T_m. Nearest Neighbors and Basic are the two methods that are discussed as follows:

$$T_m = \frac{\Delta H}{A + \Delta S + Rln\left(\frac{C}{4}\right)} - 273.15 + 16.6\log[Na^+] \quad (11.7)$$

   where,

   (a) Tm = melting temperature in °C
   (b) ΔH = enthalpy change in kcal mol^{-1} (accounts for the energy change during annealing/melting)
   (c) A = constant of −0.0108 kcal K^{-1}mol^{-1} (accounts for helix initiation during annealing/melting)
   (d) ΔS = entropy change in kcal K^{-1}mol^{-1} (accounts for energy unable to do work, i.e. disorder)
   (e) R = gas constant of 0.00199 kcal K^{-1}mol^{-1} (constant that scales energy to temperature)
   (f) C = oligonucleotide concentration in M or mol L^{-1} (use 0.0000005, i.e. 0.5 μM)
   (g) −273.15 = conversion factor to change the expected temperature in Kelvins to °C
   (h) Na+ = sodium ion concentration in M or mol L^{-1} (use 0.05, i.e. 50 mM)

   This example will demonstrate the manual calculation of the Tm for the following sequence:
   5'-AAAAACCCCCGGGGGTTTTT-3'

   This is the above sequence paired with its reverse complement:

   5'-AAAAACCCCCGGGGGTTTTT-3'

   3'-TTTTTGGGGGCCCCCAAAAA-5'

$$T_m = \frac{\Delta H}{A + \Delta S + Rln\left(\frac{C}{4}\right)} - 273.15 + 16.6\log[Na^+]$$

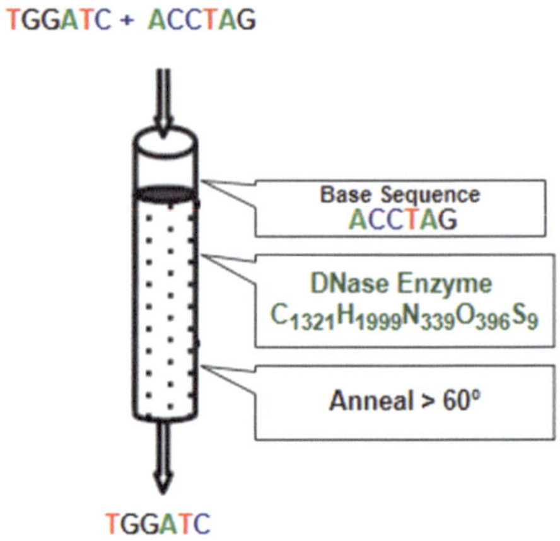

Fig. 11.8 DNA AND operational gate

$$T_m = \frac{-185.7\ kcal\ mol^{-1}}{\begin{array}{c}-0.0108\ kcal\ K^{-1}.\ mol^{-1} + -0.4672\ kcal\ K^{-1}.\ mol^{-1} \\ +0.00199\ kcal\ K^{-1}.\ mol^{-1}\ \times \ln \times\ \left(\frac{0.0000005\ mol\ L^{-1}}{4}\right) \\ -273.15 + 16.6\log[0.05\ mol\ L^{-1}]\end{array}}$$

$T_m = 69.6\ °\ C$

2. **Basic Method**

A secondary method is used to calculate T_m is the basic method of a modified Marmur Doty formula, which is used for oligonucleotides with short sequences lengths, (those that are 14 bases or less) [7, 8]. To calculate T_m the modified Marmur Doty formula is given as follows:

$\boldsymbol{T}_m = 2(A + T) + 4(C + G) - 7$

(a) T_m = melting temperature in °C
(b) A = number of adenosine nucleotides in the sequence
(c) T = number of thymidine nucleotides in the sequence
(d) C = number of cytidine nucleotides in the sequence
(e) G = number of guanosine nucleotides in the sequence
(f) *-7* = correction factor accounting for in solution.

So, for example, the melting temperature of a DNA sequence in different operational DNA gates can be calculated as AND, OR, NOT, NAND, NOR, XOR and XNOR.

1. **Heat Calculation of DNA AND Operational Gate**

Figure 11.8 shows the DNA AND operational gate. Here, ACCTAG = True and TGGATC = False. False and True inputs are given, then False output is obtained.

Here, Input 1 = TGGATC

$T_{m1} = 2(A + T) + 4(C + G) - 7$

$= 2(1 + 2) + 4(1 + 2) - 7$

$= 11.0\ °C$

Again, Input 2 = ACCTAG

$T_{m2} = 2(A + T) + 4(C + G) - 7$

$= 2(1 + 1) + 4(2 + 1) - 7$

$T_{m2} = 9.0\ °C$

Other processes should also be done for finding an output in DNA computing. That's why another generalized process has to be performed within all steps and all the following steps are applicable for each tube for performing a DNA operation.
Preparing, Mixing, and Annealing: Allosteric DNAzyme-based DNA logic circuit, described a procedure to make a DNAzyme-based logic circuit. Here, All DNA logic gates were formed by annealing twice: firstly, the mixture of the inhibitor DNA strands and E6-type DNAzymes in 1× TAE/Mg_2+ buffer (40 mM Tris, 20 mM acetic acid, 1 m MEDTA$_2$Na and 12.5 mM Mg(OAc)$_2$, pH 8.0) was heated at 95 °C for 4 min, 65 °C for 30 min, 50 °C for 30 min, 37 °C for 30 min, 22 °C for 30 min, and preserved at 20 °C; and then the substrates were added into the annealed mixture and incubated at constant temperature 20 °C for 4 h (total 6 h for preparing the DNA logic gate).
Melting, Amplifying, Separating, Extracting, Cutting, Ligating, Substituting, Marking, and Destroying sequences: After that Logic gates were triggered through displacement reaction in 1× TAE/Mg2+ buffer (40 mM Tris, 20 mM acetic acid, 1 mM EDTA$_2$Na, and 12.5 mM Mg (OAc)$_2$, pH 8.0). The input DNA strands were added to a solution containing DNA logic gates and reacted for > 2 h at 20 °C. Next, the displaced products were stored at 20 °C for native PAGE or fluorescence detection. In addition, polyacrylamide gel electrophoresis (PAGE) needs 2 h and the PCR process for fluorescence detection needs less than 2 h.
Detecting and Reading Sequences: Here to describe a specific biochemical process briefly which serves as the basis of the DNA computing approach as Polymerase Chain Reaction (PCR). Polymerases perform several functions, including the repair and duplication of DNA. PCR is a process that quickly amplifies the amount of specific DNA molecules in a given solution, using primer extension by the polymerase.
Each cycle of the reaction doubles the quantity of this molecule, leading to an exponential growth in the number of sequences. It consists of the following key processes:

(a) **Initialization:** a mixed solution of template, primer, dNTP and enzyme is heated to 94–98 °C for 1 – 9 min to ensure that most of the DNA template and primers are denatured;
(b) **Denaturation:** heat the solution to 94–98 °C for 20–30 s for separation of DNA duplexes;
(c) **Annealing:** lower the temperature enough (usually between 50 and 64 °C) for 20–40 s for primers to anneal specifically to the ssDNA template;
(d) **Elongation/Extension:** raise temperature to optimal elongation temperature of *Taq* or similar DNA polymerase (70–74°C) for the polymerase adds dNTP's from the direction of 5' to 3' that are complementary to the template;
(e) **Final Elongation/Extension:** after the last cycle, a 5–15 min elongation may be performed to ensure that any remaining ssDNA is fully extended.

Steps 2 to 4 are repeated 20–35 times; fewer cycles results in less product, and too many cycles increase the fraction of incomplete and erroneous products. PCR is a routine job in the laboratory that can be performed by an apparatus named thermal cycler. According to the PCR process, to produce an operational output of DNA computation, it needs around 2 h.
Specific steps with heat for DNA computations are as follows:

1.	Gate operation preparing	(98 °C–94 °C)
2.	Synthesizing	(98 °C–94 °C)
3.	Mixing	(95 °C–22 °C)
4.	Annealing	(70 °C–20 °C)
5.	Melting	(Depends on the sequence)
6.	Amplifying	20 °C
7.	Separating	
8.	Extracting	
9.	Cutting	
10.	Ligating	
11.	Substituting	
12.	Marking	
13.	Destroying	
14.	Detecting and Reading	(98 °C–25 °C)

So, in DNA AND gate, the overall maximum required heat is given below.
= (98+98+95+70+11+20+98) °C,
= 490 °C,

And the minimum required heat = (94+94+22+20+9+20+25) °C,

= 284 °C.

Again, in DNA AND gate, all the processes occur in the test tube after mixing is completed. Here sometimes, need to keep the temperature high, and sometimes it needs to keep the temperature low for several steps. So, it needs to keep the temperature at a maximum of 94–98 °C. When DNA gate logic operation occurs,

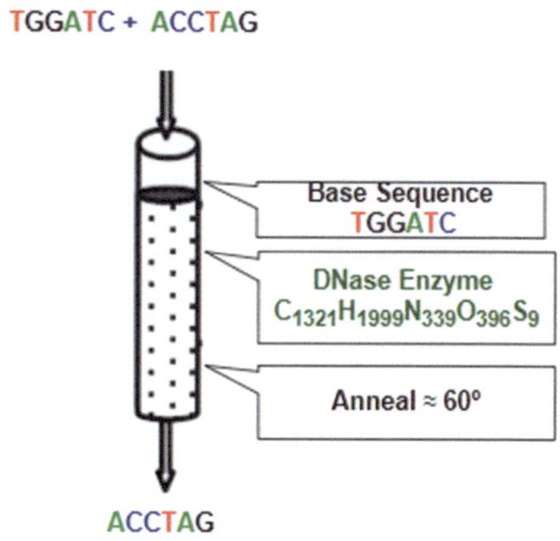

Fig. 11.9 DNA OR operational gate

it needs to keep the temperature around 20 °C and at detection time, it needs to keep it at 25 °C.

2. **Heat Calculation of DNA OR Operational Gate**
Figure 11.9 shows the DNA OR operational gate. Here, ACCTAG = True and TGGATC = False. In this case, False and True inputs are given, and the True output is obtained.

Here, Input 1 = TGGATC

$T_{m1} = 2(A + T) + 4(C + G) - 7$

$= 2(1 + 2) + 4(1 + 2) - 7$

$= 11.0\ °C$

Again,

Input 2 = ACCTAG

$= 2(A + T) + 4(C + G) - 7$

$= 2(1 + 1) + 4(2 + 1) - 7$

$T_{m2} = 9.0\ °C$

Specific steps with heat for DNA computations are given below.

1.	Gate operation preparing	(98 °C–94 °C)
2.	Synthesizing	(98 °C–94 °C)
3.	Mixing	(95 °C–22 °C)
4.	Annealing	(70 °C–20 °C)
5.	Melting	(Depends on the sequence)
6.	Amplifying	20 °C
7.	Separating	
8.	Extracting	
9.	Cutting	
10.	Ligating	
11.	Substituting	
12.	Marking	
13.	Destroying	
14.	Detecting and Reading	(98 °C–25 °C)

So, in DNA OR gate, the overall maximum required heat is
= (98+98+95+70+11+20+98) °C,

= 490 °C,

And the minimum required heat = (94+94+22+20+9+20+25) °C,

= 284 °C

Again, in DNA OR gate, all the processes happening in the test tube after mixing are completed. Here sometimes it needs to keep the temperature high and sometimes needs to keep the temperature low for several steps. So, it needs to keep the temperature at a maximum of 94–98 °C. When DNA gate logic operation occurs, it needs to keep the temperature around 20 °C and at detection time it is as 25 °C.

3. **Heat Calculation of DNA NOT Operational Gate**
Figure 11.10 shows the DNA NOT gate where one input TGGATC (False) is given and the obtained output is ACCTAG (True).

Here, Input DNA sequence = TGGATC

So, Melting temperature , $T_m = 2(A + T) + 4(C + G) - 7$

$= 2(1 + 2) + 4(1 + 2) - 7$

= 11.0 °C

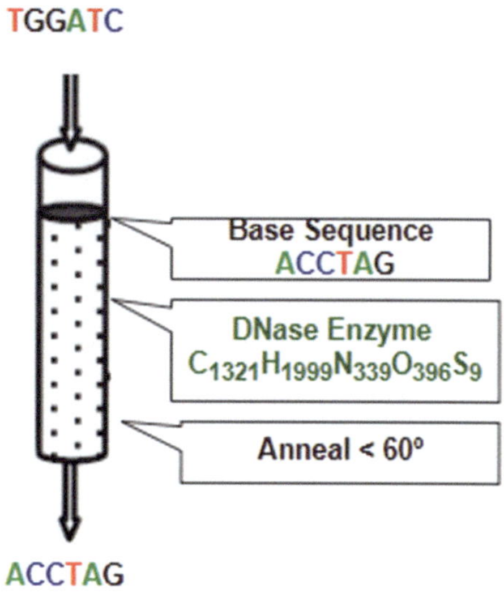

Fig. 11.10 DNA NOT operational gate

Specific steps with heat for DNA computation are given below.

1.	Gate operation preparing	(98 °C–94 °C)
2.	Synthesizing	(98 °C–94 °C)
3.	Mixing	(95 °C–22 °C)
4.	Annealing	(70 °C–20 °C)
5.	Melting	(Depends on the sequence)
6.	Amplifying	20 °C
7.	Separating	
8.	Extracting	
9.	Cutting	
10.	Ligating	
11.	Substituting	
12.	Marking	
13.	Destroying	
14.	Detecting and Reading	(98 °C–25 °C)

So, in DNA NOT gate, the overall maximum required heat is
= (98+98+95+70+11+20+98) °C,

= 490 °C,

And the minimum required heat = (94+94+22+20+11+20+25) °C,

= 286 °C.

Again, in DNA NOT operational gate, all the processes are done in the test tube after mixing are completed. Here sometimes it needs to keep the temperature

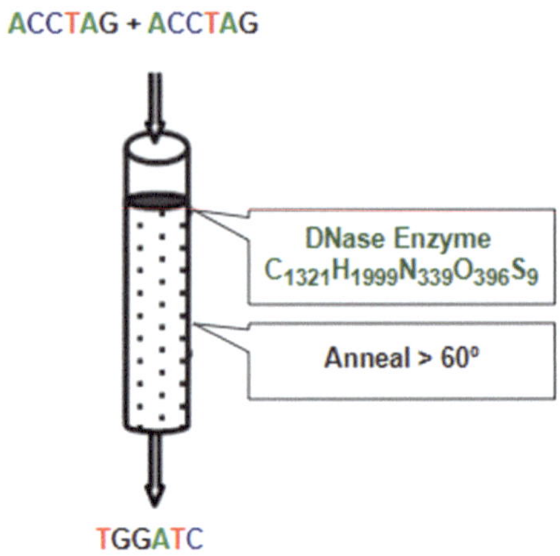

Fig. 11.11 DNA XOR gate

high and sometimes it needs to keep the temperature low for several steps. So, it needs to keep the temperature at a maximum of 94–98 °C. When DNA gate logic operation occurs, it needs to keep the temperature around 20 °C and at detection time it should be 25 °C.

4. **Heat Calculation of DNA XOR Gate**

Figure 11.11 shows the DNA XOR gate and here also two inputs are given and one output is obtained by maintaining the truth table of XOR operation.

Again, $Input_1$ = ACCTAG

$$So,\ T_{m1} = 2(A + T) + 4(C + G) - 7$$

$$= 2(1 + 1) + 4(2 + 1) - 7$$

$$= 9.0\ ^\circ C$$

Again, $Input_2$ = ACCTAG

$$So,\ T_{m2} = 2(A + T) + 4(C + G) - 7$$

$$= 2(1 + 1) + 4(2 + 1) - 7$$

$$= 9.0\ ^\circ C$$

Specific steps with heat for DNA computations are given below.

1.	Gate operation preparing	(98 °C–94 °C)
2.	Synthesising	(98 °C–94 °C)
3.	Mixing	(95 °C–22 °C)
4.	Annealing	(70 °C–20 °C)
5.	Melting	(Depends on the sequence)
6.	Amplifying	20 °C
7.	Separating	
8.	Extracting	
9.	Cutting	
10.	Ligating	
11.	Substituting	
12.	Marking	
13.	Destroying	
14.	Detecting and Reading	(98 °C–25 °C)

So, in DNA XOR gate, the overall maximum required heat is
= (98+98+95+70+11+20+98) °C,

= 490 °C,

And the minimum required heat = (94+94+22+20+9+20+25) °C,

Again, in the DNA XOR gate, all the processes happening in the test tube after mixing are completed. Here in specific cases, it is needed to keep the temperature high and sometimes the temperature should be low for several steps. So, it is needed to keep the temperature at a maximum of 94–98 °C. When DNA gate logic operation occurs, it needs to keep the temperature around 20 °C and at detection time it is as 25 °C.

11.5 Heat Calculation in DNA Circuits

This subsection is going to describe some DNA operational circuits to calculate it's performing Heat in an approximate value based on the theory described in Sect. 11.5.

11.5.1 DNA Full Subtractor

A full subtractor is a combinational circuit that performs subtraction of two bits, one is minuend and the other is subtrahend, taking into account the borrow of the previous adjacent lower minuend bit. This circuit has three inputs and two outputs. The three inputs A, B, and B_{in}, denote the minuend, subtrahend, and previous borrow, respectively. The two outputs, D and B_{out} represent the difference and output borrows, respectively. To create a full subtractor, one OR, two AND, two NOT, and two XOR gates are required.

Here, three input sequences are as follows:

1. B_{in} = TGGATC
2. A = ACCTAG
3. B = ACCTAG.

Calculation of the melting temperature of a specific DNA sequence is as follows:
For $Input_{,1,}$ B_{in} = TGGATC
$T_{m\ (Bin)} = 2(A + T) + 4(C + G) - 7$
$= 2(1 + 2) + 4(1 + 2) - 7$
= 11.0 °C
Again, $Input_2$ and $Input_3$, A = B = ACCTAG
So, $T_{m\ (A\ or\ B)} = 2(A + T) + 4(C + G) - 7$
$= 2(2 + 1) + 4(2 + 1) - 7$
= 11.0°C

Specific steps with heat for DNA full subtractor (for each tube) are given below.

1.	Gate operation preparing	(98 °C–94 °C)
2.	Synthesizing	(98 °C–94 °C)
3.	Mixing	(95 °C–22 °C)
4.	Annealing	(70 °C–20 °C)
5.	Melting	(Depends on the sequence)
6.	Amplifying	20 °C
7.	Separating	
8.	Extracting	
9.	Cutting	
10.	Ligating	
11.	Substituting	
12.	Marking	
13.	Destroying	
14.	Detecting and Reading	(98 °C–25 °C)

So, in DNA full subtractor, the overall maximum required heat is
= (98+98+95+70+11+20+98) °C,
= 490 °C,
And the minimum required heat = (94+94+22+20+11+20+25) °C,
= 286 °C.
Again, in a specific basic DNA gate, all the processes happening in the test tube after mixing are completed. Here in specific cases, it is needed to keep the temperature high and sometimes the temperature should be low for several steps. So, the temperature should be kept at a maximum of 94–98 °C. When DNA gate logic operation occurs, it needs to keep the temperature around 20 °C and at detection time it needs to keep as 25 °C. Figure 11.12 shows the DNA circuit of the full subtractor.

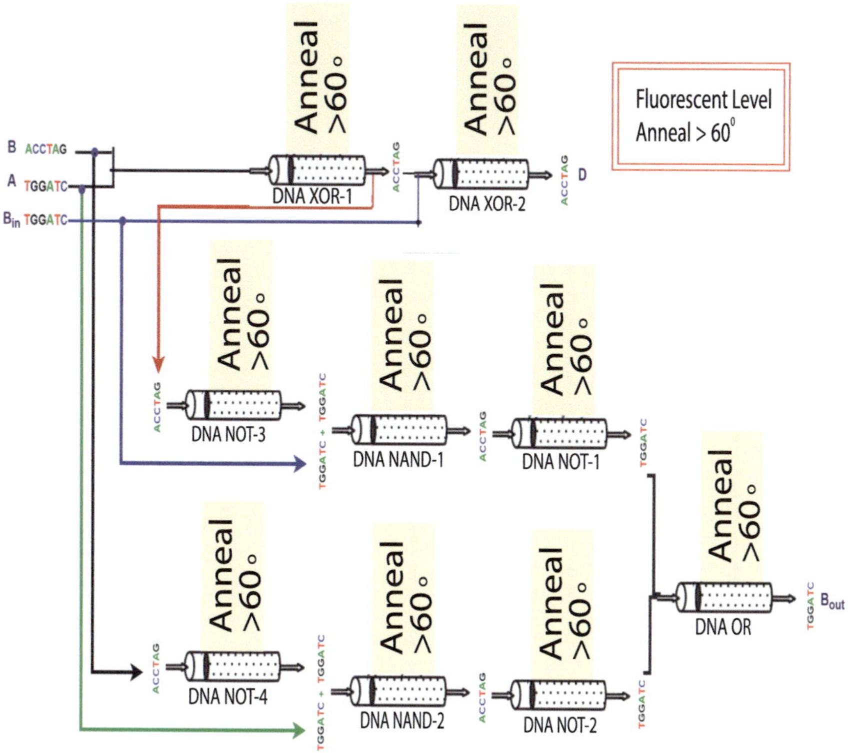

Fig. 11.12 DNA full subtractor circuit

11.5.2 DNA Full Adder

Full Adder is the adder that adds three inputs and produces two outputs. The first two inputs are A and B and the third input is an input carry as C-IN. The output carry is designated as C_{out} and the normal output is designated as S which is SUM. A full adder logic is designed in such a manner that can take eight inputs together to create a byte-wide adder and cascade the carry bit from one adder to another. To create a Full Adder, one OR, two NAND, one XOR, and two NOT gates are required. Figure 11.13 shows the DNA circuit of the Full Adder.
Here, three input sequences are as follows:

1. A_0 = TGGATC
2. A_1 = TGGATC
3. A_2 = ACCTAG.

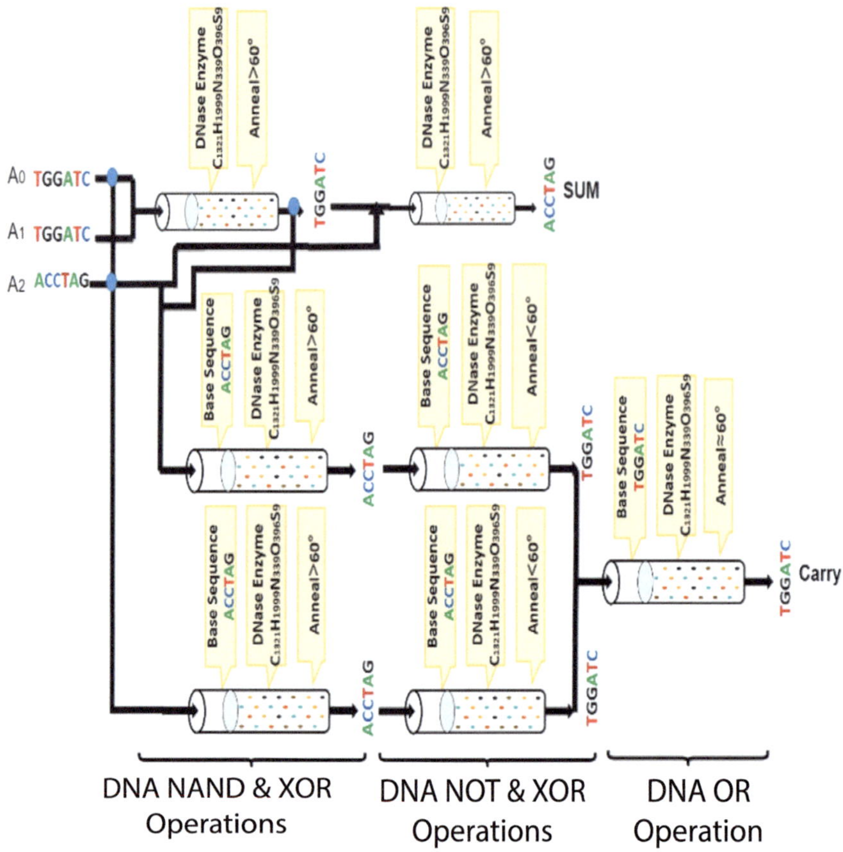

Fig. 11.13 DNA full adder circuit

Calculation of the Melting temperature of a specific DNA sequence is as follows:
For Inputs, A_0 and A_1 = TGGATC
$T_{m\ (A0\ or\ A1)} = 2(A + T) + 4(C + G) - 7$
$= 2(1 + 2) + 4(1 + 2) - 7$
$= 11.0\ °C$
Again, Input, A_3 = ACCTAG
$So,\ T_{m\ (A3)} = 2(A + T) + 4(C + G) - 7$
$= 2(2 + 1) + 4(2 + 1) - 7$
$= 11.0\ °C$

Specific steps with heat for DNA full Adder (for each tube) are as follows:

1.	Gate operation preparing	(98 °C–94 °C)
2.	Synthesizing	(98 °C–94 °C)
3.	Mixing	(95 °C–22 °C)
4.	Annealing	(70 °C–20 °C)
5.	Melting	(Depends on the sequence)
6.	Amplifying	20 °C
7.	Separating	
8.	Extracting	
9.	Cutting	
10.	Ligating	
11.	Substituting	
12.	Marking	
13.	Destroying	
14.	Detecting and Reading	(98 °C–25 °C)

So, in DNA full adder, the overall maximum required heat
= (98+98+95+70+11+20+98) °C,
= 490 °C,
And the minimum required heat = (94+94+22+20+11+20+25) °C,
= 286 °C.

Again, in a specific basic DNA gate, all the processes happening in the test tube after mixing are completed. Here in specific cases, it is needed to keep the temperature high and sometimes the temperature should be low for several steps. So, it has to keep the temperature at a maximum of 94–98 °C. When DNA gate logic operation occurs, the temperature should be around 20 °C and at detection time, it needs to keep it as 25°C.

11.5.3 DNA Multiplication Circuit

The multiplicand and multiplier can be of various bit sizes. The product's bit size depends on the bit size of the multiplicand and multiplier. The bit size of the product is equal to the sum of the bit size of the multiplier multiplicand. To create a DNA Multiplication circuit, six DNA NAND, two DNA XOR, and 6 DNA NOT gates are required. Figure 11.14 describes the DNA circuit of the 2-molecular multiplication.

Here, four input sequences are as follows:

1. A_0 = ACCTAG
2. A_1 = TGGATC
3. B_0 = TGGATC
4. B_1 = ACCTAG.

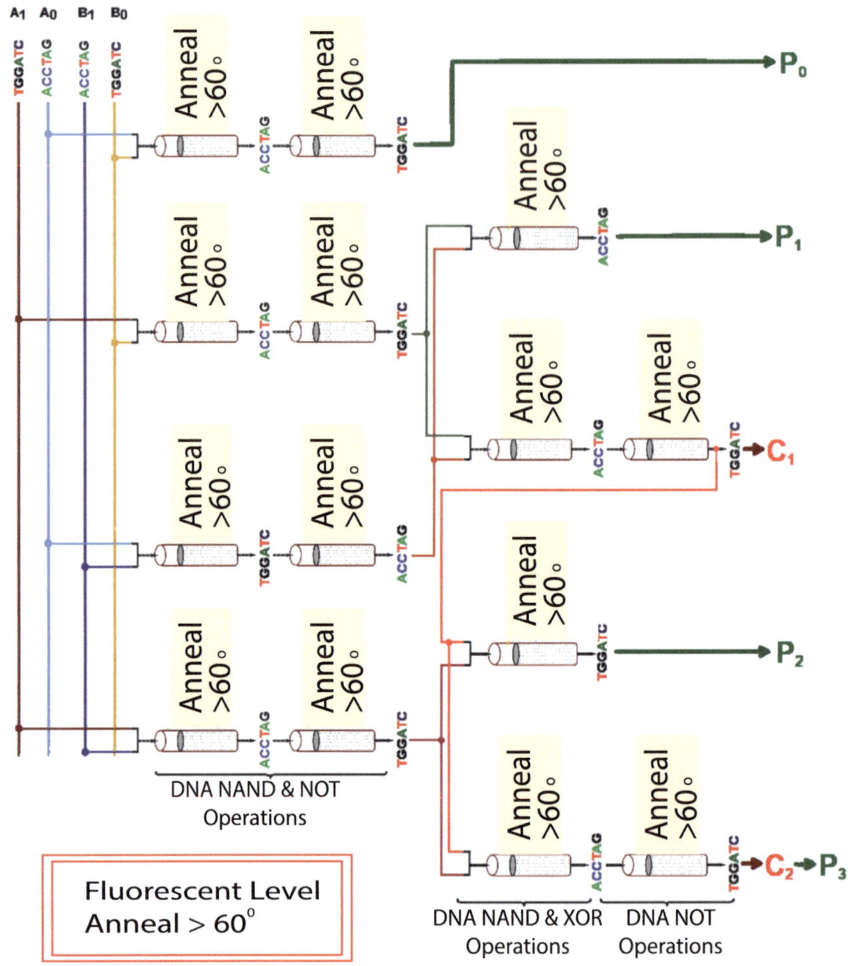

Fig. 11.14 DNA multiplier circuit

Calculation of the melting temperature of a specific DNA sequence is as follows:
For Inputs, A_0 and B_1 = ACCTAG
$T_{m\ (A0\ or\ B1)} = 2(A + T) + 4(C + G) - 7$
$= 2(2 + 1) + 4(2 + 1) - 7$
$= 11.0\ °C.$

Again, Inputs, A_1 and B_0 = TGGATC
So, $T_{m\ (A3)} = 2(A + T) + 4(C + G) - 7$
$= 2(1 + 2) + 4(1 + 2) - 7$
$= 11.0\ °C$

Specific steps with heat for DNA full Adder (for each tube) are as follows:

1.	Gate operation preparing	(98 °C–94 °C)
2.	Synthesizing	(98 °C–94 °C)
3.	Mixing	(95 °C–22 °C)
4.	Annealing	(70 °C–20 °C)
5.	Melting	(Depends on the sequence)
6.	Amplifying	20 °C
7.	Separating	
8.	Extracting	
9.	Cutting	
10.	Ligating	
11.	Substituting	
12.	Marking	
13.	Destroying	
14.	Detecting and Reading	(98 °C–25 °C)

So, in DNA multiplier, the overall maximum required heat is
= (98+98+95+70+11+20+98) °C,
= 490 °C,
And the minimum required heat = (94+94+22+20+11+20+25) °C,
= 286 °C.

Again, in a specific basic DNA gate, all the processes ocurring in the test tube after mixing are completed. Here in specific cases, the temperature should be kept high and sometimes low for several steps. So, the temperature must be at a maximum of 94–98 °C. When DNA gate logic operation occurs, it needs to keep the temperature around 20 °C and at detection time, it needs to keep it at 25 °C.

11.6 Heat Calculation in Quantum-DNA Circuits

According to quantum computing, quantum computation is faster than classical computation systems. Quantum computers are also more powerful than supercomputers in terms of computing. They are 1000 times faster than regular computers and supercomputers at processing data. Quantum computers can execute calculations that would take a regular computer 1000 years to complete in a matter of seconds. On the other hand, the use of DNA strands to compute has led to high parallel computation that makes up for the slow processing of the chip. Memory space required by DNA is around 1 molecule per cubic nanometer which is much less when compared to regular storage systems Consumption of power is almost nil as the chemical bonds in DNA produce energy to build or repair new strands. So, to find a super faster computation system with a huge memory, a Quantum-DNA computation system can be developed. This Quantum-DNA computation system can merge all advantages of quantum computing and DNA computing.

In a Quantum-DNA computing system, input will be received as a qubit and after performing in a certain number of quantum operational gates these qubits will be turned into DNA sequences by NMR relaxation.

11.6.1 Quantum-DNA Full Adder

Full Adder is the adder that adds three inputs and produces two outputs. The first two inputs are A and B and the third input is an input carry as C_{in}. The output carry is designated as C_{out} and the normal output is designated as S which is SUM. A full adder logic is designed in such a manner that can take eight inputs together to create a byte-wide adder and cascade the carry bit from one adder to another. To create a Full Adder, one OR, two AND, and two XOR gates are required. Figure 11.15 describes the Quantum-DNA circuit of the Full Adder. From the Figure, it is assumed that three quantum operational gates and two DNA operational gates are required to find the expected output from Quantum-DNA Full adder.

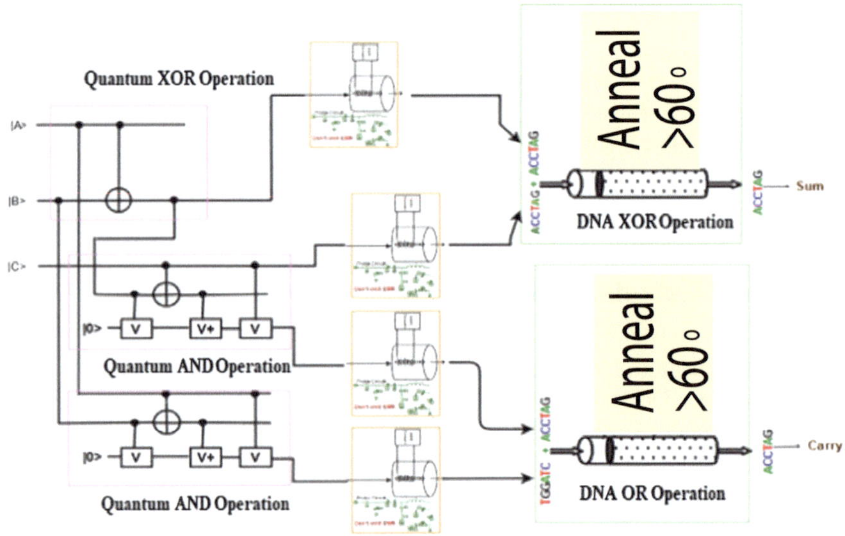

Fig. 11.15 Quantum-DNA full adder at room temperature

A quantum-DNA full adder has five qubits (considering 2 ancilla qubits and 3 input qubits) quantum operation. The output qubit from the quantum operation will be converted into a DNA sequence by using NMR relaxation later. The following formula will be used to compute the amount of heat generated by the quantum process.

$$\frac{dS_{th}(E)}{dE} = \frac{n}{T}$$

It is known that, for N-qubit gate, S_{th} (E, N) = N (Kb ln 2) S (E/N)

$$= -k_B \frac{\epsilon - E/N}{\epsilon} ln \frac{\epsilon - E/N}{\epsilon} - k_B \frac{E/N}{\epsilon} ln \frac{E/N}{\epsilon}$$

The quantum-DNA full adder has five qubits as shown in the figure. So, N = 5.

Thus, $S_{th}(E, N) = 1.984 \times 10^{-18}$

$T = \frac{dE \ x \ N}{dS_{th}(E)}$

$$= \frac{5.7051 \times 10^{-16} \times 5}{1.984 \times 10^{-18}}$$

$$= 1436.49 \text{ K}$$

Thus, the produced heat from Quantum-DNA full adder is 1436.49 K.

Here, at most, the two DNA sequences for the DNA operational gate in the Quantum-DNA full adder are as follows:

1. TGGATC
2. ACCTAG

To calculate the melting temperature of a specific DNA sequence, it is as follows:
For TGGATC
$T_m = 2(A + T) + 4(C + G) - 7$
$= 2(1 + 2) + 4(1 + 2) - 7$
= 11.0 °C
Again, For ACCTAG
So, $T_m = 2(A + T) + 4(C + G) - 7$
$= 2(2 + 1) + 4(2 + 1) - 7$
= 11.0 °C

Specific steps with heat for Quantum-DNA Full Adder in DNA operation (for each tube) are as follows:

1.	Gate operation preparing	(98 °C–94 °C)
2.	Synthesizing	(98 °C–94 °C)
3.	Mixing	(95 °C–22 °C)
4.	Annealing	(70 °C–20 °C)
5.	Melting	(Depends on the sequence)
6.	Amplifying	20 °C
7.	Separating	
8.	Extracting	
9.	Cutting	
10.	Ligating	
11.	Substituting	
12.	Marking	
13.	Destroying	
14.	Detecting and Reading	(98 °C–25 °C)

So, in DNA operation, the overall maximum required heat is
= (98+98+95+70+11+20+98) °C,
= 490 °C,
And the minimum required heat = (94+94+22+20+11+20+25) °C,
= 286 °C.

Again, in a specific basic DNA gate, all the processes ocurring in the test tube after mixing are completed. Here in specific cases, the temperature is high and low for several steps. So, it is needed to keep the temperature at a maximum of 94–98 °C. When DNA gate logic operation occurs, the temperature should be around 20 °C and at the detection time, it needs to keep it as 25 °C.

11.7 Heat Calculation in DNA-Quantum Circuits

Quantum computation, according to quantum computing, is faster than traditional processing systems. Quantum computers are also more powerful in terms of computation than supercomputers. They process data 1000 times faster than normal computers and supercomputers. Quantum computers can perform computations in a fraction of a second that would take a traditional computer 1000 years to finish. The usage of DNA strands to calculate, on the other hand, has resulted in high parallel computation, which compensates for the chip's slow processing. When compared to traditional storage systems, DNA requires just about 1 bit per cubic nanometer of memory space. The chemical interactions in DNA provide energy to make or repair new strands, therefore there is essentially no power use. As a result, a Quantum-DNA computation system can be created to find a super-fast computation system with a lot of memory. This quantum-DNA computation device combines the benefits of both quantum and DNA computing.

In a DNA-Quantum computing system, input will be received in DNA sequences and after performing in a certain number of DNA operational gates these DNA sequences will be turned into quantum qubits by the NMR process.

11.7.1 DNA-Quantum Full Adder

Full Adder is the adder that adds three inputs and produces two outputs. The first two inputs are A and B and the third input is an input carry as C_{in}. The output carry is designated as C_{out} and the normal output is designated as S which is SUM. A full adder logic is designed in such a manner that can take eight inputs together to create a byte-wide adder and cascade the carry bit from one adder to another. To create a Full Adder, one OR, two AND, and two XOR gates are required.

Figure 11.16 illustrates the DNA-Quantum circuit of the Full Adder and from the figure, it is assumed that two DNA operational gates and three quantum operational gates are required to find the expected output from the DNA-Quantum Full adder.

Here, at most, the two DNA sequences for the DNA operational gate in the DNA-Quantum Full adder are as follows:

1. TGGATC
2. ACCTAG

To calculate the melting temperature of a specific DNA sequence is as follows:
For TGGATC
$T_m = 2(A + T) + 4(C + G) - 7$
$= 2(1 + 2) + 4(1 + 2) - 7$
= 11.0 °C
Again, For ACCTAG
So, $T_m = 2(A + T) + 4(C + G) - 7$
$= 2(2 + 1) + 4(2 + 1) - 7$
= 11.0 °C

Specific steps with heat for DNA-Quantum Full Adder in DNA operation (for each tube) are as follows:

1.	Gate operation preparing	(98 °C–94 °C)
2.	Synthesizing	(98 °C–94 °C)
3.	Mixing	(95 °C–22 °C)
4.	Annealing	(70 °C–20 °C)
5.	Melting	(Depends on the sequence)
6.	Amplifying	20 °C
7.	Separating	
8.	Extracting	
9.	Cutting	
10.	Ligating	
11.	Substituting	
12.	Marking	
13.	Destroying	
14.	Detecting and Reading	(98 °C–25 °C)

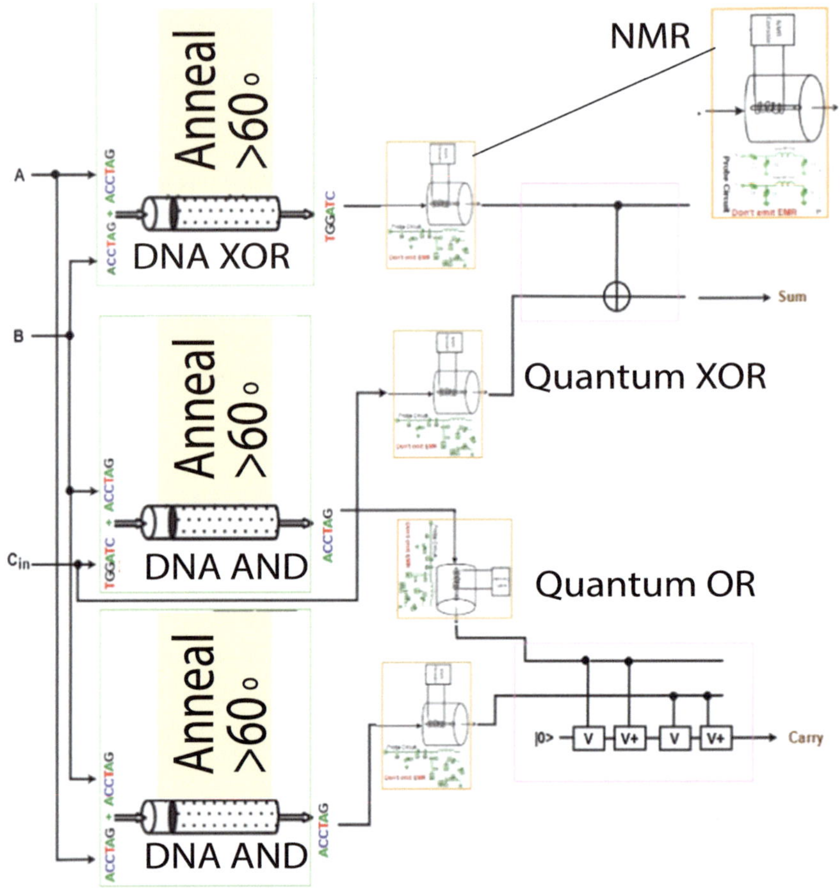

Fig. 11.16 DNA-Quantum full adder at room temperature

So, in DNA Full subtractor, the overall maximum required heat is
= (98+98+95+70+11+20+98) °C,
= 490 °C,
And the minimum required heat = (94+94+22+20+11+20+25) °C,
= 286 °C.
Again, in a specific basic DNA gate, all the processes ocurring in the test tube after mixing are completed. Here in specific cases, it is needed to keep the temperature high and sometimes the temperature should be low for several steps. So, the temperature must be at a maximum of 94–98 °C. When DNA gate logic operation occurs, it is needed to keep the temperature around 20 °C and at detection time and it needs to keep as 25 °C.

Further, a DNA-Quantum Full Adder is 5 qubits (considering 1 ancilla qubit and 4 input qubits) quantum operation. The output of a DNA operational gate is a DNA sequence, which is converted into a quantum qubit after processing through the NMR at room temperature.

The produced heat generated from quantum operation in the DNA-Quantum Full Adder will be calculated using the following formula:

$$\frac{dS_{th}(E)}{dE} = \frac{n}{T}$$

It is known that, for N-qubit gate, S_{th} (E, N) = N (Kb ln 2) S (E/N)

$$= -k_B \frac{\epsilon - E/N}{\epsilon} ln \frac{\epsilon - E/N}{\epsilon} - k_B \frac{E/N}{\epsilon} ln \frac{E/N}{\epsilon}$$

The DNA-quantum full adder has five qubits as shown in the figure. So, N = 5.

Thus, $S_{th}(E, N) = 1.984 \times 10^{-18}$

T = $\frac{dE\ x\ N}{dS_{th}(E)}$

$$= \frac{5.7051 \times 10^{-16} \times 5}{1.984 \times 10^{-18}}$$
$$= 1436.49\ \text{K}$$

Thus, the produced heat from DNA-Quantum full adder is 1436.49 K.
= 11.63 °C,
and the minimum required heat = (94+94+22+20+11+20+25) °C,
= 286 °C.

Again, in a specific basic DNA gate, all the processes are done in the test tube after mixing are completed. Here in specific cases, the temperature needs to keep high and sometimes low for several steps. So, it is needed to keep the temperature at a maximum of 94–98 °C. When DNA gate logic operation occurs, it is needed need to keep the temperature around 20 °C and at detection time and it needs to keep at 25 °C.

11.8 Applications

This section describes some real-life applications of DNA and Quantum operation where it is used and provides superior performance against classical computing systems. In the case of different applications, classical computing systems might fail but Quantum and DNA computing systems can show their capability.

Traveling salesman problem: The first theory of DNA computation was proposed by Leonard Adleman in 1994. He put his experimental theory to the test with a seven-point Hamiltonian path problem also called the traveling salesman problem. The salesman in this problem needs to find the shortest path between seven cities whose distances are known in such a way that he crosses no city twice and returns to the original city. Adleman represented each city with a short DNA sequence with about 20 bases and a complementary strand with 20 bases as the street between the cities. All the fragments are capable of linking with each other. When the fragments were put in a tube and mixed, the natural bonding tendencies of the DNA created 109 formations or solutions in less than a second. Not all were correct and he took a week to extrapolate and filter out the shortest path through various techniques. Though this solution was not ideal, this demonstration opened floodgates to a wide range of possibilities and applications.

Security: Deploying DNA algorithms in cryptography to build an intrusion detection model is the most recent development. The ability to store 108 terabytes of data in 1 g of DNA has led to the potential of holding a huge one-time pad. Another example is DNA steganography, in which a novel method was used to hide the messages in a microdot. Instead of the traditional binary encoding, each letter was denoted by three chemical bases i.e. the letter A was encoded by CGA. These messages are then encoded into DNA sequences and concealed by mixing them in a tube with a large amount of sonicated random human DNA. This led to the formation of microdots, which were then decoded by the receiver with appropriate primers (short sequence with complementary bases). However, such encryption techniques have posed problems. The lack of a theoretical basis to explain the implementation and come up with good schemes seem to be a challenge. These are also expensive to apply, and analyze, and it requires modern infrastructure.

Artificial intelligence and machine learning: Artificial intelligence and machine learning are some of the prominent areas right now, as the emerging technologies have penetrated almost every aspect of humans' lives. However, as the number of applications increases, it becomes a challenging task for traditional computers to match up the accuracy and speed. And that's where quantum computing can help in processing complex problems in very little time, which would have taken traditional computers thousands of years.

11.9 Summary

Quantum computing focuses on speedy technology based on quantum-theoretical principles, which is the behavior of energy and matter of a qubit. A combination of qubits is used to perform any specific task in quantum computing. Quantum computers represent a significant advancement in computing capability, with enormous performance benefits for specific use cases. The ability of qubits to be in several states at the same time gives the quantum computer a lot of computing capabilities and it is much faster than classical bitwise computing.

Furthermore, DNA computing uses biological molecules to do computations. The four-character genetic alphabet (A-adenine, G-guanine, C-cytosine, and T-thymine) is used in DNA computing. The input of any DNA operation can be represented by

DNA molecules with specific sequences. The instructions are carried out by laboratory operations on the molecules, and the result is defined as some property of the final set of molecules. DNA computing promises meaningful linkages between computers and life systems, as well as massively parallel computations. DNA computing can carry out millions of operations at the same time.

To the advanced computations, it is possible to make or use DNA-Quantum computing and Quantum-DNA computing systems, which will merge all the advantages of both DNA computing and Quantum computing.

Heat is an important property of any operation for computation. It is found that, in quantum computing operation, much heat is produced by circuits, which is dependent on the number of qubits in the operational circuit. On the other hand, DNA computing needs heat in the test tube to execute the operation. Its different stage needs different amounts of time and heat, which are highlighted in this chapter. In addition, the produced heat of quantum computing can be used in the DNA computing operation by using some heat transfer nanotubes.

Chapter 12
Speed Calculation

12.1 Introduction

DNA computing is a kind of natural computing that uses the molecular characteristics of DNA to conduct logical and arithmetic operations instead of typical carbon/silicon chips. The four-character genetic alphabets (A-adenine, G-guanine, C-cytosine, and T-thymine) are used in DNA computing instead of the binary digits (1 and 0) utilized by standard computers. This enables massively parallel computation, making it possible to answer difficult mathematical equations or problems in a fraction of the time. As a result, computation is far more efficient with a large volume of self-replicating DNA than with a standard computer, which would require a lot more hardware. Information or data will now be kept in the form of the bases A, T, G, and C, rather than binary digits. The capacity to generate short DNA sequences artificially allows these sequences to be used as inputs for algorithms. This is possible due to the ability to create small DNA molecules with any arbitrary sequence. The input of any DNA operation can be represented by DNA molecules with specific sequences. The instructions are carried out by laboratory operations on the molecules, and the result is defined as some property of the final set of molecules. DNA computing makes significant and meaningful linkages between computers and life systems, as well as massively parallel computations. DNA computing can also carry out millions of operations at the same time.

On the other hand, Quantum computing is a field of study that focuses on the creation of computer-based technologies based on quantum-theoretical principles. On the quantum (atomic and subatomic) level, quantum theory describes the nature and the behavior of energy and matter. To execute certain computational tasks, quantum computing employs a combination of qubits. All of these are done at a far higher rate than their traditional computing equipment. Quantum computers represent a significant advancement in computing capability, with enormous performance benefits for specific use cases.

H. Md. Hasan Babu, *Quantum Biocomputing in Quantum Biology Volume II*,
https://doi.org/10.1007/978-981-97-5349-9_12

Besides the accuracy of a system, time or speed is a metric that can be used to measure the performance of a system. This chapter will describe how to calculate speed or consumed time for different operations in Quantum, DNA, Quantum-DNA, and DNA-Quantum computation systems.

12.2 Speed Calculation for Quantum Operations

Researchers have proposed the theory and formula to calculate the average required operational time in any quantum computation. Numerous researchers have also applied it in their work and research studies. The average computation time needed for an operation is calculated by Eq. 12.1.

$$\tau = \frac{h}{4E} \tag{12.1}$$

where τ is the required operational time, h is Planck's constant, and E is the quantum mechanical average energy of the performing system. It has also shown that the minimum operation time of any digital logic gate in quantum computation is given below:

$$\tau = \frac{h}{4E}\left(1 + 2\frac{\theta}{\pi}\right) \tag{12.2}$$

Here, τ is the required operational time, h is the plank's constant, E is the quantum mechanical average energy of the performing system, and θ is the phase shift modulo π. It considers any basic quantum gate that complements the state of a qubit and then adds to it an arbitrary phase shift.

Numerous researchers have noted that any basic quantum operation needs some basic amount of time which is tabulated in Table 12.1. It describes that a single gate operation as NOT needs 1 μs and a double qubit gate as CNOT, V, and V+ needs approximately 10 μs. Sometimes, one output of a gate operation in quantum computting can be needed in another gate operation to perform. In that case, one needs a movement time of approximately 20 μs, and in the future, it can be required more or less than 10 μs.

As a basic quantum operational gate is composed of a single or double qubit gate, it is easy to calculate the required operational time for a quantum operation by using the information provided in Table 12.1.

Now, basic quantum operational time will be calculated in their operational gates like AND, OR, XOR, NAND, NOR, and XNOR quantum operational gates. These quantum operational gates are prepared with quantum gates as NOT, CNOT, V+, and V.

Table 12.1 Execution times for basic quantum gate operations

Operations	Time (μs) now (future)
Single qubit gate	1 (1)
Double qubit gate	10 (10)
Movement	20 (10)

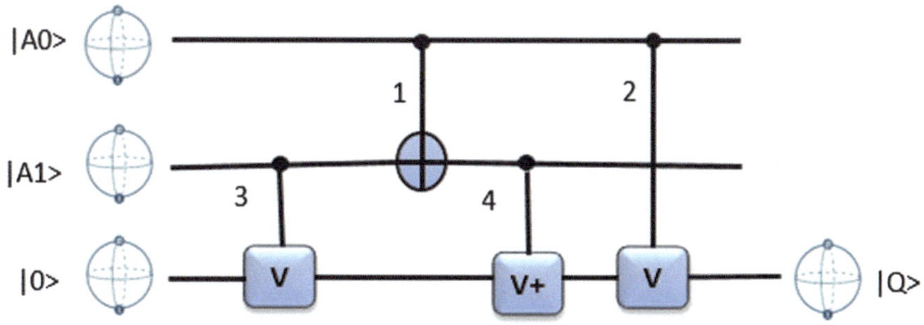

Fig. 12.1 Quantum AND operation circuit

1. **Speed Calculation for Quantum AND Operation**

   Figure 12.1 shows the quantum AND gate operation, which is nothing but a circuit that is composed of four quantum gates (CNOT, V, and V+). It has three lines to the output, where only the last one can provide the expected output (represented as |Q> in Fig. 12.1 as AND operations do.
   The CNOT (1) and V (3) quantum gates can operate in parallel in this operational circuit. So, the required operational time for both of them is 10 μs.
   The V+ (4) gate operates sequentially and its input depends on the output of CNOT (1) and V (2) quantum gates. So, the required operational time for V+ (4) gate is 10 μs.
   Again, the V (2) gate operates sequentially and its input depends on the input A0 and output of the V+ (4) quantum gate. So, the required operational time for V (2) gate is 10 μs.
   So, the total required time for AND operation = (Operational time for CNOT (1) or V (3) gate + Operational time for V+ (4) gate + Operational time for V (2) gate)
   = (10+10+10) μs
   = 30 μs.
   Thus, the total required operational time for Quantum AND operation is 30 μs.

2. **Speed Calculation for Quantum OR Operation**

   Figure 12.2 represents the quantum OR gate operation, which is a circuit that is composed of four quantum gates (CNOT and V). It has three lines to the output,

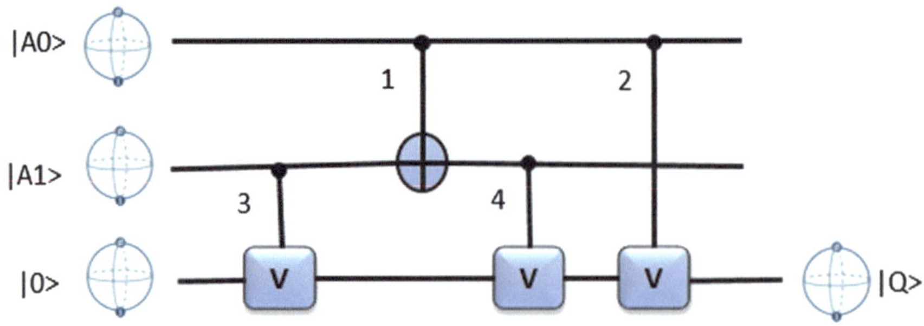

Fig. 12.2 Quantum OR operation circuit

but only the final one, like OR operations do, can deliver the desired result.
The CNOT (1) and V (3) quantum gates can operate in parallel in this operational circuit. So, the required operational time for both of them is 10 μs.
The V (4) gate operates sequentially and its input depends on the output of CNOT (1) and V (2) quantum gates. So, the required operational time for the V (4) gate is 10 μs.
Again, the V (2) gate operates sequentially and its input depends on the input A0 and output of the V (4) quantum gate. So, the required operational time for the V (2) gate is 10 μs.
So, the total required time for OR operation = (Operational time for CNOT (1) or V (3) gate + Operational time for V (4) gate + Operational time for V (2) gate)
= (10 + 10 + 10) μs
= 30 μs.
Thus, the total required operational time for Quantum OR operation is 30 μs.

3. **Speed Calculation for Quantum XOR Operation**

Figure 12.3 shows the quantum XOR or Exclusive OR or Ex-OR gate operation, which is a circuit that is composed of one basic Quantum gate (CNOT). It has two lines to the output, but only the final one with XOR operations, will deliver the anticipated result (as shown in Fig. 12.3).
The total required time for XOR operation depends on the operational time of the double qubits CNOT gate.
So, the required time for XOR operation is 10 μs.

4. **Speed Calculation for Quantum NAND Operation**

Figure 12.4 describes Quantum NAND gate operation, which is nothing but a circuit that is composed of four Quantum gates (CNOT, V+, and V). It has three lines to the output, but only the final one with NAND operations, can provide the expected result as shown in Fig. 12.4.
The CNOT (1) and V+ (3) quantum gates can operate in parallel in these oper-

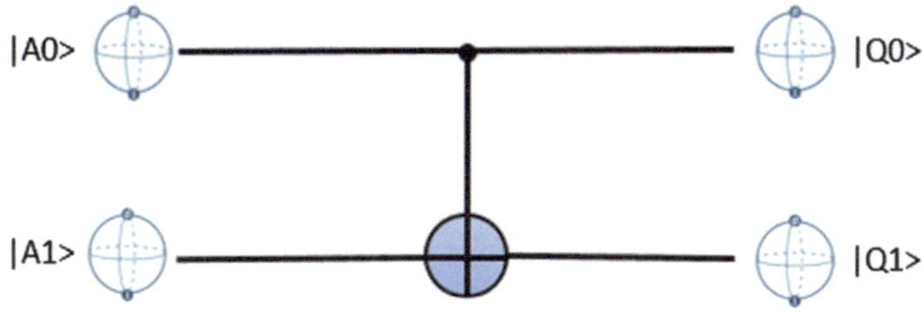

Fig. 12.3 Quantum XOR operation circuit

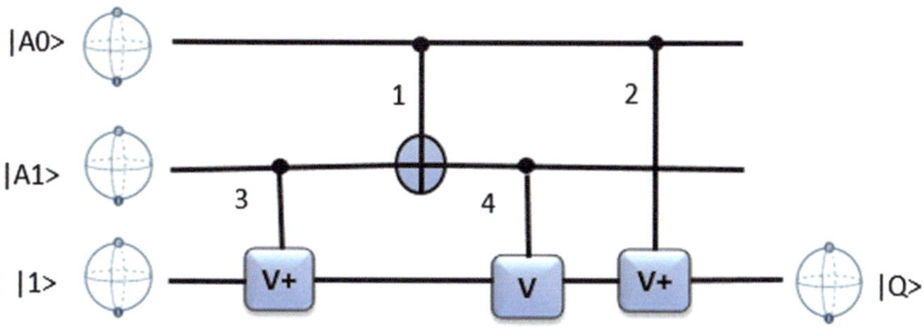

Fig. 12.4 Quantum NAND operation circuit

ational circuits. So, the required operational time for both of them is = 10 μs.
The V (4) gate operates sequentially and its input depends on the output of CNOT (1) and V+ (3) quantum gates. So, the required operational time for the V (4) gate is = 10 μs.
Again, the V+ (2) gate operates sequentially and its input depends on the input A0 and output of V (4) quantum gate. So, the required operational time for the V+ (2) gate is = 10 μs.
So, the total required time for NAND operation = (Operational time for CNOT (1) or V+ (2) gate + Operational time for V (4) gate + Operational time for V+ (2) gate)
= (10 + 10 + 10) μs
= 30 μs.
Thus, the total required operational time for Quantum NAND operation is 30 μs.

5. **Speed Calculation for Quantum NOR Operation**

Figure 12.5 describes Quantum NOR gate operation, which is a circuit that is composed of four quantum gates (CNOT and V). It has three lines to the output, but only the final one, like with NOR operations, it will deliver the desired result shown as |Q> in Fig. 12.5.
The CNOT (1) and V (3) quantum gates can operate in parallel in this operational

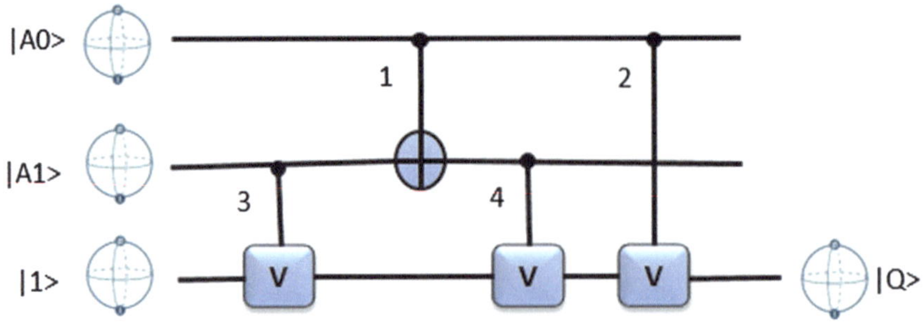

Fig. 12.5 Quantum NOR operation circuit

circuit. So, the required operational time for both of them is = 10 μs.
The V (4) gate operates sequentially and its input depends on the output of CNOT (1) and V (3) quantum gates. So, the required operational time for the V (4) gate is = 10 μs.
Again, the V (2) gate operates sequentially and its input depends on the input A0 and output of the V (4) quantum gate. So, the required operational time for the V (2) gate is = 10 μs.
So, the total required time for NOR operation = (Operational time for CNOT (1) or V (3) gate + Operational time for V (4) gate + Operational time for V (2) gate)
= (10 + 10 + 10) μs
= 30 μs.
Thus, the total required operational time for Quantum NOR operation is 30 μs.

6. **Speed Calculation for Quantum XNOR Operation**

Figure 12.6 shows the quantum XNOR operation, which is a circuit that is composed of two quantum gates (CNOT and NOT). It has two lines to the output, but only the final one, like with XNOR operations, will yield the desired output shown as |Q> in Fig. 12.6.
The CNOT gate operates sequentially and its input depends on the input of A1 and A0. So, the required operational time for the CNOT gate is = 10 μs.
Again, the NOT gate operates sequentially and its input depends on the output of the CNOT quantum gate. So, the required operational time for the NOT gate is 1 μs.
So, the total required time for XNOR operation = (Operational time for CNOT gate + Operational time for NOT gate)
= (10 + 1) μs
= 11 μs.
Thus, the total required operational time for Quantum XNOR operation is 11 μs.

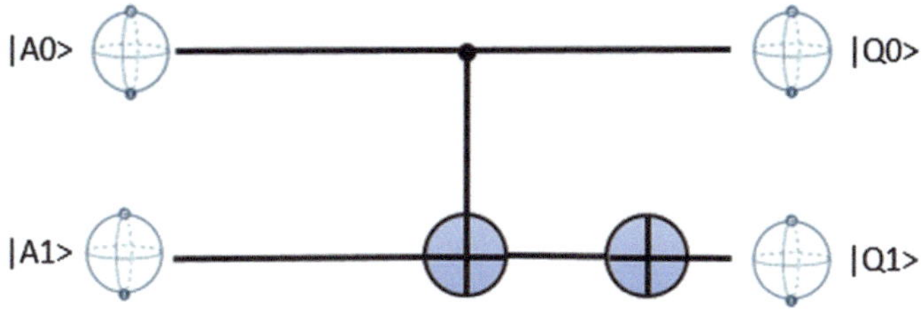

Fig. 12.6 Quantum XNOR operation circuit

12.2.1 Speed Calculation in Quantum Operational Circuits

In this section, the total operational time is going to be calculated for a quantum operational circuit. The basic information for calculating the operational time of a circuit is provided in the previous section.

12.2.1.1 Quantum Full Subtractor

A full subtractor is a combinational circuit which is developed to overcome the drawback of the half subtractor circuit. It can take three inputs and after subtracting them creates two outputs. Here, A full subtractor is implemented using the quantum circuit. To create a full subtractor, two NOT, two XOR, two AND, and an OR gate are required. Figure 12.7 depicts the quantum circuit of a full subtractor. A full subtractor receives three inputs and produces two outputs containing "D" and "B_{out}". To find the required performing time of a quantum full subtractor, it is divided into three

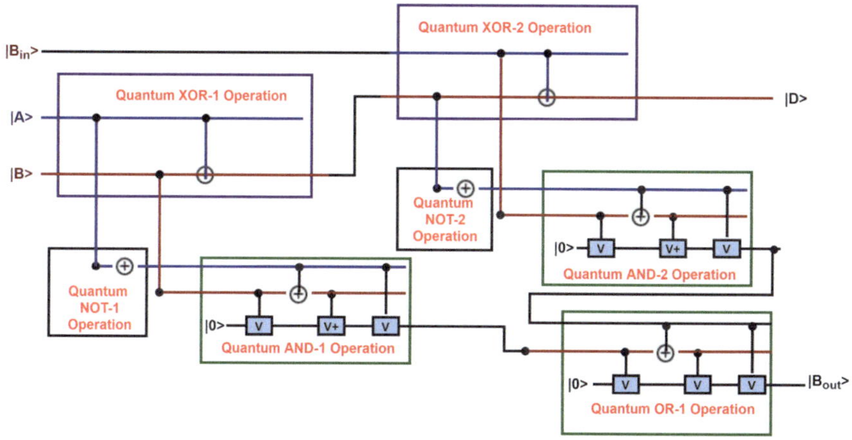

Fig. 12.7 Quantum full subtractor circuit

pipelines as some of the basic quantum gate operations are performed in parallel. Three pipelines are as follows:

1. XOR, XOR
2. XOR, NOT, AND, OR
3. NOT, AND, OR.

It is already perceived that XOR, AND, and OR quantum gate operation needs 10, 30, and 30 μs, respectively. And, a single qubit quantum operation as NOT need 1 μs. As the second pipeline is the largest pipeline for providing an output of the full subtractor, taking it for measuring the total required performing time. Other's quantum basic gate operations will be performed within this time in parallel.

The required time for a full subtractor is = (XOR+ NOT+ AND+ OR) μs.
where, required time for basic quantum NOT gate = 1 μs,
the required time for basic quantum XOR gate = 10 μs,
the required time for basic quantum AND gate = 30 μs.
the required time for basic quantum OR gate = 30 μs.
The required time for a full subtractor is = (10 + 1 + 30 + 30) μs = 71 μs.

12.2.1.2 Quantum Three-Qubit Even Parity Qubit Checker

A circuit that checks the parity in the receiver is called a Parity Checker. A combined circuit or device of parity generators and parity checkers are commonly used in digital systems to detect single-qubit errors in the transmitted data. To create a three-qubit even parity qubit-checker, three XOR gates are required. Figure 12.8 describes the digital and quantum circuits of a three-qubit even parity qubit checker. A three-qubit even parity qubit checker receives four inputs and produces one output containing "E".

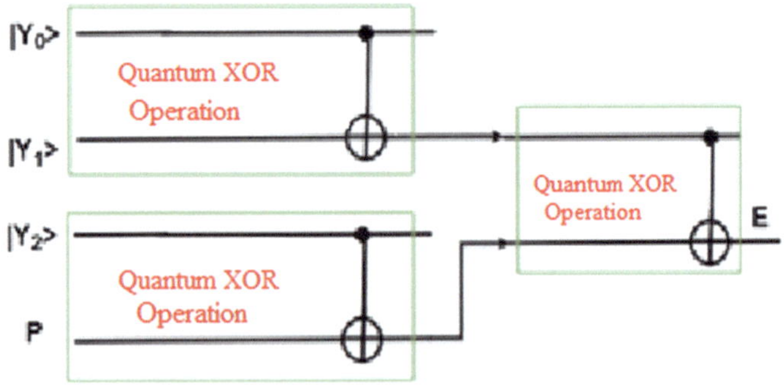

Fig. 12.8 Quantum three-qubit even parity qubit checker

To find the required performing time of Quantum three-qubit even parity qubit checker, it is divided into two pipelines as some of the basic quantum gate operations are performed in parallel. Two pipelines are as follows:

1. XOR, XOR
2. XOR, XOR

It could be perceived that the quantum XOR gate operation needs 10μs. As both pipelines are the equal pipeline for providing an output of the three-qubit even parity-qubit checker, any of them is taken for measuring the total required performing time. Other quantum basic gate operations will be performed within this time in parallel.

12.2.1.3 Quantum 2-to-1 Multiplexer

A multiplexer (MUX) is a device that can receive multiple input signals and synthesize a single output signal in a recoverable manner for each input signal. It is also an integrated system that usually contains a certain number of data inputs and a single output. To create a multiplexer, one NOT, two AND, and an OR gate are required. Figure 12.9 describes the quantum multiplexer circuit. A multiplexer receives three inputs and produces one output containing "Y".

To find the required performing time of a quantum multiplexer, it is divided into two pipelines as some of the basic quantum gate operations are performed in parallel. Two pipelines are as follows:

1. NOT, AND, OR
2. AND, OR

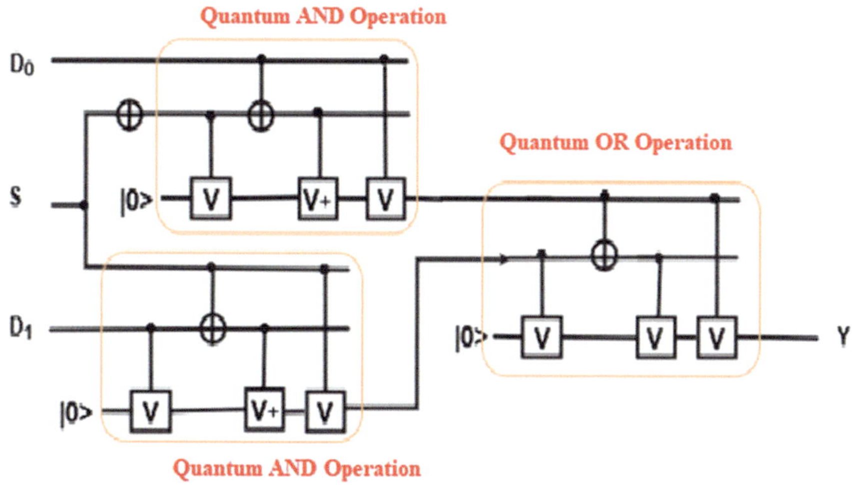

Fig. 12.9 Quantum multiplexer circuit

It is perceived that AND and OR quantum gate operation needs 30 μs, respectively. And, a single qubit quantum operation as NOT need 1 μs. As the first pipeline is the largest pipeline for providing an output of the Multiplexer, let's take it for measuring the total required performing time. Other quantum basic gate operations will be performed within this time in parallel.

So, the performing time for full multiplexer will be (NOT+ AND+ OR) μs,

where the required time for basic quantum NOT gate = 1 μs,
the required time for basic quantum AND gate = 30 μs.
the required time for basic quantum OR gate = 30 μs.
So, the required time for Multiplexer is = (1 + 30 + 30) = 61 μs.

12.3 Speed Calculation for DNA Operations

DNA molecules can be used as information storage media. Usually, DNA sequences of around 8–20 base pairs are used to represent bits, and numerous methods have been developed to manipulate and evaluate them. To manipulate a wet technology to perform computations, one or more of the following techniques are used as computational operators for copying, sorting, splitting, or concatenating the information contained within DNA molecules as ligation, hybridization, polymerase chain reaction (PCR), gel electrophoresis, and enzyme reaction.

Allosteric DNAzyme-based DNA logic circuit, described a procedure to make a DNAzyme-based logic circuit. Here, All DNA logic gates were formed by annealing twice: firstly, the mixture of the inhibitor DNA strands and E6-type DNAzymes in 1× TAE/Mg_2+ buffer (40 mM Tris, 20 mM acetic acid, 1 m $MEDTA_2Na$ and 12.5 mM $Mg(OAc)_2$, pH 8.0) was heated at 95 °C for 4 min, 65 °C for 30 min, 50 °C for 30 min, 37 °C for 30 min, 22 °C for 30 min, and preserved at 20 °C; and then the substrates were added into the annealed mixture and incubated at constant temperature 20 °C for 4 h (total 6 h for preparing the DNA logic gate).

After that logic gates were triggered through displacement reaction in 1× TAE/Mg2+ buffer (40 mM Tris, 20 mM acetic acid, 1 mM $EDTA_2Na$, and 12.5 mM Mg $(OAc)_2$, pH 8.0). The input DNA strands were added to a solution containing DNA logic gates and reacted for >2 h at 20 °C. Next, the displaced products were stored at 20 °C for native PAGE or fluorescence detection. In addition, polyacrylamide gel electrophoresis (PAGE) needs 2 h and the PCR process for fluorescence detection needs less than 2 h.

A specific biochemical process described briefly serves as the basis of the DNA computing approach as Polymerase Chain Reaction (PCR). Polymerases perform several functions, including the repair and duplication of DNA. PCR is a process that quickly amplifies the amount of specific DNA molecules in a given solution, using primer extension by the polymerase. Each cycle of the reaction doubles the quantity of this molecule, leading to an exponential growth in the number of sequences. It consists of the following key processes:

1. **Initialization:** a mixed solution of template, primer, dNTP, and the enzyme is heated to 94–98 °C for 1–9 min to ensure that most of the DNA template and primers are denatured;
2. **Denaturation:** heat the solution to 94–98 °C for 20–30 s for separation of DNA duplexes;
3. **Annealing:** lower the temperature enough (usually between 50–64 °C) for 20–40 s for primers to anneal specifically to the ssDNA template;
4. **Elongation/Extension:** raise temperature to optimal elongation temperature of *Taq* or similar DNA polymerase (70–74 °C) for the polymerase adds dNTP's from the direction of 5' to 3' that are complementary to the template;
5. **Final Elongation/Extension:** after the last cycle, a 5–15 min elongation may be performed to ensure that any remaining ssDNA is fully extended.

Steps 2–4 are repeated 20–35 times; fewer cycles results in less product, too many cycles increase the fraction of incomplete and erroneous products. PCR is a routine job in the laboratory that can be performed by an apparatus named a thermal cycler. According to the PCR process, to produce an operational output of DNA computation, it needs around 2 h.

So, except for initial preparation and the last phase of fluorescence detection for each operation in a particular test tube, it needs more or less than 2 h.

12.4 Speed Calculation in DNA Operational Circuits

This subsection is going to describe some DNA circuits to calculate their performing time or speed in an approximate value based on the theory described in Sect. 12.3.

12.4.1 DNA Full Subtractor

A full subtractor is a combinational circuit that performs two-molecular subtraction, one is minuend and the other is subtrahend, taking into account the borrow of the previous adjacent lower minuend bit. This circuit has three inputs and two outputs. The three inputs A, B, and B_{in}, denote the minuend, subtrahend, and previous borrow, respectively. The two outputs, D and B_{out} represent the difference and output borrows, respectively. To create a full subtractor, one OR, two AND, two NOT, and two XOR gates are required. Figure 12.10 describes the DNA circuit of a full subtractor. Here, two inputs will be DNA sequences and it also provides two outputs that contains DNA sequence. In DNA computing, DNA AND operation will be easy to represent by DNA NOT and DNA NAND operation.

According to Sect. 12.2, it is perceived that any DNA basic gate (i.e. AND, OR, NOT, and XOR) operation needs more or less than 2 h to perform. In addition, 6

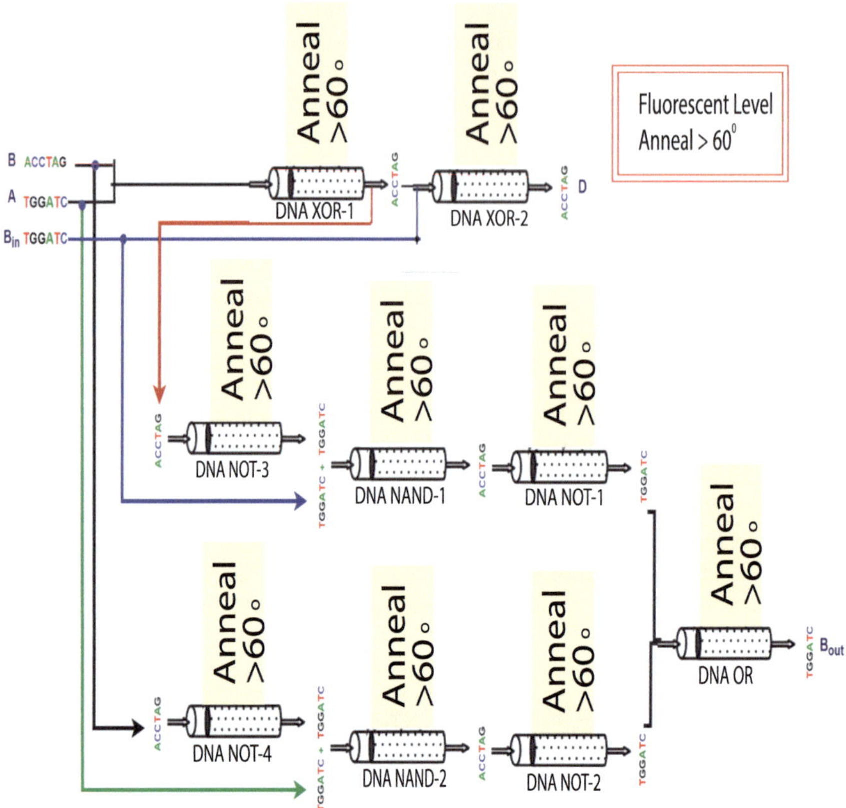

Fig. 12.10 DNA full subtractor circuit

h are needed for preparing any DNA basic gate operation and 2 h for fluorescence detection which is fixed for any multi-(basic) gate operation.

As the second pipeline is the largest pipeline for processing input to the output of the Full Subtractor, it will be considered to measure the total required operational time. The other DNA basic gate operations will be performed within this time in parallel.

So, the required time for four basic gate operations in DNA is

$= (2 + 2 + 2 + 2)$

$= 8\,\text{h},$

where the required time for DNA gate (AND, OR, NOT, and XOR) operation needs more or less than 2 h.

The total required time for performing DNA full subtractor the summation of initial preparation time, fluorescence detection time, and DNA gate operational time.

So, the required time for DNA multiplexer operation is
= (Basic gate preparation time + four basic DNA gate operational time + Fluorescence detection)
= (6 + 8 + 2)
= 16 h (approximately).

12.4.2 DNA Full Adder

A full adder is an adder that adds three inputs and produces two outputs. The first two inputs are A and B and the third input is an input carry as C_{in}. The output carry is designated as C_{out} and the normal output is designated as S which is SUM. A full-adder logic is designed in such a manner that can take eight inputs together to create a byte-wide adder and cascade the carry bit from one adder to another. To create a full adder, one OR, two NAND, one XOR, and two NOT gates are required. Figure 12.11 depicts the DNA circuit of the full adder.

To find the required operational time of DNA full adder, it is divided into three pipelines as some of the basic DNA gate operations are performed in parallel. Three pipelines are as follows:

1. XOR, XOR
2. XOR, NAND, NOT, OR
3. NAND, NOT, OR.

According to Sect. 12.3, it can be perceived that any DNA basic gate (i.e., AND, OR, NOT, and XOR) operation needs more or less than 2 h to perform. In addition, 6 h are needed for preparing any DNA basic gate operation and 2 h for fluorescence detection which is fixed for any multi-(basic) gate operation.

As the second pipeline is the largest pipeline for processing input to the output of the Full Adder, it is taken for measuring the total required operational time. Other DNA basic gate operations will be performed within this time in parallel.

So, the required time for four basic gate operations in DNA is = (2 + 2 + 2 + 2) = 8 h,

where the required time for DNA gate (NAND, OR, NOT, and XOR) operation needs more or less than 2 h.

The total required time for performing DNA full adder operation is a summation of initial preparation time, fluorescence detection time, and DNA gate operational time.

So, the required time for DNA full adder operation is
= (Basic gate preparation time + Four basic DNA gate operational time + Fluorescence detection)
= (6 + 8 + 2) h
= 16 h (approximately).

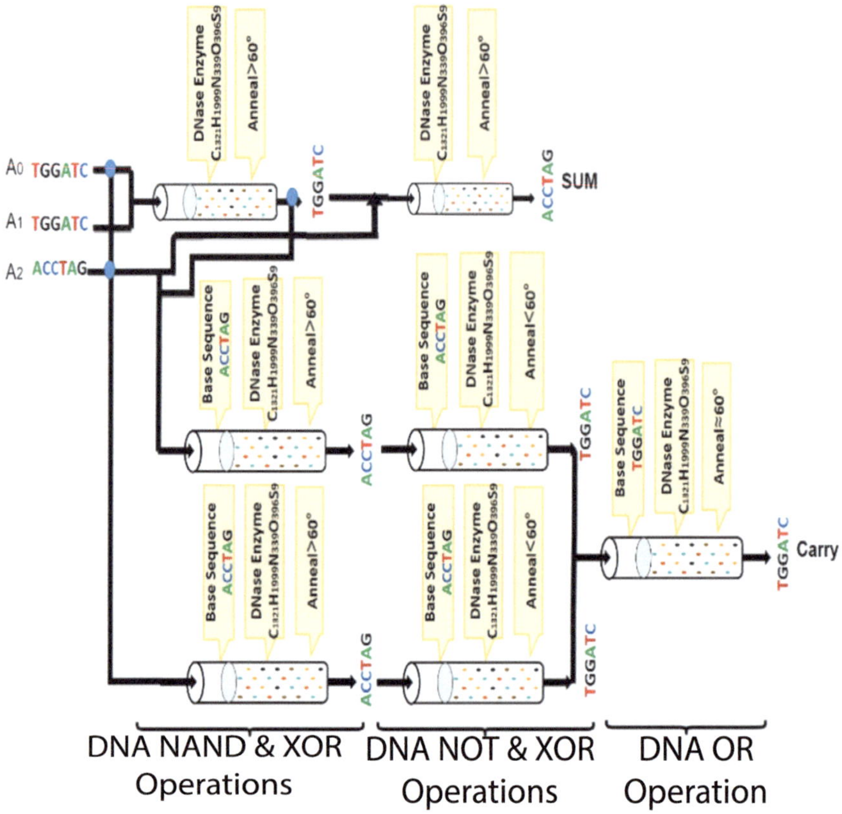

Fig. 12.11 DNA full adder circuit

12.4.3 DNA Multiplication Circuit

A binary multiplier is a combinational logic circuit or digital device used for multiplying two binary numbers. The two numbers are more specifically known as multiplicand and multiplier and the result is known as a product. The multiplicand and multiplier can be of various bit sizes. The product's bit size depends on the bit size of the multiplicand and multiplier. The bit size of the product is equal to the sum of the bit size of multiplier and multiplicand. To create a multiplication circuit, six NAND, two XOR, and six NOT gates are required. Figure 12.12 describes the DNA circuit of the two-molecular multiplication.

To find the required performing time of a DNA multiplication circuit, it is divided into multiple pipelines as some of the basic DNA gate operations are performed in parallel. Among all of the pipelines, five are as follows:

1. NAND, NOT
2. NAND, NOT, XOR

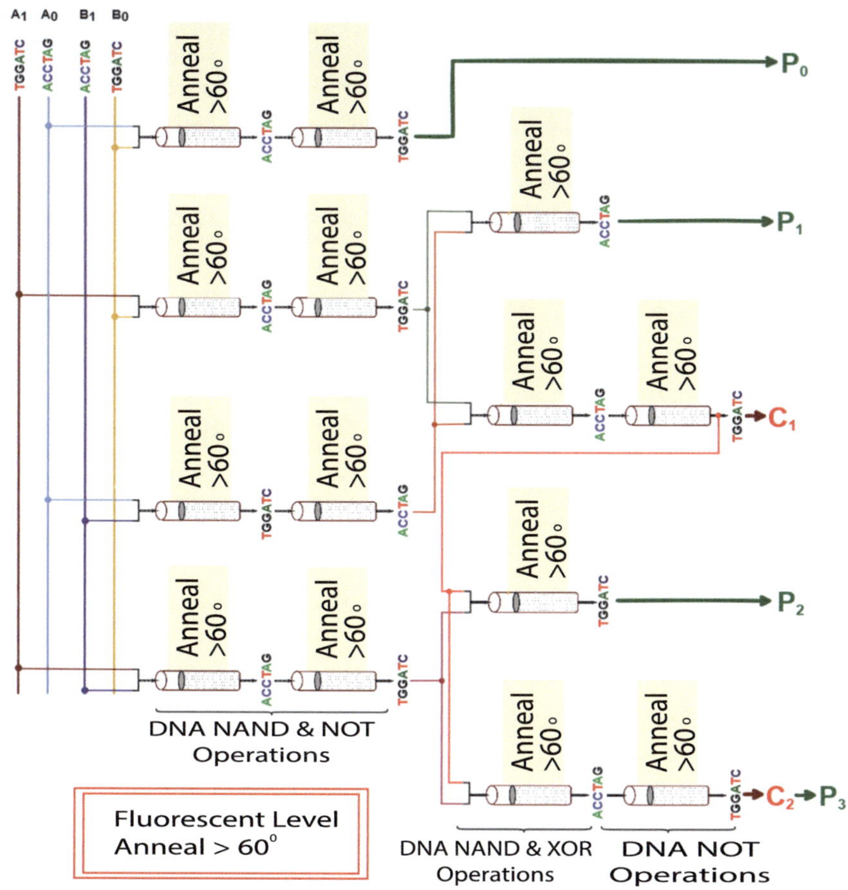

Fig. 12.12 DNA multiplier circuit

3. NAND, NOT, NAND, NOT, NAND, NOT
4. NAND, NOT, NAND, NOT; and
5. NAND, NOT, XOR.

According to Sect. 12.2, it can be perceived that any DNA basic gate (i.e., AND, OR, NOT, and XOR) operation needs more or less than 2 h to perform. In addition, 6 h are needed for preparing any DNA basic gate operation and 2 h for fluorescence detection which is fixed for any multi-(basic) gate operation.

As the third pipeline is the longest pipeline for processing input to the output of the two-molecular multiplication circuit, it is taken for measuring the total required performing time. Other DNA basic gate operations will be performed within this time in parallel.

So, the required time for four basic gate operations in DNA is $= (2 \times 6) = 12$ h, where the performing time for DNA gate (NAND and NOT) operation needs more or less than 2 h.

The total required time for performing DNA two-molecular multiplication circuit operation operation is the summation of initial preparation time, fluorescence detection time, and DNA gate operational time.

So, the required time for DNA two-molecular multiplication circuit operation is
= (Basic gate preparation time + six basic DNA gate operational time + fluorescence detection)
= (6 + 12 + 2) h
= 20 h (approximately).

12.5 Speed Calculation in Quantum-DNA Circuits

A Quantum-DNA computation system can merge all advantages of quantum computing and DNA computing. In a Quantum-DNA computing system, the input will be received as a qubit and after performing in a certain number of quantum operational gates these qubits will be turned into DNA sequences by NMR relaxation.

12.5.1 Full Subtractor at 0 K

A full subtractor is a combinational circuit that is developed to overcome the drawback of the half subtractor circuit. It can take inputs and after subtracting them creates two outputs. Here the full subtractor is implemented using the Quantum-DNA Circuit. To create a full subtractor, two Not, two XOR, two AND, and an OR operational circuit are required. Figure 12.13 describes the Quantum-DNA circuit for full subtractor. A full subtractor receives three inputs and produces two outputs containing "D" and "B_{out}".

From the Quantum-DNA circuit in Fig. 12.13, it is found that 1 XOR, 2 AND quantum gate operational circuits perform with quantum qubits. The output of all quantum operational gates goes through NMR relaxation at 0 K. It produces DNA sequence as an output and it can be used as an input in all DNA operational circuits.

Here, two DNA operational circuits (one XOR and one OR) are used to find the expected output. Four DNA sequences are the outputs from the NMR relaxation process and are used as input sequence in the XOR and OR DNA operational gate.

To find the required performing time of a Quantum-DNA full subtractor, it can be divided into three pipelines as some of the basic quantum and DNA operational gates are performing in parallel.

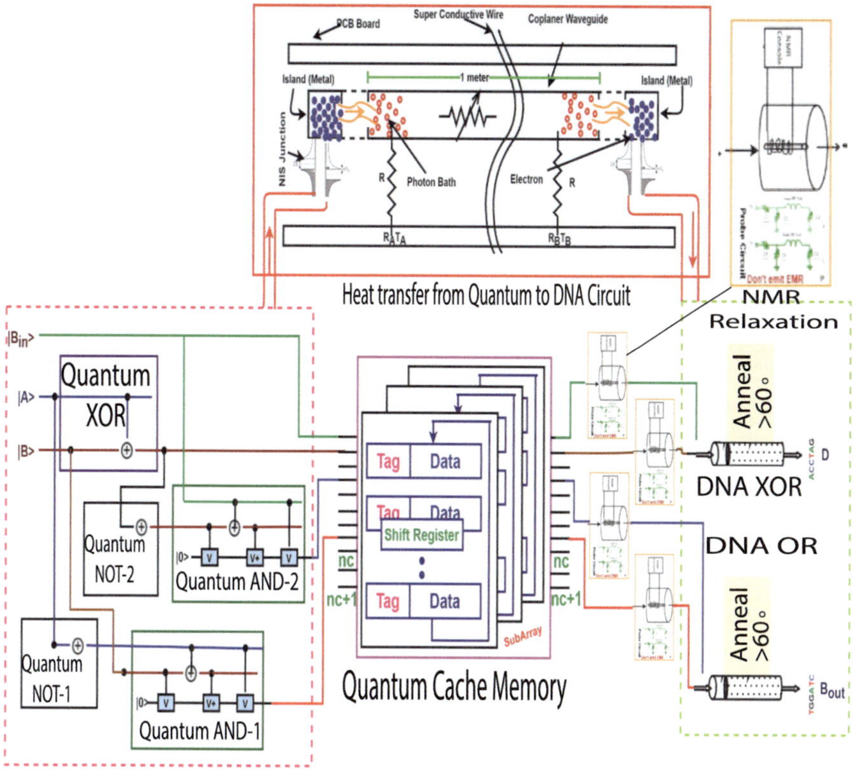

Fig. 12.13 A quantum-DNA full subtractor

Three pipelines are as follows:

(1) Quantum XOR, DNA XOR
(2) Quantum (XOR, NOT, AND), DNA OR
(3) Quantum (NOT, AND), DNA OR.

As the second pipeline is the largest pipeline for providing an output of the full subtractor, it is taken for measuring the total required performing time. Other quantum and DNA basic operational gates will be performed in parallel within this time. Three pipelines for extracting the output, one can understand that last one operational gate will perform with DNA sequence and others will perform with qubit quantum operational gate. According to Sect. 12.2, it can be perceived that XOR and AND quantum operational gates need 10 μs and 30 μs, respectively. And a single qubit quantum operation as NOT needs 1 μs.

So, the required time for three basic quantum operational gates is $(10 + 1 + 30)$ μs.

$= 41$ μs.

Furthermore, from Sect. 12.3, it is found that any DNA basic gate (i.e., AND, OR, NOT, and XOR) operation needs more or less than 2 h to perform. In addition, 6 h are needed for preparing any DNA basic gate operation and 2 h for fluorescence detection which is fixed for any multi-(basic) gate operation.

The total operational time required for DNA OR operation is the summation of initial preparation time, fluorescence detection time, and DNA gate operational time.

So, the required time for DNA OR operation is
= (Basic gate preparation time + OR gate operational time + Fluorescence detection)
= (6 + 2 + 2)
= 10 h (approximately).

Thus, to find the expected output of the Quantum-DNA full subtractor, the required time will be the summation of quantum operation and DNA operation.

So, the total required time for Quantum-DNA full subtractor operation is
= (Required time for quantum operation + Required time for DNA operation)
= (41 μs + 10 h)
= 10 h (approximately).

12.5.2 *Full Adder at 0 K*

Full adder is the adder that adds three inputs and produces two outputs. The first two inputs are A and B and the third input is an input carry as C_{in}. The output carry is designated as C_{out} and the normal output is designated as S which is sum. A full adder logic is designated in such a manner that it can take eight inputs together to create a qubit adder and cascade the carry qubit from one adder to another. Figure 12.14 describes the Quantum-DNA circuit of the full adder.

From the Quantum-DNA circuit in Fig. 12.14, it is found that one XOR and two AND quantum gate operational circuit performs with quantum qubits. The output of all quantum operational gates goes through NMR relaxation at 0 K. It produces the DNA sequence as an output and it can be used as an input in all DNA operational circuits.

Here, two DNA operational circuits (one XOR and one OR) are used to find the expected output. Four DNA sequences are the outputs from the NMR relaxation process and are used as input sequence in the XOR and OR DNA operations.

To find the required performing time of Quantum-DNA full adder, it can be divided into four pipelines as some of the basic quantum and DNA operational gates are performing in parallel. Three pipelines are as follows:

(1) Quantum XOR, DNA XOR
(2) Quantum (XOR, AND), DNA OR
(3) Quantum (AND), DNA XOR
(4) Quantum AND, DNA OR.

As the second pipeline is the largest pipeline for providing an output of the full adder, it is taken for measuring the total required performing time. Other quantum

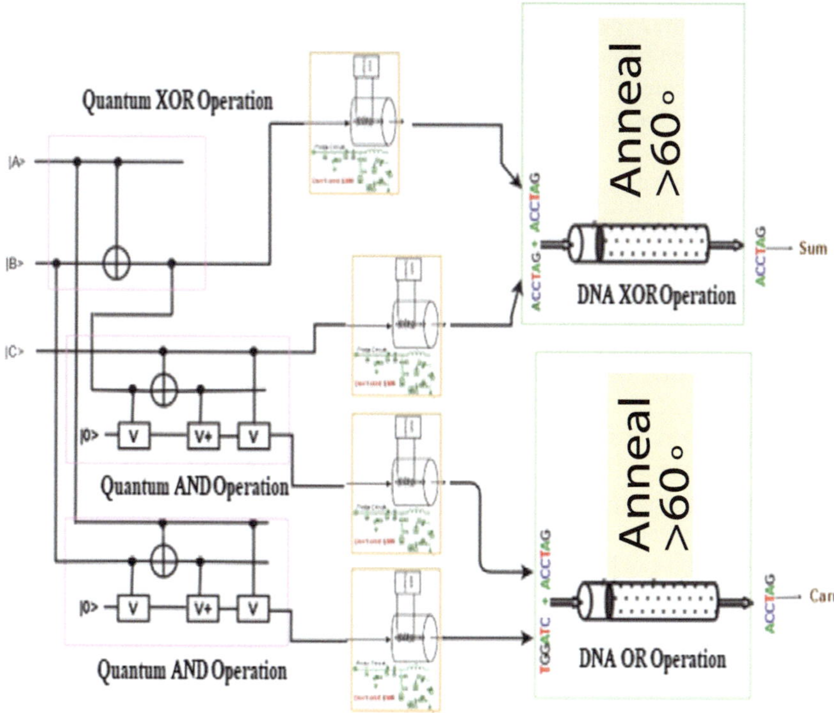

Fig. 12.14 A quantum-DNA full adder

and DNA basic operational gates will be performed in parallel within this time. Four pipelines for extracting the output, one can understand that the last operation will be performed with DNA sequence and others will perform with qubit in the quantum operational gate. According to Sect. 12.2, it can be perceived that XOR and AND quantum operational gates need 10 μs and 30 μs, respectively.

So, the required time for two basic quantum operational gates is (10 + 30) μs. = 40 μs.

Furthermore, from Sect. 12.3, it is found that any DNA basic operational gate (i.e., AND, OR, NOT, and XOR) needs more or less than 2 h to perform. In addition, 6 h are needed for preparing any DNA basic gate operation and 2 h for fluorescence detection which is fixed for any multi-(basic) gate operation.

The total operational time required for DNA OR operation is the summation of initial preparation time, fluorescence detection time and DNA gate operational time.

So, the required time for DNA OR operation is

= (Basic gate preparation time + OR gate operational time + Fluorescence detection)
= (6 + 2 + 2)
= 10 h (approximately).

Thus, to find the expected output of Quantum-DNA full subtractor, the required time will be the summation of Quantum operation and DNA operation.

So, the total required time for Quantum-DNA full subtractor is
= (Required time for quantum operation + Required time for DNA operation)
= (40 μs + 10 h)
= 10 h (approximately).

12.5.3 *Multiplier at 0 K*

To create a Quantum-DNA multiplier circuit, four AND operations in quantum, two XOR, and two AND DNA operational gates are required. Figure 12.15 describes the Quantum-DNA circuit of the two-bit multiplication.

From the Quantum-DNA circuit in Fig. 12.15, it is found that four AND quantum gate operational circuits perform with quantum qubits. The output of all quantum operational gates goes through NMR relaxation at 0 K. It produces DNA sequence as an output and it can be used as an input in all DNA operational circuits.

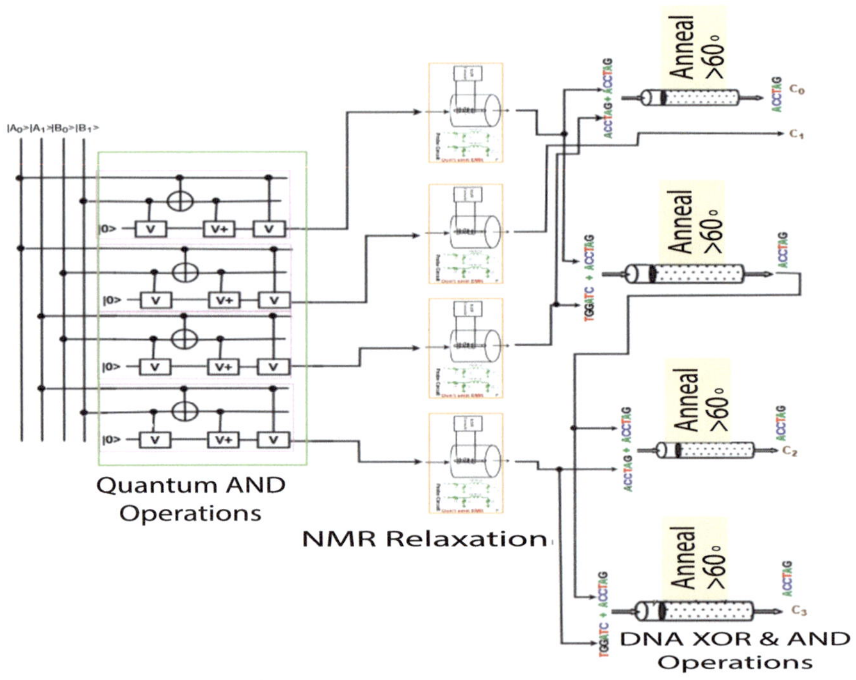

Fig. 12.15 Quantum-DNA multiplier circuit

To find the required operational time of a Quantum-DNA multiplier, it can be divided into five pipelines as some of the basic quantum and DNA operational gates are performing in parallel. Five pipelines are as follows:

(1) Quantum AND, DNA XOR
(2) Quantum AND, DNA (AND, XOR)
(3) Quantum AND, DNA (AND, AND)
(4) Quantum AND
(5) Quantum AND, DNA AND.

As the second or third pipeline is the largest pipeline for providing an output of the multiplier, any of these two can be taken for measuring the total required performing time. Other quantum and DNA basic operational gates will perform in parallel within this time. There are five pipelines for extracting the output, where one can understand that the last two operational gates will perform with DNA sequence and others will perform with qubit using quantum gates. According to Sect. 12.2, it can be perceived that the quantum AND gate needs 30 μs for providing the output.

So, the required time for the basic quantum operational gate is 30 μs.

Furthermore, from Sect. 12.3, it is found that any DNA basic operational gate (i.e., AND, OR, NOT and XOR) needs more or less than 2 h to perform. In addition, it need 6 h for preparing any DNA basic gate operation and 2 h for fluorescence detection which is fixed for any multi-(basic) gate operation.

So, the required time for DNA XOR and AND operations are $(2 + 2)$ h
$= 4$ h

The total operational time required for DNA OR is the summation of initial preparation time, fluorescence detection time and DNA gate operational time.

So, the required time for DNA OR operation is
= (Basic gate preparation time + DNA gate operational time + Fluorescence detection)
$= (6 + 4 + 2)$
$= 12$ h (approximately).

Thus, to find the expected output of Quantum-DNA multiplier, the required time will be the summation of quantum operation and DNA operation.

So, the total required time for Quantum-DNA multiplier operation is
= (Required time for Quantum operation + Required time for DNA operation)
= (30 μs + 12 h)
$= 12$ h (approximately).

12.5.4 Multiplexer at 0 K

A multiplexer (MUX) is a device that can receive multiple input signals and synthesize a single output signal in a recoverable manner for each input signal. It is also an integrated system that usually contains a certain number of data inputs and a single output. To create a multiplexer, one NOT, two AND, and an OR gates are required.

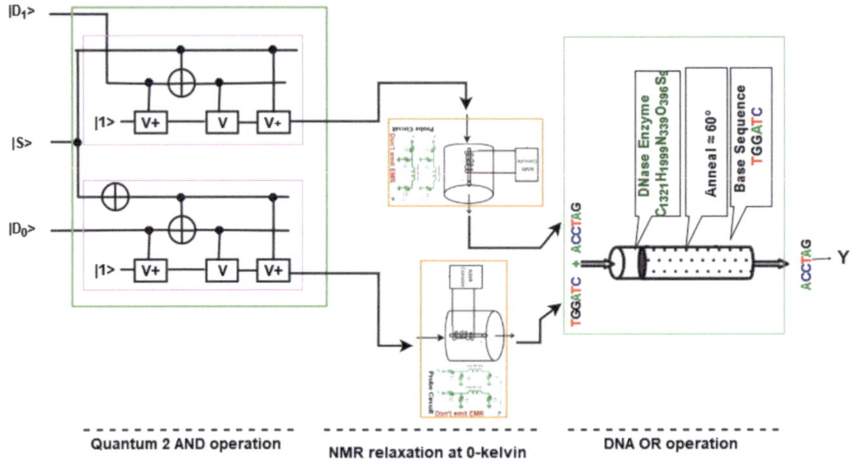

Fig. 12.16 Quantum-DNA 2-to-1 multiplexer circuit

Figure 12.16 describes the Quantum-DNA circuit of the multiplexer. A multiplexer receives three inputs and produces one output containing "Y".

From the Quantum-DNA multiplexer circuit in Fig. 12.16, two AND quantum gate operational circuits performs with quantum qubits. The output of all quantum operational gates goes through NMR relaxation at 0 K. It produces a DNA sequence as an output and it can be used as an input in all DNA operational circuits. Here, a DNA operational circuit (OR) is used to find the expected output. Two DNA sequences are the outputs from the NMR relaxation process and are used as an input sequences in OR DNA operational gate.

To find the required operational time of Quantum-DNA Multiplexer, divide it into five pipelines as some of the basic quantum and DNA operations are performing in parallel. Five pipelines are as follows:

1. Quantum AND, DNA OR
2. Quantum AND, DNA OR.

As both pipelines are equal for providing an output of the Multiplexer, taking any of these two for measuring the total required performing time. Other quantum and DNA basic operations will be performed in parallel within this time. Two pipelines for extracting the output, one can understand that the last operation will be performed with DNA sequence and others will perform with qubit in the quantum operational gate. According to Sect. 12.2, AND quantum, operational gate needs 30 μs for providing the output.

So, the required operational time for a basic quantum gate is 30 μs.

Furthermore, from Sect. 12.3, any DNA basic gate (i.e., AND, OR, NOT and XOR) needs more or less than 2 h to perform. In addition, it needs 6 h for preparing any DNA basic gate operation and 2 h for fluorescence detection which is fixed for any basic gate operation.

So, the required time for DNA OR operations is 2 h.

The total performing time required for DNA OR operation is the summation of initial preparation time, fluorescence detection time and DNA gate operational time.

So, the required time for DNA OR operation is
= (Basic gate preparation time + DNA gate operational time + Fluorescence detection)
= (6 + 2 + 2)
= 10 h (approximately).

Thus, to find the expected output of Quantum-DNA Multiplier, the required time will be summation of Quantum operation and DNA operation.

So, total required time for Quantum-DNA Multiplier is
= (Required time for Quantum operation + Required time for DNA operation)
= (30 μs + 10 h)
= 10 h (approximately).

12.6 Speed Calculation in DNA-Quantum Circuits

A DNA-Quantum computation system is introduced to find a super-fast computation system with a lot of memory. This DNA-Quantum computation device combines the benefits of both quantum and DNA computing.

In a DNA-Quantum computing system, input will be received in DNA sequences and after performing in a certain number of DNA operational gates these DNA sequences will be turned into quantum qubits by the NMR process.

12.6.1 3-Qubit Parity Qubit Checker at 0 K

A circuit that checks the parity in the receiver is called a Parity Checker. A combined circuit or device of parity generators and parity checkers are commonly used in digital systems to detect single-bit errors in the transmitted data. To create a 3-qubit even parity bit checker, three quantum XOR gates are required.

Figure 12.17 describes the DNA-Quantum circuit of 3-qubit even parity-qubit checker. A 3-qubit even parity qubit checker receives four inputs and produces one output containing “E”.

From the DNA-Quantum circuit in Fig. 12.17, two XOR DNA operational gate performs with DNA sequence. The output of all DNA operational gates goes through NMR at 0 K. It produces a quantum qubit as an output and it can be used as an input in all quantum operational circuits. Here a DNA operational circuit (XOR) is used to find the expected output. Two quantum qubit is the output from the NMR process and is used as input qubit in Quantum OR operation. To find the required performing time of DNA-Quantum 3-molecular Even parity molecular checker, divide it into

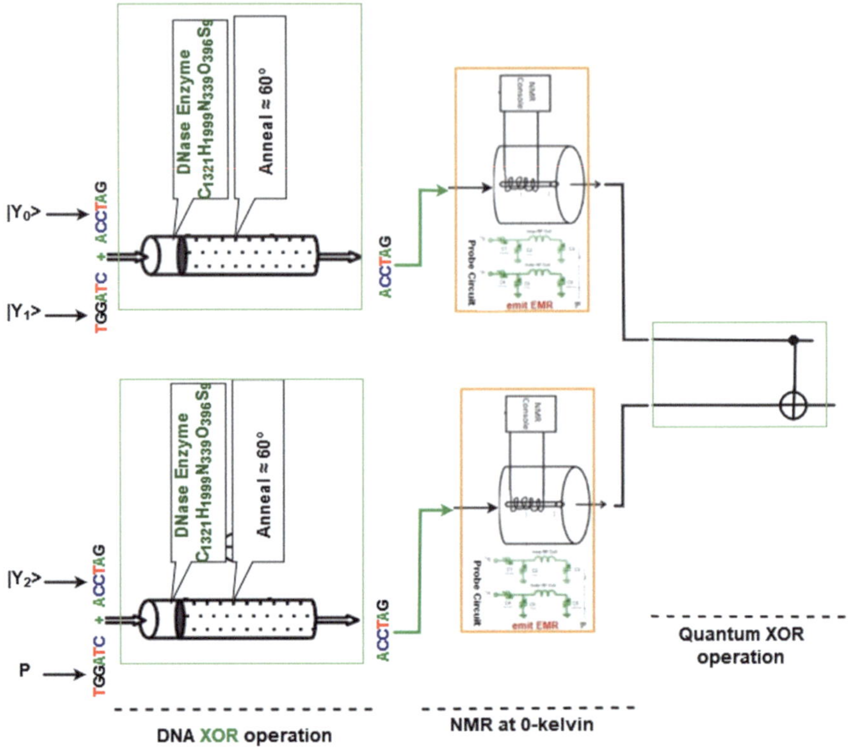

Fig. 12.17 DNA-quantum circuit of 3-molecular even parity molecular checker

two pipelines as some of the basic quantum and DNA operational gate is performing in parallel. Two pipelines are as follows:

1. DNA XOR, Quantum XOR
2. DNA XOR, Quantum XOR.

As both pipelines are equal for providing an output of the 3-molecular even parity molecular checker, taking any of these two for measuring the total required performing time. Other DNA and Quantum operations will be performed in parallel within this time. DNA basic operational gate (i.e., AND, OR, NOT and XOR) needs more or less than 2 h to perform. In addition, it needs 6 h for preparing any DNA basic gate operation and 2 h for fluorescence detection which is fixed for any multi-(basic) gate operation.

So, the required time for DNA XOR operations is 2 h.

The total performing time required for DNA XOR operation is the summation of initial preparation time, fluorescence detection time and DNA gate operational time.

So, the required time for DNA OR operation is
= (Basic gate preparation time + DNA gate operational time + Fluorescence detection)
= (6 + 2 + 2)
= 10 h (approximately).

XOR quantum operation needs 10 μs to obtain the output.

So, the required time for basic quantum operational gate is = (10) μs.

Thus, to find the expected output of the DNA-Quantum 3-molecular even parity molecular checker, the required time will be summation of Quantum operation and DNA operation.

So, the total required time for DNA-Quantum 3-molecular even parity molecular checker is
= (The required time for DNA operation + The required time for Quantum operation)
= (10 h + 10 μs)
= 10 h (approximately).

12.6.2 Full Subtractor at 0 K

A Full Subtractor is a combinational circuit that is developed to overcome the drawback of the half subtractor circuit. It can take inputs and after subtracting them creates two outputs. Here full Subtractor is implemented using the DNA- Quantum Circuit. To create a Full Subtractor, two NOT, two XOR, two AND, and an OR operational circuit are required. Figure 12.18 describes the DNA-Quantum circuit for Full Subtractor. A Full Subtractor receives three inputs and produces two outputs with "D" and "B_{out}".

From the Quantum-DNA circuit in Fig. 12.18, 1 XOR, 2 NOT, and 2 AND DNA gate operational circuit performs with DNA sequences. The output of all DNA operational gates goes through NMR at 0 K. It produces Quantum qubit as an output and it can be used as an input in all quantum operational circuits.

Here 2 quantum operational circuit (1 XOR and 1 OR) is used to find the expected output. Four quantum qubits are the output from the NMR process and is used as input qubit in XOR and OR quantum operational gate.

To find the required operational time of DNA-Quantum Full Subtractor, divide it into three pipelines as some of the basic DNA and Quantum operations are performed in parallel. Three pipelines are as follows:

1. DNA (XOR, NOT), Quantum XOR
2. DNA (XOR, NOT, AND), Quantum OR
3. DNA (NOT, AND), Quantum OR.

As the second pipeline is the largest pipeline for providing an output of the Full Subtractor, it is considered it for measuring the total required operational time. The

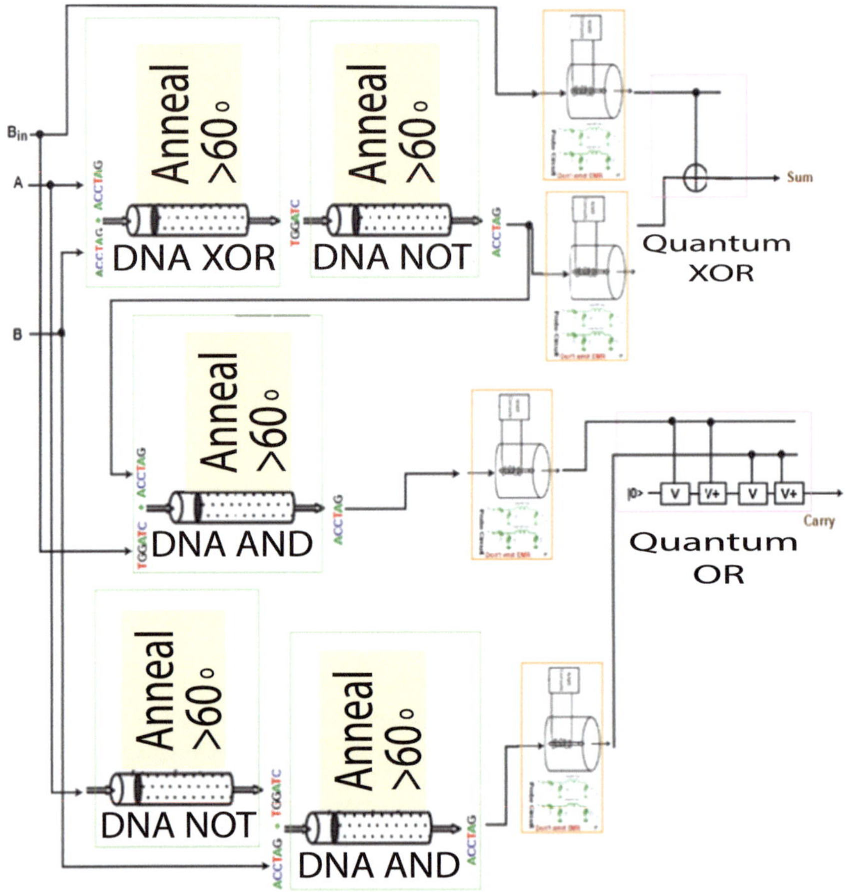

Fig. 12.18 DNA-quantum circuit of full subtractor at 0 K operation

other quantum and DNA basic gate operations will perform in parallel within this time. Three pipelines for extracting the output, one can understand that the last operation will be performed with quantum qubit and others will perform with DNA sequence.

From Sect. 12.3, any DNA basic gate (i.e., AND, OR, NOT, and XOR) operation needs more or less than 2 h to perform. In addition, it needs 6 h for preparing any DNA basic gate operation and 2 h for fluorescence detection which is fixed for any multi-(basic) gate operation.

The total operational time required for DNA operations is the summation of the initial preparation time, fluorescence detection time, and DNA gate operational time.

So, the required time for DNA operations is
= (Basic gate preparation time + DNA gate operational time + Fluorescence detection)
= [6 + (2 + 2 + 2) + 2)
= 14 h (approximately).

According to Sect. 12.2, OR quantum gate operation needs 30 μs, respectively.

So, the required time OR quantum gate operation is 30 μs.

Thus, to find the expected output of DNA-Quantum Full Subtractor, the required time will be summation of quantum operation and DNA operation.

So, the total required time for Quantum-DNA Full Subtractor is
= (The required time for DNA operation + The required time for quantum operation)
= (30 μs + 14 h)
= 14 h (approximately).

12.6.3 Full Adder at 0 K

Full Adder is the adder that adds three inputs and produces two outputs. The first two inputs are A and B and the third input is an input carry as C_{in}. The output carry is designated as C_{out} and the normal output is designated as S which is Sum. A Full Adder logic is designated in such a manner that can create a qubit adder and cascade the carry qubit from one adder to another. To create a Full Adder, one OR, two NAND one XOR, and two NOT gates are required. Figure 12.19 describes the DNA-Quantum circuit of the Full Adder. From the DNA-Quantum circuit in Fig. 12.19, one XOR and two AND DNA gate operational circuits perform operations with DNA sequence. The output of all DNA operations through NMR at 0 K. It produces Quantum qubit as an output which can be used as an input in all quantum circuit. Here 2 Quantum operational circuit (1 XOR and 1 OR) is used to find the expected output. Four quantum qubits are the output from the NMR process and are used as input in XOR and OR quantum operations.

To find the required operational time of DNA-Quantum Full Adder, it is needed to divide it into four pipelines as some of the basic quantum and DNA operations are performed in parallel. Four pipelines are as follows:

1. DNA XOR, Quantum XOR
2. DNA AND, Quantum OR
3. Quantum XOR
4. DNA AND, Quantum OR.

As the first, second, and fourth pipeline is the equal and largest pipeline for providing an output of the Full Adder, any of them for measuring the total required operational time. Others DNA and quantum basic is considered for gates will perform in parallel within this time. Four pipelines for extracting the output, one can understand that last one operational gate will perform with quantum qubit and others will perform with DNA gate operation.

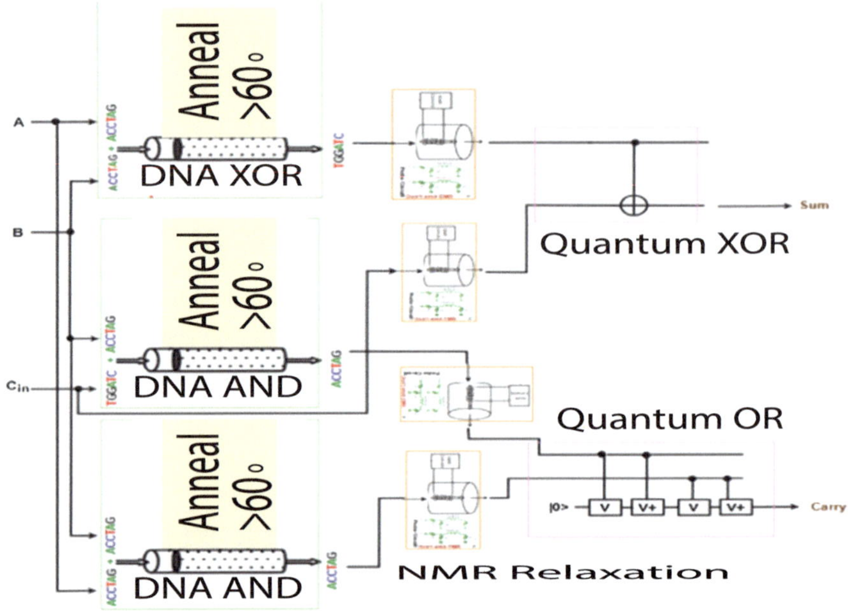

Fig. 12.19 DNA-quantum operational circuit for full adder at 0 K

From Sect. 12.3, any DNA basic gate operations (i.e., AND, OR, NOT, and XOR) needs more or less than 2 h to perform. In addition, it needs 6 h for preparing any DNA basic gate operations and 2 h for fluorescence detection which is fixed for any operation.

The total operational time required for DNA XOR operation is the summation of the initial preparation time, fluorescence detection time, and DNA operation time.

So, the required time for DNA XOR operation is
= (Basic operation time + OR operation time + Fluorescence detection)
= (6 + 2 + 2)
= 10 h (approximately).

According to Sect. 12.2, the XOR quantum operation needs 10 μs.

So, the required time for XOR quantum operation is 10 μs.

Thus, to find the expected output of the DNA-Quantum Full Adder, the required time will be the summation of DNA operation and Quantum operation.

So, the total required time for Quantum-DNA full Subtractor is
= (The required time for DNA operation + The required time for quantum operation)
= (20 μs + 10 h)
= 10 h (approximately).

12.7 Applications

This section describes some real-life applications of DNA and Quantum operation, where it is used and provides superior performance against classical computing systems. In the case of different applications, classical computing systems might fail but Quantum and DNA computing systems can show their capability.

Encryption: The Data Encryption Standard (DES) employs a 56-bit key to encrypt 64-bit information. Breaking DES means finding a key that maps the plain-text to the cipher-text given a (plain-text, cipher-text) pair. A traditional DES assault would need an exhaustive search of all 256 DES keys, which would take 10,000 years at a rate of 100,000 operations per second. As a result, molecular programs were developed, which needed around 4 months of laboratory work instead.

DNA computation is expected to provide significant benefits in terms of speed, energy efficiency, and cost-effective information storage. The number of operations per second in Adleman's model might be as high as 1.2×10^{18}. This is around 1,200,000 times quicker than the most powerful supercomputer. Additionally, DNA computers have the potential to be extremely energy efficient. In theory, one joule is enough to do about 2×10^{19} ligation operations. This is surprising given that the second law of thermodynamics states that there may only be 34×10^{19} (irreversible) operations per joule in theory (at room temperature). Existing supercomputers are significantly less efficient, with 10^9 operations per joule at most. Finally, storing information in DNA molecules could allow for an information density of around 1 bit per cubic nanometer, compared to 1 bit per 10^{12} nm^3 in current storage media.

Weather Forecasting: Weather affects over 30% of US GDP ($6 trillion) directly or indirectly, affecting food production, transportation, and retail trade, among other things. The capacity to better predict the weather would be extremely beneficial in a variety of sectors, not to mention giving people more time to prepare for disasters. While scientists have a long desire to do this, the equations controlling such processes involve a large number of variables, making traditional simulation time-consuming. "Using a classical computer to undertake such analysis might take longer than it takes for the actual weather to evolve!" said quantum researcher Seth Lloyd. This prompted Lloyd and colleagues at MIT to demonstrate that the weather equations have a hidden wave nature that can be solved by a quantum computer. Quantum computers could aid in the development of improved climate models, allowing us to gain a better understanding of how humans affect the ecosystem. These models form the foundation for the projections of future warming, and they assist us in determining what steps need to be taken now to avoid calamities.

12.8 Summary

Speed calculation is an important part in quantum-DNA computing and DNA-quantum computing. Quantum computing is itself a speedy process. It is faster than any other supercomputer in the world. Even it is thousand times faster than the most powerful computer till now. DNA computing process is another way of computing where the storage capacity is huge and can work in a parallel way. This chapter has presented the way to calculate the speed and time of quantum-DNA computing and DNA-quantum computing in detail. Some examples have been presented to clear the concept of speed calculation. The necessary figures and explanations have also been shown in this chapter.

Chapter 13
Heat Transfer

13.1 Introduction

Combinatorics, a type of calculation that traditional computers have trouble with, is one such challenge. These calculations involve arranging elements in a way that achieves a particular objective. As the number of things increases, so does the number of possible combinations. To find the ideal arrangement, today's digital computers must loop through each permutation until an outcome is found, and then determine which is the most effective at achieving the goal. In many circumstances, this will demand a large number of calculations. Combinatorics calculations prove problematic in a variety of sectors, ranging from banking to pharmaceuticals. Quantum computers come into play here. Quantum computing reduces the cost of calculating difficult combinatorial problems in the same way as classical computers have reduced the cost of arithmetic.

According to the theory of thermodynamics, huge amounts of produced heat from a quantum circuit can deteriorate its performance. On the other hand, in the DNA operational circuit, to complete each step, it needs a certain amount of heat. If these two findings are merged, can provide a novel procedure to provide heat in DNA operational circuits by taking away the produced heat in Quantum operational circuits. The main objectives of chapter two are to transfer produced heat from Quantum operational circuit to DNA operational circuits by heat conduction circuits in Quantum-DNA operational circuit and multivalued Quantum-DNA operational circuits. Heat transfer is important in protecting the environment by reducing emissions and pollutants. The heat conduction circuit receives heat from the Quantum operational gate. The NIS junctions produce heat flows between the normal-metal islands and the superconducting leads. Then, the electrons in the normal metal exchange heat with the phonon bath. Next, the islands exchange heat with each other by photons traveling in the transmission line. Finally, the model takes into account the geometrical properties of the samples as well as properties specific to the measurement setup. Basically, heat passes through the junction into metal then electrons are heated and

H. Md. Hasan Babu, *Quantum Biocomputing in Quantum Biology Volume II*,
https://doi.org/10.1007/978-981-97-5349-9_13

this electron transfers the heat to photon bath on the coplanar waveguide channel, and then it goes to one 1-meter distance.

13.2 Quantum Heat Conductance Circuit

It is possible to calculate the heat of the Quantum circuit by using the laws of thermodynamics. In addition, this heat can transfer into DNA circuits using some basic operations of heat transfer. This section will discuss how to transfer the produced heat from the quantum operation to DNA circuits. Figure 13.1 depicts the Quantum heat conduction circuit using photon and nanotube.

13.2.1 Design Procedure

In this design, nanotube photons are used for heat conduction. Here heat conduction will happen from one island to another island. The length of the coplanar waveguide is either 20 cm or 1 m, and it has a double-spiral structure on the a silicon chip with a size of 1×1 cm^2 or 2×2 cm^2, respectively. In all samples, the normal-metal islands terminating the waveguide have two galvanic contacts to superconducting lines: one to the center conductor of the waveguide and the other to the ground plane.

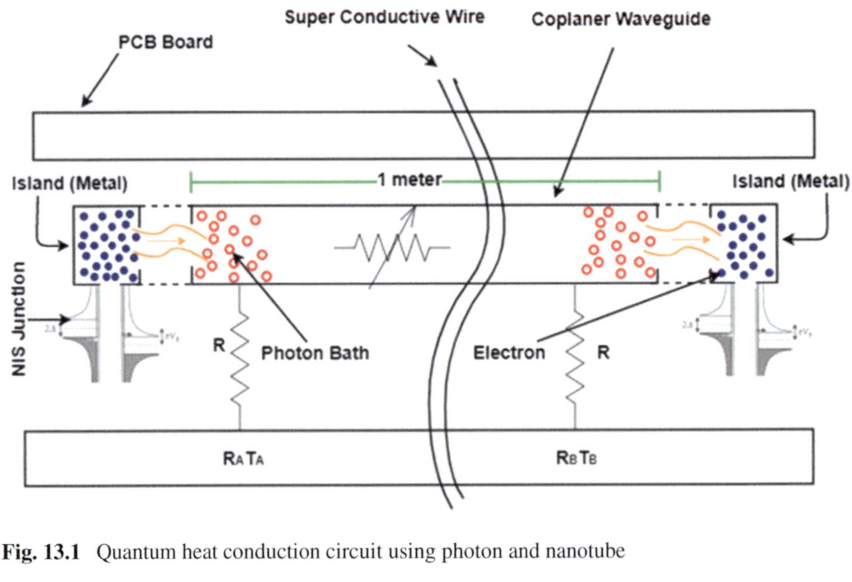

Fig. 13.1 Quantum heat conduction circuit using photon and nanotube

In practice, the two resistors are connected symmetrically by two aluminum superconducting lines, interrupted by a DC-SQUID (superconducting quantum interference device) in each line. These nanostructures are fabricated with electron beam lithography. The two AuPd resistors are nominally identical—6.6 mm long, 0.8 mm wide, and 15 nm thick. Their resistances are Ri $<$ 200 V each. The register is also connected with the downstairs aluminum beam for thermometry and joule heating. On the upside, the chip is attached to a sample holder containing a printed circuit board (PCB), to which the sample is connected by A1 bond wires. The PCB is connected to a room-temperature measurement set-up with glossy coaxial cables.

13.2.2 Working Principle

The quantum of thermal conductance, GQ $= \pi k_B 2T/6\hbar$, provides the fundamental upper limit for heat conduction through a single channel. Here, T is the temperature, k_B denotes the Boltzmann constant, and $\hbar$ is the reduced Planck's constant. From that equation, a measurement of quantum conductance can be taken.

With the help of the circuit using photons, the heat conduction limit will be 1 m. In this circuit in the nanotube, photons are used because photons, unlike many other carriers of heat, can travel macroscopic distances without significant scattering. For example, in the case of optical fibers or superconducting waveguides.

Here heat conduction happens through a single channel formed by photons traveling in a long superconducting waveguide in a single transverse mode. This heat transport does not depend on only photon temperature rather it depends on control of the temperature. One island transports the heat to another island or the temperature of each island is increasing when another island's temperature decreases.

The NIS superconducting junction is connected to islands, which are metal. NIS junction is sometimes used to cool or to heat the normal metal.

Firstly, the NIS junctions produce heat flows between the normal-metal islands and the superconducting leads. Secondly, the electrons in the normal metal exchange heat with the phonon bath. Then, the islands exchange heat between each other by photons traveling in the transmission line. Finally, the model takes into account the geometrical properties of the samples as well as properties specific to the measurement setup.

Here, in this circuit heat passes through the junction into metal then the electron is heated and this electron transfers the heat to the photon bath on the coplanar waveguide channel, and then it goes to one 1-meter distance.

13.3 Heat Transfer in Quantum-DNA Logic Operations

Now, this section will show how to transfer heat between basic quantum and DNA operational gates like AND, OR, XOR, and NOT operational gates. These quantum operational gates are prepared with Quantum basic gates as NOT, CNOT, V+, and

V. On the other hand, DNA operational gates are prepared with base DNA sequence and enzymes.

13.3.1 Heat Transfer from Quantum AND Operation to DNA NOT Operation

Figure 13.2 shows the heat transfer circuit of quantum-DNA NAND operation. Quantum AND operation is transferring excessive heat to DNA NOT operation.

13.3.1.1 Design Procedure

The given circuit design describes a quantum operational gate as AND and a DNA operational gate as NOT. In the AND gate, three v+ and two NOT gates are used It is a three-qubit gate and two qubits are input and one qubit is an ancilla qubit. From the quantum AND gate, connect the junction to transfer heat to metal. The island is metal in a heat conduction circuit. There are two resistors and a photon bath working as a heat conductance at a distance of 1 m. Then heat is transferred by junction into the DNA circuit. The output qubit produced from AND operational gate being relaxed using the NMR relaxation process and becomes a DNA sequence which is the input in DNA NOT operation. Quantum AND gate and DNA NOT gate make together a NAND operation.

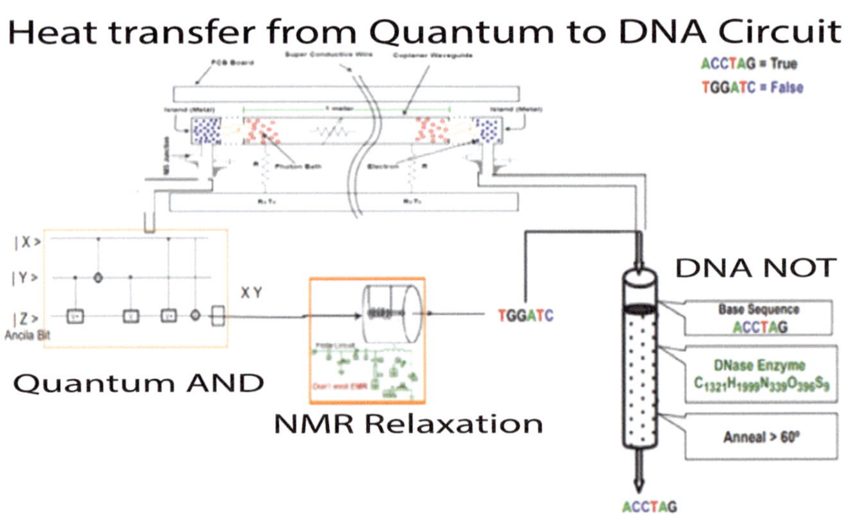

Fig. 13.2 Quantum-DNA NAND operation circuit

13.3.1.2 Working Procedure

In the quantum AND operational circuit, 2 qubits as input and an ancilla bit are used. The Ancilla bit is used to reverse the operation. The ancilla qubit is also used for error correction and it is called check bit. Here in Quantum AND operation, qubit is used as |0> because without this qubit, the gate cannot be reversed and an error will occur in the output without an ancilla qubit.

When qubits are started to operate as a gate operation, they create a huge temperature but quantum computers need an absolute zero temperature environment. With the junction heat transfer to metal molecules. Then molecules transfer the heat into the photon bath and then photons transfer heat through the channel. And finally, getting sequence by NMR relaxation from the first quantum operation and heat into the DNA NOT operation. In addition, for DNA operation, it needs some basic amount of heat such as mixing, melting, and amplifying.

13.3.2 Heat Transfer from Quantum OR Operation to DNA NOT Operation

Figure 13.3 shows the Quantum-DNA NOR operation heat transfer circuit. In this NOR operation circuit OR operation is performed using quantum OR operation and NOT operation is performed using DNA NOT operation. The output qubit of OR operation is converted into DNA corresponding to DNA sequence using NMR relaxation. The design procedure and working procedure of Quantum-DNA NOR operation are explained in the Sects. 13.3.2.1 and 13.3.2.2.

13.3.2.1 Design Procedure

The given circuit design describes a Quantum operational gate as OR and a DNA operational gate as NOT. In the quantum OR operation, two v+ and two v gates are used. It is a three-qubit gate and two qubits are input and one qubit is an ancilla qubit. From the quantum OR operation, the junction is connected to transfer heat to metal. The island is metal in a heat conduction circuit. There are two resistors and a photon bath working as a heat conductance at a distance of 1 m. Then heat is transferred by junction into the DNA circuit. The output qubit produced from quantum OR operation being relaxed using NMR relaxation process and becomes a DNA sequence which is the input in DNA NOT operation. Quantum OR operation and DNA NOT operation make together a NOR operation.

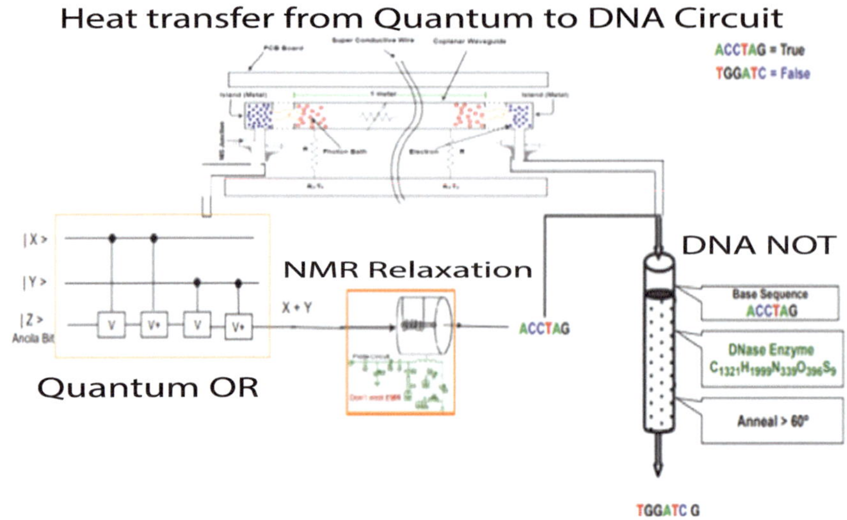

Fig. 13.3 Quantum-DNA NOR operation circuit

13.3.2.2 Working Procedure

In the quantum OR operational circuit, 2 qubits as input and an ancilla qubit are used. The ancilla qubit is used to reverse the operational gate. Ancilla qubit is also used for error correction and it is called check qubit. Here in Quantum OR operation, qubit |0> is used because without this qubit it is not possible to reverse the gate and an error in output without an ancilla qubit will be produced.

When qubits are started to operate as a gate operation, they create a huge temperature but quantum computers need an absolute zero temperature environment. With the junction heat transfer to metal molecules. Then molecules transfer the heat into the photon bath and then photons transfer heat through the channel. Finally, getting sequence by NMR relaxation from the first quantum operation and heat into the DNA NOT operational gate. In addition, for DNA operation, it needs some basic amount of heat such as for mixing, melting, and amplifying.

13.3.3 Heat Transfer from Quantum XOR Operation to DNA NOT Operation

Figure 13.4 shows the heat transfer circuit of Quantum-DNA XNOR operation. In this circuit, XOR operation is performed using quantum XOR operation and NOT

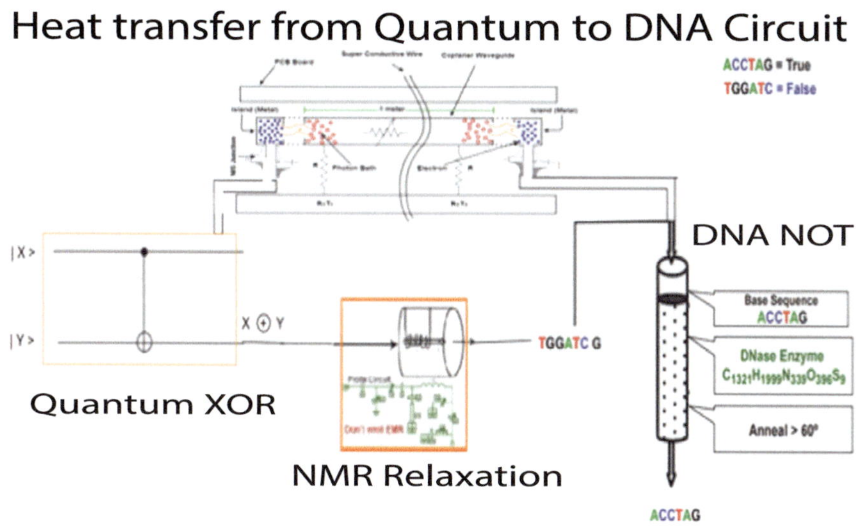

Fig. 13.4 Quantum-DNA XNOR operation heat transfer circuit

operation is performed using DNA NOT operation. The output qubit of XOR operation is converted into DNA corresponding to DNA sequence using NMR relaxation. The design procedure and working procedure of the heat transfer circuit of Quantum-DNA XNOR operation are explained in the Sects. 13.3.3.1 and 13.3.3.2.

13.3.3.1 Design Procedure

The given circuit design describes a quantum operational gate as XOR and a DNA operational gate as NOT. In the XOR gate, one CNOT gate is used. It is a two-qubit gate and both are inputs. From the quantum XOR gate, the junction is connected to transfer heat to metal. The island is metal in a heat conduction circuit. There are two resistors and a photon bath working as a heat conductance at a distance of 1 m. Heat is transferred by junction into the DNA circuit. The output qubit produced from XOR operational gate is relaxed using the NMR relaxation process and becomes a DNA sequence which is the input in DNA NOT the operational gate. Quantum XOR gate and DNA NOT gate make together an XNOR operational gate.

13.3.3.2 Working Procedure

In the quantum XOR operational circuit, two qubits are used as input. When qubits are started to operate as a gate operation, they create a huge temperature but quantum computers need an absolute zero temperature environment. With the junction heat transfer to metal molecules. Then molecules transfer the heat into the photon bath and next, photons transfer heat through the channel. And finally, getting sequence by NMR relaxation from the first quantum operational gate and heat into the DNA NOT the operation. In addition, for DNA operation, the basic amount of heat is needed such as mixing, melting, and amplifying.

13.4 Heat Transfer in Quantum-DNA Circuits

According to quantum computing, quantum computation is faster than classical computation systems. Quantum computers are also more powerful than supercomputers in terms of computing. They are 1000 times faster than regular computers and supercomputers at processing data. Quantum computers can execute calculations that would take a regular computer 1000 years to complete in a matter of seconds. On the other hand, the use of DNA strands to compute has led to high parallel computation that makes up for the slow processing of the chip. Memory space required by DNA is around 1 bit per cubic nanometer which is much less when compared to regular storage systems Consumption of power is almost nil as the chemical bonds in DNA produce energy to build or repair new strands. So, To find a super faster computation system with huge memory, a Quantum-DNA computation system can be developed. This Quantum-DNA computation system can merge all the advantages of quantum computing and DNA computing.

In a Quantum-DNA computing system, input will be received as a qubit and after performing in a certain number of quantum operations, these qubits will be turned into DNA sequences by NMR relaxation.

13.4.1 Heat Transfer in Quantum-DNA Full Adder Circuit

Full Adder is the adder that adds three inputs and produces two outputs. The first two inputs are A and B and the third input is an input carry as C_{in}. The output carry is designated as C_{out} and the normal output is designated as S which is Sum. A Full Adder logic is designated in such a manner that can take eight inputs together to create a qubit adder and cascade the carry qubit from one adder to another. To create a Full Adder, one OR, two AND, and two XOR operational gates are required. Figure 13.5 describes the Quantum-DNA circuit of the Full Adder.

13.4.1.1 Design Procedure

To design a Quantum-DNA Full Adder for heat transfer, Quantum and DNA operations are used to operate the input qubit for their corresponding outputs. The Quantum operational gates will be used for receiving the input qubits and the DNA operations will be used to produce the final output against the corresponding set of inputs. Each

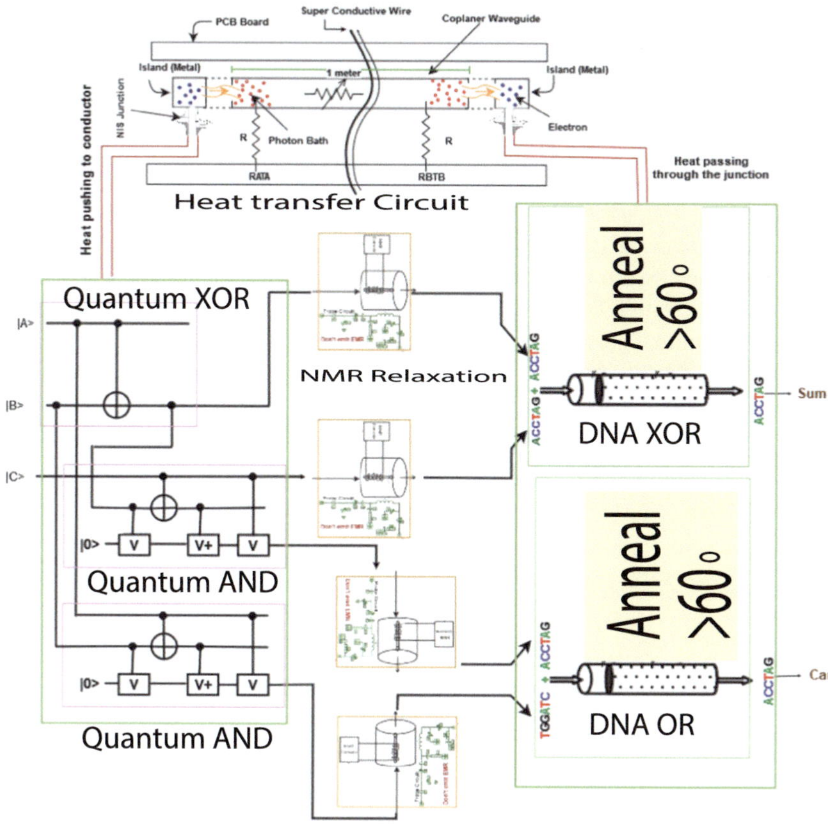

Fig. 13.5 Quantum-DNA full Adder for heat transfer

time, the Quantum-DNA Full adder will receive three qubits as input. After operating in a certain number of Quantum operations, the qubit will be turned into a corresponding DNA sequence by using an NMR relaxation room temperature probe. By using a room temperature probe and corresponding components of NMR relaxation, the excited qubit turns into a ground state and produces a DNA sequence. Then the DNA sequence is processed through DNA operations and outputs are received.

From the quantum operational gate, the junction is connected to metal for transferring heat. The island is metal in a heat conduction circuit. There are two resistors and a photon bath working as a heat conductance at a distance of 1 m. Heat is transferred by junction into the DNA circuit. Here, Fig. 13.5 describes Quantum-DNA Full adder using Quantum, DNA operations, and heat conduction nanotubes.

From the figure, it is found that the Quantum-DNA Full Adder consists of three Quantum operations and two DNA operations. Here, two AND and one XOR are used as Quantum operations and further one XOR and one OR operations is used as DNA operations.

13.4.1.2 Working Procedure

Here, using DNA sequence ACCTAG = TRUE for Quantum qubit | 1 > and DNA sequence TGGATC = FLASE for Quantum qubit | 0 >.

The heat conduction circuit receives heat from the Quantum operations. Firstly, the NIS junctions produce heat flows between the normal-metal islands and the superconducting leads. Secondly, the electrons in the normal metal exchange heat with the phonon bath. Then, the islands exchange heat between each other by photons traveling in the transmission line. Finally, the model takes into account the geometrical properties of the samples as well as properties specific to the measurement setup.

Here in this circuit heat passes through the junction into metal then the electron is heated and this electron transfers the heat to photon bath on the coplanar waveguide channel and then it goes to one 1-m distance.

Using NMR relaxation quantum qubits turned into DNA sequences. Further, these DNA sequences and supplied heat operate DNA operation of the Full Adder.

13.4.2 Heat Transfer in Quantum-DNA Multiplier Circuit

To create a multiplication circuit, six AND and two XOR gates are required. Figure 13.6 describes the Quantum-DNA circuit of the 2-qubit multiplication.

13.4.2.1 Design Procedure

To design a Quantum-DNA multiplier for heat transfer, Quantum and DNA operations are used to operate the input qubit for their corresponding outputs.

The Quantum operations will be used for receiving the input qubits and the DNA operations will be used to produce the final output against the corresponding set of inputs. Each time, the Quantum-DNA Full adder will receive three qubits as input. After operating in a certain number of Quantum operations, the qubit will be turned into a corresponding DNA sequence by using the NMR relaxation room temperature probe. By using a room temperature probe and corresponding components of NMR relaxation, the excited qubit turns into a ground state and produces a DNA sequence. Then the DNA sequence is processed through DNA operations; and outputs are received.

From the quantum operation, the junction is connected to metal for transferring heat. The island is metal in a heat conduction circuit. There are two resistors and a photon bath working as a heat conductance at a distance of 1 m. Heat is transferred by junction into the DNA circuit. Here, Fig. 13.6 describes Quantum-DNA Multiplier using Quantum, DNA operations, and heat conduction nanotubes.

From the figure, the Quantum-DNA Full Adder consists of four Quantum operations and four DNA operations. Here, four AND operations are used as Quantum operations and further two XOR and two AND DNA operations are used.

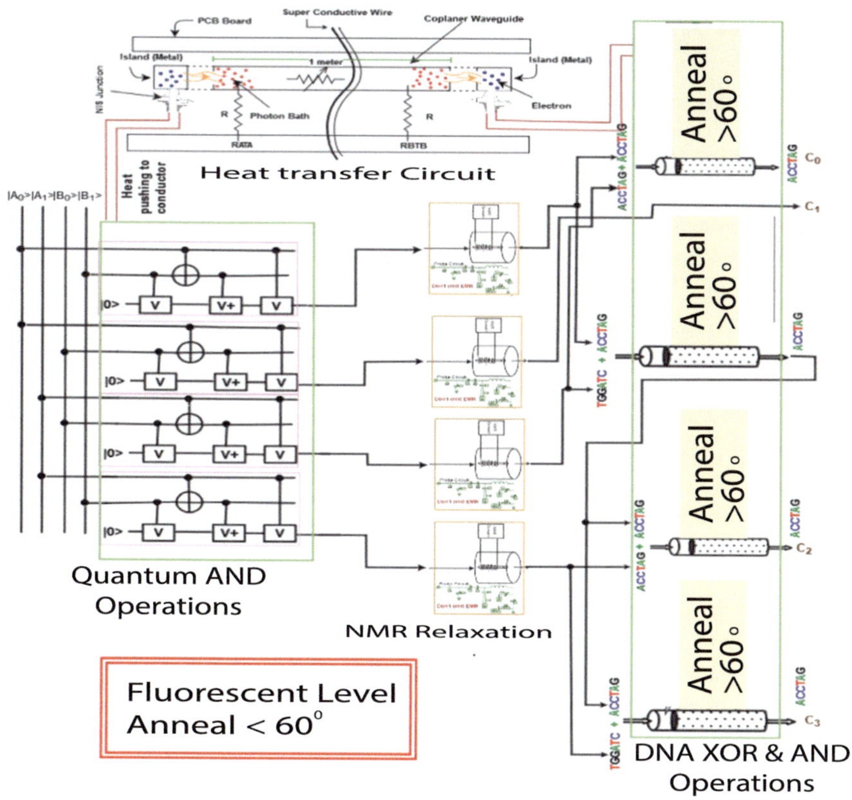

Fig. 13.6 Quantum-DNA multiplier for heat transfer using nanotubes

13.4.2.2 Working Procedure

Here, using DNA sequence ACCTAG = TRUE for Quantum qubit | 1 > and DNA sequence TGGATC = FLASE for Quantum qubit | 0 >.

The heat conduction circuit receives heat from the Quantum operations. Firstly, the NIS junctions produce heat flows between the normal-metal islands and the superconducting leads. Secondly, the electrons in the normal metal exchange heat with the phonon bath. Then, the islands exchange heat between each other by photons traveling in the transmission line. Finally, the model takes into account the geometrical properties of the samples as well as properties specific to the measurement setup.

Here in this circuit heat passes through the junction into metal then the electron is heated and this electron transfers the heat to the photon bath on the coplanar waveguide channel and then it goes to one 1-meter distance.

Using NMR relaxation quantum qubits turned into DNA sequences. Further, these DNA sequences and supplied heat operate the DNA operation of the multiplier.

13.5 Heat Transfer in DNA-Quantum Circuits

DNA computing uses biological components such as DNA, biochemistry, and molecular biology instead of standard silicon-based computing technology. When applied to problems that can be broken down into independent, discontinuous tasks, DNA computers have obvious advantages over traditional computers. This is because strands of DNA can store large amounts of data and perform many operations at the same time, thus solving degradable problems significantly faster. On the other hand, the fastest computing system can be defined as a quantum computing system that processes qubits. 1>, | 0>, and both | 1> and | 0> at the same time. In addition, quantum computer computations are particularly promising for analyzing or simulating highly complex processes that use large amounts of data.

So, a DNA-Quantum computation system is created to find a super-fast computation system with a lot of memory. This DNA-Quantum computation system combines the benefits of both quantum and DNA computing. It will be able to do super-fast parallel operations.

In a DNA-Quantum circuit, it will use DNA operational gates to control inputs and Quantum qubits to control output. In this scenario, NMR is applied for transforming DNA sequences to quantum qubits. A cache memory is used to store DNA sequence temporarily which will be discussed in later chapter.

In this regard, there is no need of any circuit for transferring heat from DNA operational circuit to the Quantum operational circuit. The reason behind not using any heat transferring circuit is, DNA operational circuit does not produce any heat rather it needs heat, which can be supplied from different outside sources. On the other hand, the Quantum operational circuit produces heat, which cannot be used in the DNA operational circuit because DNA operational circuit operates before the Quantum operational circuit. Here heat transfer from outside sources to DNA circuit to perform computations and the excessive heat produced by quantum circuit will be transferred to a cooler to make the quantum part cool.

Figure 13.7 shows DNA-quantum Full Adder operational circuit, which clearly states that the required heat for the DNA operational circuit does not depend on the produced heat from the Quantum operational circuit.

For DNA-Quantum Full Adder, three DNA operations as XOR and AND DNA operational circuit are required and for Quantum operations, two Quantum operations as XOR and OR are required.

The DNA operations produced four outputs as DNA sequence and converted through the NMR process to produce quantum qubits. The Quantum qubits are then processed through OR and XOR quantum operations and produce the expected output.

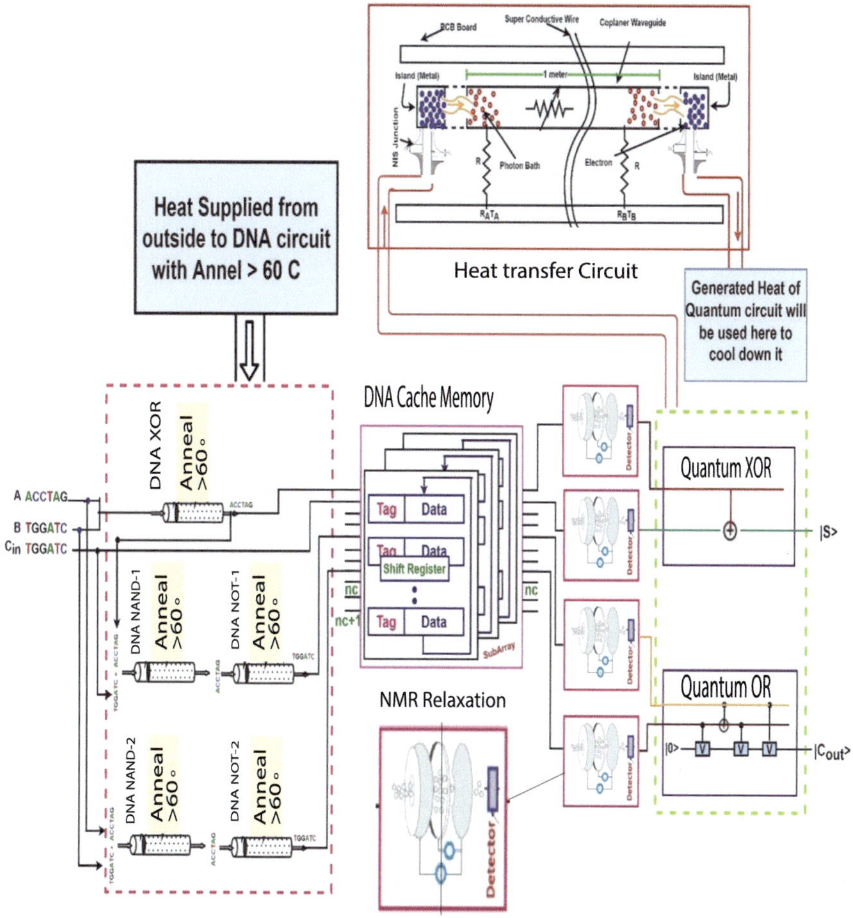

Fig. 13.7 DNA-quantum Full Adder at room temperature

13.6 Applications

This section describes some real-life applications of DNA and Quantum operation, where it is used and provides superior performance against classical computing systems. In the case of different applications, classical computing systems might fail but Quantum and DNA computing systems can show their capability.

Scheduling: A DNA computing-based algorithm was presented to solve the job scheduling problem. To explain the model with six tasks can be demonstrated the working operations, mimicking the method used for the Hamiltonian Path problem. This however was not the first time, Watada in early 2000 used DNA algorithms to work out elevator schedule systems and rearrangement of Flexible Manufacturing

System. However, due to a lack of a theoretical base, only medium-sized tasks were taken into consideration.

Clustering: Clustering deals with deriving highly meaningful relationships in a complex collection of data by creating a structure using various concepts and algorithms. DNA- based clustering involves using strands to assign edges and vertices. Iterative calculations are performed for every produced cluster to improve quality. This method is of particular interest when dealing with large heterogeneous data with an unknown number of clusters. It helps in reducing the time complexity by high parallelism features of DNA.

13.7 Summary

Quantum computing focuses on speedy technology based on quantum-theoretical principles, which is the behavior of energy and matter of a qubit. A combination of qubits is used to perform any specific task in quantum computing. Quantum computers represent a significant advancement in computing capability, with enormous performance benefits for specific use cases. The ability of bits to be in several states at the same time gives the quantum computer a lot of computing capability and it is much faster than classical bitwise computing.

Furthermore, DNA computing uses biological molecules to do computations. The four-character genetic alphabet (A-adenine, G-guanine, C-cytosine, and T-thymine) is used in DNA computing. The input of any DNA operation can be represented by DNA molecules with specific sequences. The instructions are carried out by laboratory operations on the molecules, and the result is defined as some property of the final set of molecules. DNA computing promises significant and meaningful linkages between computers and life systems, as well as massively parallel computations. DNA computing can actually carry out millions of operations at the same time.

To advance computation, it is easy to make or use DNA-Quantum computing systems, which will merge all the advantages of both DNA computing and Quantum computing.

Heat is an important property of any operation for computation, in quantum computing operation, much heat is produced by circuits, which is dependent on the number of qubits in the operational circuit. On the other hand, in DNA computing, it provides heat in the test tube to execute the operation. Here, the heat transfer circuit is used to pass the produced heat of quantum computing operations in the DNA computing operations. The heat conduction circuit receives heat from the Quantum operational gate. The NIS junctions produce heat flows between the normal-metal islands and the superconducting leads. The electrons of normal metal exchange heat with the phonon bath. Then, the islands exchange heat between each other by photons traveling in the transmission line. In every circuit heat passes through the junction into metal then electrons are heated and this electron transfers the heat to the photon bath on the coplanar waveguide channel and then it goes to one 1-m distance.

Chapter 14
Data Conversion Mechanisms

14.1 Introduction

DNA-quantum computing and quantum-DNA computing systems can be used to combine the benefits of both DNA and quantum computation systems. It will be able to compute parallel operations at super-fast speed. In this, it needs to convert data from quantum qubits to DNA sequence and further DNA sequence to Quantum qubits. In a Quantum-DNA circuit, it operates inputs in quantum operation and provides outputs in DNA sequence. For providing output, it needs to use NMR relaxation or trap ion for converting quantum qubit to the DNA sequence. After getting a qubit from a quantum operation, it needs to operate NMR relaxation or trap ion. In NMR relaxation , a room temperature probe or a cryogenic probe can be used at 0 K temperature. By NMR relaxation qubit of superposition state turns into a normal ground state and gets the equivalent DNA sequence. In NMR relaxation at room temperature probes, decoherence problems, polarization, and scaling problems can occur thus cryogenic probes at 0 K temperature can be preferred for Quantum-DNA computation systems.

In a DNA-Quantum circuit, the inputs are in DNA sequence and provide output in Quantum qubits. For providing output, NMR operation or Quadrupole trap ion is used for converting DNA sequence to quantum qubits. After getting the DNA sequence from the DNA operation, it needs to be operated by the NMR process to convert it into Quantum qubits. In the NMR process, a room temperature probe or a cryogenic probe can be used at 0 K temperature. By applying the NMR process for the DNA sequence, the qubit of the normal ground state turns into a superposition state or ground state according to the input sequence in NMR. The output of the NMR process is a quantum qubit. In NMR at room temperature probes, decoherence problems, polarization, and scaling problems can occur, thus cryogenic probes at 0 K temperature can also be preferred for the DNA-Quantum computation system.

H. Md. Hasan Babu, *Quantum Biocomputing in Quantum Biology Volume II*,
https://doi.org/10.1007/978-981-97-5349-9_14

14.2 Data Conversion in Quantum-DNA Circuits

The fastest computation system can be defined as a Quantum computation system, which works with qubits |1 >, |0 >, and both |1 > and |0 > at the same time. On the other hand, in place of standard silicon-based computer technology, DNA computing uses biological components such as DNA, biochemistry, and molecular biology. When applied to issues that can be separated into independent, non-sequential tasks, the DNA computer has demonstrable benefits over conventional computers. The reason for this is that DNA strands can store a lot of data and do numerous operations at the same time, allowing them to solve decomposable issues considerably faster. Again, Quantum computer calculations are especially promising for analyzing or simulating extremely complicated processes involving large volumes of data.

So, to find a super faster computation system with huge memory, a Quantum-DNA computation system can be developed. This Quantum-DNA computation system can merge all the advantages of quantum computing and DNA computing. It will be able to compute parallel operations at super-fast speed.

In a Quantum-DNA circuit, inputs are operated in quantum operation and provide output in DNA sequence. In this case, NMR relaxation is used for converting quantum qubit to a DNA sequence. Section 14.2.1 will describe the procedure for converting quantum qubits to the DNA sequences.

14.2.1 NMR Relaxation at Room Temperature

NMR relaxation is the process by which an excited magnetic state returns to its equilibrium distribution. When a molecule drops into the NMR probe as a sample then it goes to an excited state with the help of a magnetic field. When the electromagnetic resonance is not emitted then the magnetic field becomes weak then the superposition state molecule loses its energy and comes into the ground state and this process is called NMR relaxation. Different components of NMR relaxation are as follows:

1. *Magnet*: In magnet, superconducting magnets such as shim coils, liquid helium, and nitrogen containers can be used.
2. *Probe*: One important part of the probe is the RF coil. It controls temperature and molecules become superpositioned by the impact of this component. Figure 14.2 describes the circuit of the probe. There are two types of NMR probes which are going to be described in the following subsection.
3. *Console*: Electronics for generating RF pulse, power and gradient amplifiers, lock system, temperature control with almost all the components of NMR controlled by the console.
4. *Computer*: When all the data from the console are received then think the computer as a spectrum. Computers are used for data storage, processing, analysis, and communicating with other components.

14.2.1.1 Structure of NMR Relaxation at Room Temperature

NMR relaxation is one part of the NMR process. Figure 14.1 describes the circuit structure of NMR relaxation and represents the basic schematic setup of the NMR relaxation process. The box is shown here in the schematic Fig. 14.1 holds superconducting coil generations or a magnet. In this NMR relaxation process, data or qubit are provided into the tube where RF coils exist. RF coil is transmitting a signal into a sample. After collecting the signal from the components, the console digitalized the data and the computer visualized this data as a spectrum.

The probe contains the radiofrequency (RF) coils, tuned at specific frequencies for specific nuclei in a given magnetic field.

There are two types of probes. Those are as follows:

1. Room temperature Probe
2. Cryogenic Probe.

Figure 14.2 shows the circuit of a room temperature probe.

There is a little difference in the room temperature probe in NMR relaxation. In NMR, it needs strong magnetic fields to give more energy to the NMR solvent and push them into an excited state that's why emit EMR. But in relaxation, the EMR should not be emitted and the reason behind it will be explained in the working principle section in detail.

Same as like room temperature probe, cryogenic probes don't need EMR in the NMR relaxation process. Figure 14.3 describes the inner RF coil and outer RF coil for the Cryogenic probe.

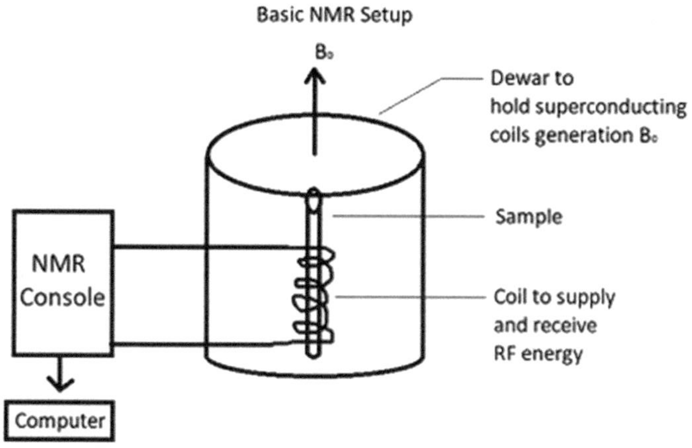

Fig. 14.1 Circuit structure of NMR relaxation

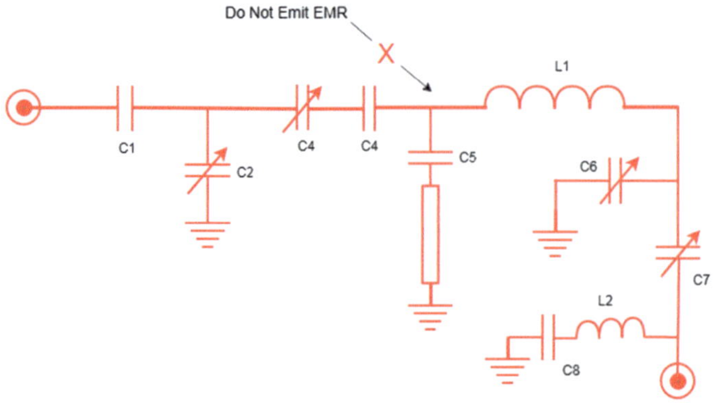

Fig. 14.2 Circuit of a room temperature probe

Fig. 14.3 Circuit of **a** Inner RF coil **b** Outer RF coil in cryogenic probe

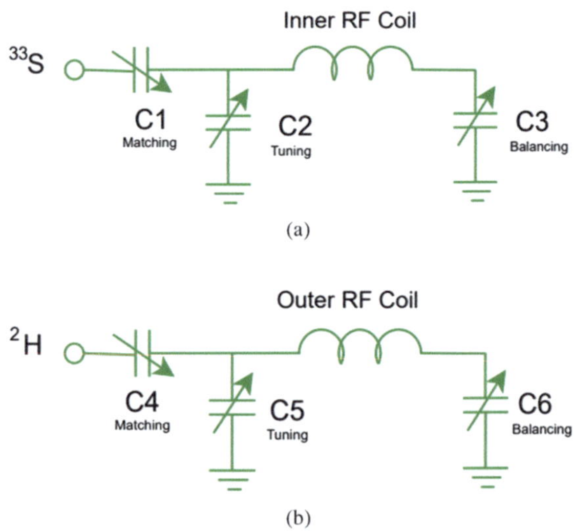

14.2.1.2 Working Principle of NMR Relaxation at Room Temperature

NMR is an analytical chemistry technique that creates a strong magnetic field and makes molecules nucleus excited then molecules exist in superposition. From the Fig. 14.4 (Spin State Realization), it is clear that at first molecules have no spin. When molecules are provided into the NMR relaxation Probe as a sample then molecules are bound by a magnetic field. Molecules' nuclear spin started spinning but they exist in the ground state still because the magnetic field didn't create a strong magnetic field. So, EMR need not be emitted.

But in NMR relaxation, the reverse process of EMR should be done. Molecules are needed to be back to the ground state. Therefore, if the EMR is not emitted and

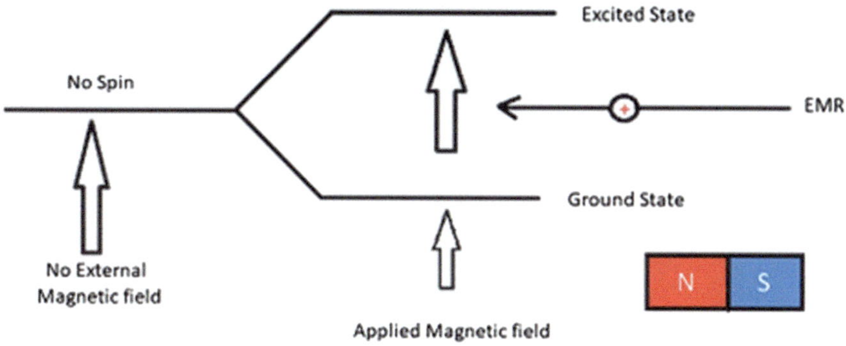

Fig. 14.4 Spin state realization

the magnetic field does not get strong then molecules start to lose their energy and come to the ground state position.

There are two procedures for NMR relaxation. They are as follows.

1. *Spin lattice Relaxation*: If the excited molecule transfers energy to the vector present molecule in the surrounding environment, then the excited molecule will lose the energy nearby solvent molecules.
2. *Spin Relaxation*: In this system, superposition state molecules transfer energy to other molecule's nucleus and then molecules come into the ground state but other molecules stay in superposition.

Shortly, it is understood that from the above discussion, if it does not produce a strong magnetic field using EMR then find molecules in the ground state. Figure 14.5 describes the overall conversion of a quantum qubit into a DNA sequence.

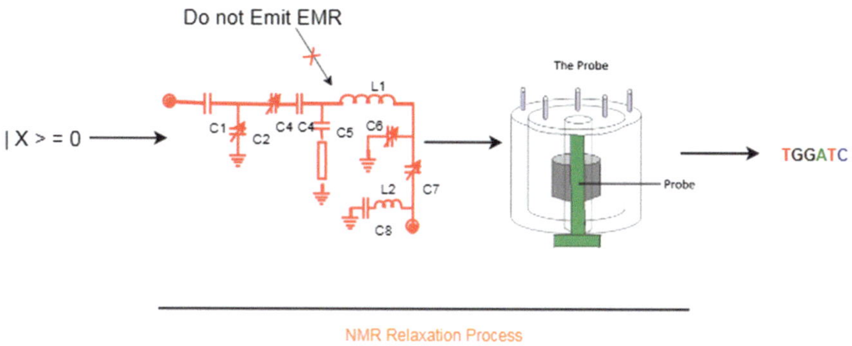

Fig. 14.5 Conversion of qubit into DNA sequence

14.2.1.3 Quantum-DNA AND Operation at Room Temperature

Figure 14.6 shows the conversion of the qubit to a DNA sequence from a quantum AND operation at room temperature. The qubit output of AND gate passes through the NMR relaxation process to create a corresponding DNA sequence. The design and working procedure of the conversion of the qubit into DNA sequence from an AND operation are explained in this sections.

1. **Design Procedure**
   Quantum AND operation circuit is prepared using two V+ two NOT and one V operations. Here are three qubits |X> and |Y> and a constant qubit (ancilla) |0>. From |X> and |Y> inputs, two lines to constant output and the target output line (|Q>). After getting qubit from quantum AND operation, NMR relaxation is needed. In NMR relaxation, a room temperature probe is used. By NMR relaxation, qubit of the superposition state turns into a normal ground state and generates the DNA sequence.
2. **Working Procedure**
   At first, in this Quantum AND operational circuit, there are three inputs as qubits, two qubits are input and the other is the ancilla qubit. Ancilla qubit is used here to reverse the gate. Ancilla qubit is also used for error correction and it is called check bit. Here in Quantum AND gate, 1 is used because without this qubit cannot reverse the gate and will get an error in output without an ancilla qubit. After processing input going through the Quantum AND operational gate, it's getting some output as a qubit. The provided output from Quantum AND operation will be used as an

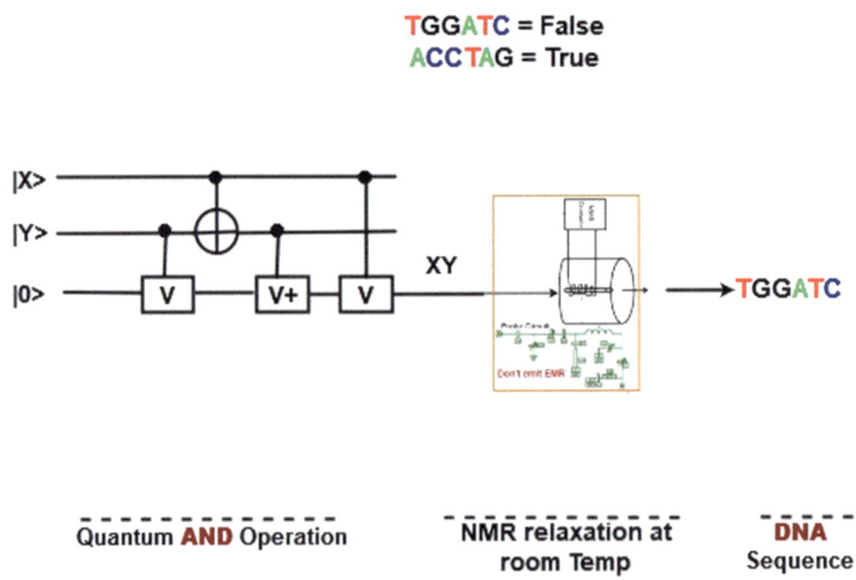

Fig. 14.6 Conversion of qubit into DNA sequence from an AND operation

Table 14.1 Outputs of data conversion for quantum AND operation

|X>	|Y>	XY	DNA sequence
|0>	|0>	|0>	TGGATC
|0>	|1>	|0>	TGGATC
|1>	|0>	|0>	TGGATC
|1>	|1>	|1>	ACCTAG

input in NMR relaxation and found the ground state. When any molecule comes to the ground state then a real sequence will be found. The DNA sequence is used for DNA logic operation. Here NMR relaxation at room temperature; and in relaxation, EMR is not to be emitted.

So, DNA sequence ACCTAG = TRUE for Quantum qubit |1 > and DNA sequence TGGATC = FALSE for qubit |0 >. The outputs for different combinations of inputs are given in Table 14.8.

14.2.1.4 Quantum-DNA OR Operation at Room Temperature

Figure 14.7 shows the conversion of the qubit to a DNA sequence from a quantum OR operation at room temperature. The qubit output of the OR gate passes through

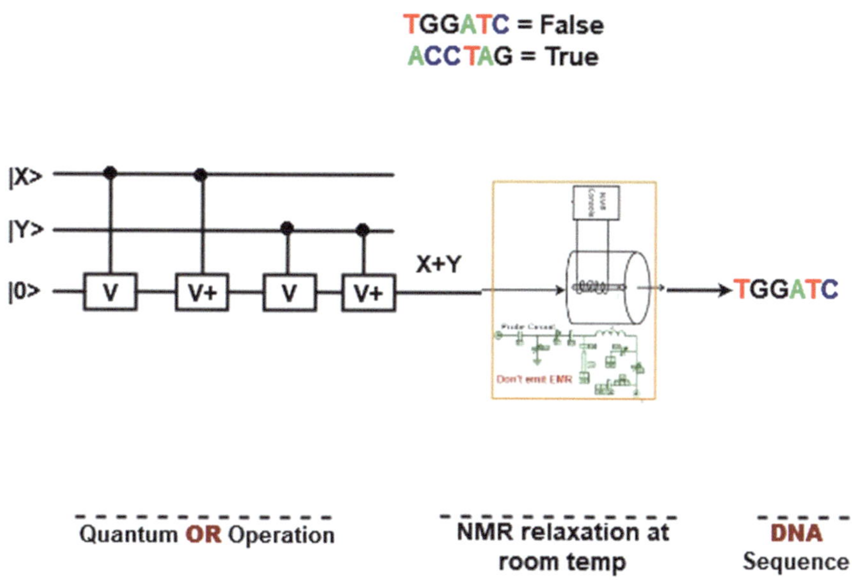

Fig. 14.7 Conversion of qubit into DNA sequence from a quantum OR operation

Table 14.2 Outputs of data conversion for quantum OR operation

$\vert X\rangle$	$\vert Y\rangle$	X+Y	DNA sequence
$\vert 0\rangle$	$\vert 0\rangle$	$\vert 0\rangle$	TGGATC
$\vert 0\rangle$	$\vert 1\rangle$	$\vert 1\rangle$	ACCTAG
$\vert 1\rangle$	$\vert 0\rangle$	$\vert 1\rangle$	ACCTAG
$\vert 1\rangle$	$\vert 1\rangle$	$\vert 1\rangle$	ACCTAG

the NMR relaxation process to create a corresponding DNA sequence. The design and the working procedure of the conversion of the qubit into DNA sequence from an OR operation are explained in this sections.

1. **Design Procedure**
   Quantum OR operation circuit is prepared using two V+ and two V gates. Here are three qubits |X> and |Y> and a constant qubit (ancilla) |0 >. From |X > and |Y > inputs, two lines to constant output, and here the target output line (|Q>) can be expressed as X+Y. After getting a qubit from quantum OR operation, NMR relaxation should be done. In NMR relaxation, a room temperature probe is used. By NMR relaxation qubit of the superposition state turns into a normal ground state and gets the DNA sequence.
2. **Working Procedure**
   At first, in this Quantum OR operational circuit, there are three inputs as qubits, two qubits are inputs and the other is the ancilla qubit. Here in Quantum OR gate, |0> as an ancilla is used qubit because without this qubit it is not possible to reverse the gate and an error will come in output without an ancilla qubit. After processing input going through the Quantum OR operational gate, it's getting some output as a qubit. The provided output from Quantum OR operation will be used as an input in NMR relaxation and find the ground state. When any molecule comes to the ground state then the real sequence will be produced. The DNA sequence is used for DNA logic gate operation. Here while applying NMR relaxation at room temperature and in relaxation, it doesn't need to emit EMR. Here, using DNA sequence ACCTAG = TRUE for qubit |1 > and DNA sequence TGGATC = FALSE for Quantum qubit |0 >. The outputs for different combinations of inputs are given in Table 14.2.

14.2.1.5 Quantum-DNA NOT Operation at Room Temperature

Figure 14.8 shows the conversion of the qubit to a DNA sequence from a quantum NOT operation at room temperature. The qubit output of the NOT gate passes through the NMR relaxation process to create a corresponding DNA sequence. The design and the working procedures of the conversion of the qubit into DNA sequence from a quantum NOT operation are explained in this section.

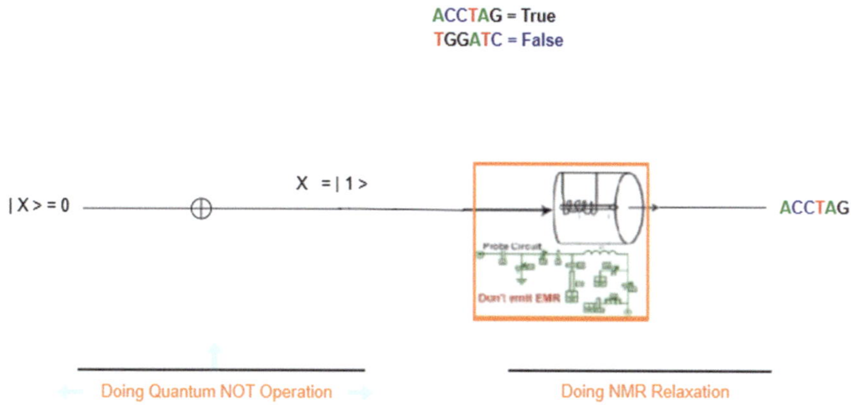

Fig. 14.8 Conversion of qubit into DNA sequence from a quantum NOT operation

1. **Design Procedure**
   Quantum NOT operation circuit is prepared using one NOT gate. Here is one qubit |X> as an input. From the input, the target output line (|X>) can be expressed as the inverse of the input. After a getting qubit from the quantum NOT operation, NMR relaxation is used. In NMR relaxation, a room temperature probe is used. By NMR relaxation qubit of the superposition state turns into the normal ground state and the DNA sequence is produced.
2. **Working Procedure**
   At first, in this Quantum NOT operational circuit, here is one input as a qubit. Here in Quantum NOT gate, no ancilla qubit is needed.
   After processing input going through the Quantum NOT operational gate, it's getting some output as a qubit. The provided output from Quantum NOT operation will be used as an input in NMR relaxation and the ground state is found. When any molecule comes to the ground state then the real sequence produces. The DNA sequence is used for DNA logic gate operation. Here while doing NMR relaxation at room temperature and in relaxation, it doesn't need to emit EMR.
   Here, using DNA sequence ACCTAG = TRUE for Quantum qubit |1 > and DNA sequence TGGATC = FALSE for Quantum qubit |0 >. The outputs for different combinations of inputs are given in Table 14.3.

Table 14.3 Outputs of data conversion for quantum NOT operation

|X>	|X>	DNA sequence
|0>	|1>	ACCTAG
|1>	|0>	TGGATC

14.2.1.6 Quantum-DNA XOR Operation at Room Temperature

Figure 14.9 shows the conversion of the qubit to a DNA sequence from a quantum XOR operation at room temperature. The qubit output of the XOR gate passes through the NMR relaxation process to create a corresponding DNA sequence. The design and working procedure of the conversion of the qubit into DNA sequence from a quantum XOR operation are explained in this section.

1. **Design Procedure**
   Quantum XOR operation circuit is prepared using one Controlled NOT or CNOT gate. There are two qubits |X> and |Y>. From |Y > inputs, one line goes through the output and the target output X XOR Y. After getting qubit from quantum XOR operation, NMR relaxation. In NMR, relaxation is applied and a room temperature probe is used. By NMR relaxation qubit of superposition state turns into the normal ground state and the DNA sequence is produced.
2. **Working Procedure**
   At first, in this Quantum XOR operational circuit, two inputs as qubits in which no need of any ancilla qubits. After processing input going through the Quantum XOR operational gate, it's getting some output as a qubit. The provided output from Quantum XOR operation will be used as an input in NMR relaxation and the ground state is found. When any molecule comes to the ground state then real sequence has come. The DNA sequence is used for DNA logic gate operation. Here while doing NMR relaxation at room temperature and in relaxation, it doesn't need to emit EMR.
   Here, DNA sequence ACCTAG = TRUE for Quantum qubit |1 > and DNA sequence TGGATC = FALSE for Quantum qubit |0 >. The outputs for different combinations of inputs are given in Table 14.4.

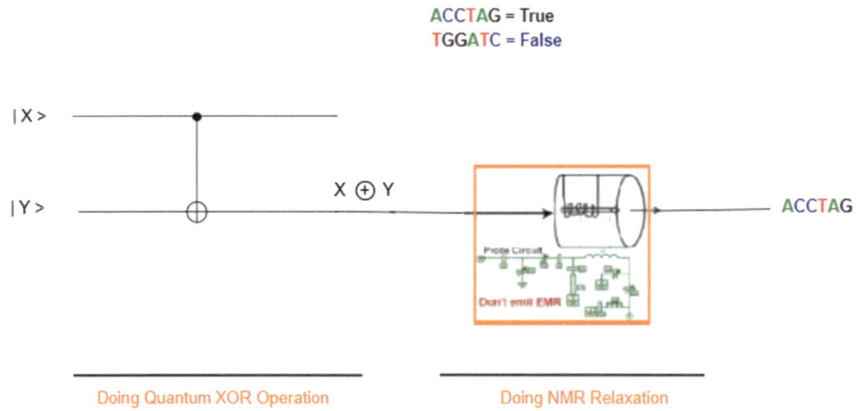

Fig. 14.9 Conversion of qubit into DNA sequence from a quantum XOR operation

Table 14.4 Outputs of data conversion for quantum XOR operation

|X>	|Y>	X XOR Y	DNA sequence
|0>	|0>	|0>	TGGATC
|0>	|1>	|1>	ACCTAG
|1>	|0>	|1>	ACCTAG
|1>	|1>	|0>	TGGATC

14.2.1.7 Quantum-DNA Full Adder at Room Temperature

Full Adder is the adder that adds three inputs and produces two outputs. The first two inputs are A and B and the third input is an input carry as C_{in}. The output carry is designated as C_{out} and the normal output is designated as S which is Sum. A Full Adder logic is designated in such a manner that can take eight inputs together to create a qubit adder and cascade the carry qubit from one adder to another. To create a Full Adder, one OR, two AND, and two XOR operational gates are required. Figure 14.10 shows the Quantum-DNA circuit of the Full Adder.

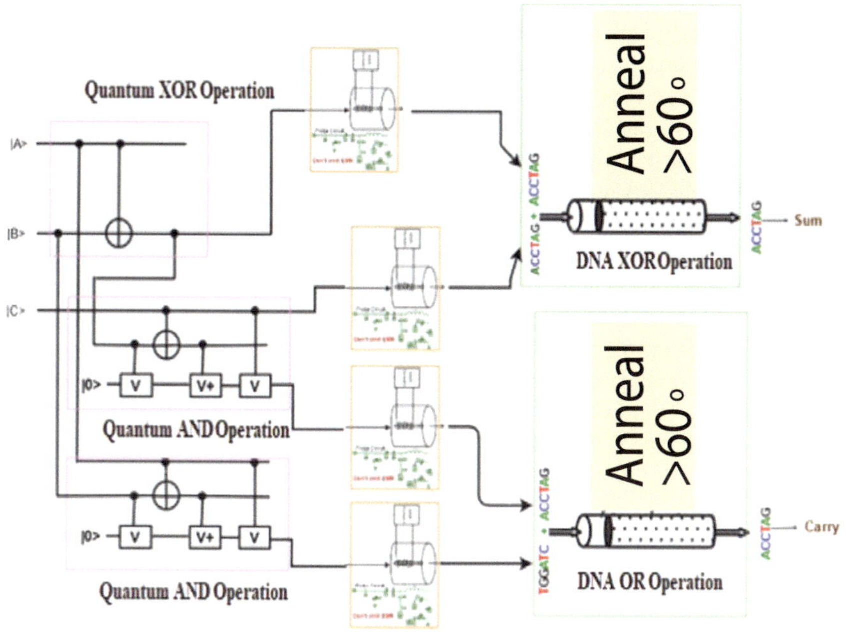

Fig. 14.10 Quantum-DNA full adder

Table 14.5 Outputs of quantum-DNA full adder operation

|A>	|B>	|C>	**SUM**	**Carry**
|0>	|0>	|0>	TGGATC	TGGATC
|0>	|0>	|1>	ACCTAG	TGGATC
|0>	|1>	|0>	ACCTAG	TGGATC
|0>	|1>	|1>	TGGATC	ACCTAG
|1>	|0>	|0>	ACCTAG	TGGATC
|1>	|0>	|1>	TGGATC	ACCTAG
|1>	|1>	|0>	TGGATC	ACCTAG
|1>	|1>	|1>	ACCTAG	ACCTAG

1. **Design Procedure**
   To design a Quantum-DNA Full Adder, Quantum and DNA operations are used to operate the input qubit for their corresponding outputs. The Quantum operations will be used for receiving the input qubits and the DNA operations will be used to produce the final output against the corresponding set of inputs. Each time, the Quantum-DNA Full adder will receive three qubits as input. After operating in a certain number of Quantum operations, the qubit will be turned into a corresponding DNA sequence by using an NMR relaxation room temperature probe. By using a room temperature probe and corresponding components of NMR relaxation, the excited qubit turns into a ground state and produces a DNA sequence. Then the DNA sequence is processed through DNA operations and outputs are received. Here, Fig. 14.10 describes Quantum-DNA Full adder using Quantum and DNA operational gates.
   From the Fig. 14.10, it is obvious that the Quantum-DNA Full Adder consists of three Quantum operations and two DNA operations. Here, two AND and one XOR are used as Quantum operations and further one XOR and one OR DNA operations are used.
2. **Working Procedure**
   The working procedure of the Quantum-DNA Full adder is given below for each pattern of input qubits. Here, the DNA sequence, ACCTAG = TRUE for Quantum qubit |1 >; and the DNA sequence TGGATC = FLASE for Quantum qubit |0 >. The working procedure of the Quantum-DNA Full Adder is given below in Table 14.5.

1. For inputs A, B, C = 0, **Sum** = DNA XOR (Quantum XOR (A, B), C)
   = DNA XOR (Quantum XOR (0, 0), 0)
   = DNA XOR (0, 0)
   = DNA XOR (TGGATC, TGGATC) [Using NMR relaxation]
   = TGGATC
   Carry = DNA OR (Quantum AND (C, XOR (A, B)), Quantum AND (A, B))
   = DNA OR (Quantum AND (0, XOR (0,0)), Quantum AND (0, 0))

= DNA OR (Quantum AND (0, XOR (0, 0)), Quantum AND (0, 0))
= DNA OR (0, 0)
= DNA OR (TGGATC, TGGATC) [Using NMR relaxation]
= TGGATC

2. For inputs A, B, C = 0, 0, 1 **Sum** = DNA XOR (Quantum XOR (A, B), C)
= DNA XOR (Quantum XOR (0, 0), 1)
= DNA XOR (0, 1)
= DNA XOR (TGGATC, ACCTAG) [Using NMR relaxation]
= ACCTAG
Carry = DNA OR (Quantum AND (C, XOR (A, B)), Quantum AND (A, B))
= DNA OR (Quantum AND (1, XOR (0, 0)), Quantum AND (0, 0))
= DNA OR (Quantum AND (1, 0), 0)
= DNA OR (0, 0)
= DNA OR (TGGATC, TGGATC) [Using NMR relaxation]
= TGGATC
3. For inputs A, B, C = 0, 1, 0 **Sum** = DNA XOR (Quantum XOR (A, B), C)
= DNA XOR (Quantum XOR (0, 1), 0)
= DNA XOR (1, 0)
= DNA XOR (ACCTAG, TGGATC) [Using NMR relaxation]
= ACCTAG
Carry = DNA OR (Quantum AND (C, XOR (A, B)), Quantum AND (A, B))
= DNA OR (Quantum AND (0, XOR (0,1)), Quantum AND (0,1))
= DNA OR (Quantum AND (0, 1), 0)
= DNA OR (0, 0)
= DNA OR (TGGATC, TGGATC) [Using NMR relaxation]
= TGGATC
4. For inputs A, B, C = 0, 1, 1 **Sum** = DNA XOR (Quantum XOR (A, B), C)
= DNA XOR (Quantum XOR (0, 1), 1)
= DNA XOR (1, 1)
= DNA XOR (ACCTAG, ACCTAG) [Using NMR relaxation]
= TGGATC
Carry = DNA OR (Quantum AND (C, XOR (A, B)), Quantum AND (A, B))
= DNA OR (Quantum AND (1, XOR (0, 1)), Quantum AND (0, 1))
= DNA OR (Quantum AND (1, 1), 0)
= DNA OR (1, 0)
= DNA OR (ACCTAG, TGGATC) [Using NMR relaxation]
= ACCTAG
5. For inputs A, B, C = 1, 0, 0 **Sum** = DNA XOR (Quantum XOR (A, B), C)
= DNA XOR (Quantum XOR (1, 0), 0)
= DNA XOR (1, 0)
= DNA XOR (ACCTAG, TGGATC) [Using NMR relaxation]
= ACCTAG
Carry = DNA OR (Quantum AND (C, XOR (A, B)), Quantum AND (A, B))
= DNA OR (Quantum AND (0, XOR (1, 0)), Quantum AND (1, 0))
= DNA OR (Quantum AND (0, 1), 0)

= DNA OR (0, 0)
= DNA OR (TGGATC, TGGATC) [Using NMR relaxation]
= TGGATC

6. For inputs A, B, C = 1, 0, 1 **Sum** = DNA XOR (Quantum XOR (A, B), C)
   = DNA XOR (Quantum XOR (1, 0), 1)
   = DNA XOR (1, 0)
   = DNA XOR (ACCTAG, TGGATC) [Using NMR relaxation]
   = ACCTAG
   Carry = DNA OR (Quantum AND (C, XOR (A, B)), Quantum AND (A, B))
   = DNA OR (Quantum AND (1, XOR (1,0)), Quantum AND (1,0))
   = DNA OR (Quantum AND (1, 1), 0)
   = DNA OR (1, 0)
   = DNA OR (ACCTAG, TGGATC) [Using NMR relaxation]
   = ACCTAG

14.2.1.8 Quantum-DNA Full Subtractor at Room Temperature

A full subtractor is a combinational circuit that performs subtraction of two qubits, one is minuend and the other is subtrahend, taking into account the borrow of the previous adjacent lower minuend qubit. This circuit has three inputs and two outputs. The three inputs A, B, and B_{in}, denote the minuend, subtrahend, and previous borrow respectively. The two outputs, D and B_{out} represent the difference and output borrows, respectively. To create a full subtractor, one OR, two AND, two OR, and 2 XOR gates are required. Figure 14.11 describes the Quantum-DNA circuit of the Full Subtractor.

1. **Design Procedure**
   To design a Quantum-DNA Full Subtractor, Quantum and DNA operations are used to operate the input qubit for their corresponding outputs. The Quantum operations will be used for receiving the input qubits and the DNA operation will be used to produce the final output against the corresponding set of inputs. Each time, the Quantum-DNA Full Subtractor will receive three qubits as input. After operating in a certain number of Quantum operations, the qubit will be turned into a corresponding DNA sequence by using an NMR relaxation room temperature probe. The process of producing a DNA sequence against a qubit state is explained in Sect. 14.2.1. By using a room temperature probe and corresponding components of NMR relaxation, the excited qubit turns into the ground state and produces a DNA sequence. Then the DNA sequence is processed through DNA operations and outputs are received. Here, Fig. 14.11 describes Quantum-DNA Full Subtractor using Quantum and DNA operations.
   From the Figure, it is obvious that the Quantum-DNA Full Subtractor consists of three Quantum operations and two DNA operations. Here, two AND, and one XOR are used as Quantum operations and further one XOR and one OR DNA operations are used.

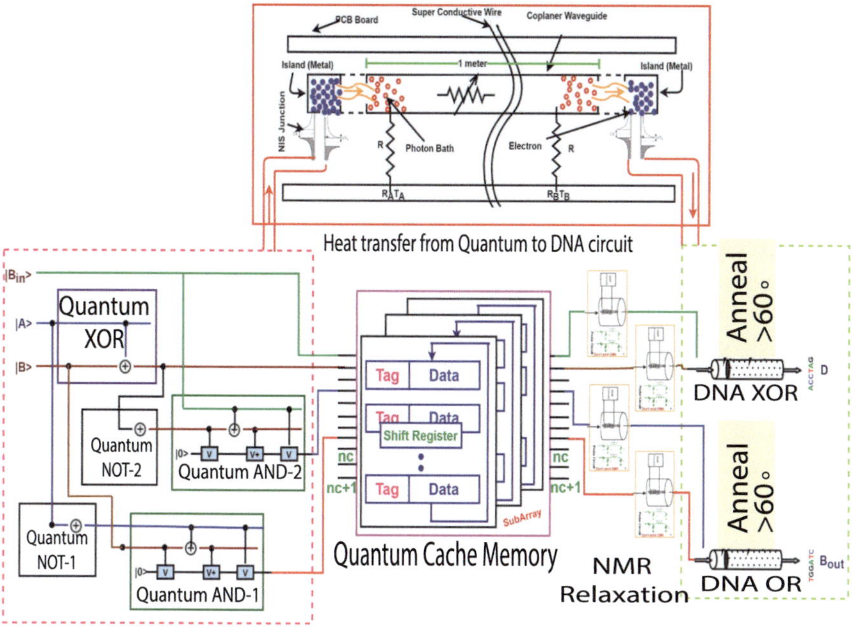

Fig. 14.11 Quantum-DNA full subtractor

2. **Working Procedure**
   The working procedure of the Quantum-DNA Full Subtractor is given below for each pattern of input qubits. Here, DNA sequence ACCTAG = TRUE for qubit |1 > and DNA sequence TGGATC = FALSE for qubit |0 >.
   The working procedure of the Quantum-DNA Full Subtractor is given below for 4 patterns of input qubits and the outputs for different combinations of inputs are given in Table 14.9.

1. For inputs A, B, B_{in} = 0, **Sum** = DNA XOR (Quantum XOR (A, B), B_{in})
   = DNA XOR (Quantum XOR (0, 0), 0)
   = DNA XOR (0, 0)
   = DNA XOR (TGGATC, TGGATC) [Using NMR relaxation]
   = TGGATC
   Carry = DNA OR (Quantum AND (B_{in}, NOT (XOR (A,B))), Quantum AND (NOT (A), B))
   = DNA OR (Quantum AND (0, 1), Quantum AND (1, 0))
   = DNA OR (0, 0)
   = DNA OR (TGGATC, TGGATC) [Using NMR relaxation]
   = TGGATC
2. For inputs A, B, B_{in} = 0, 0, 1 **Sum** = DNA XOR (Quantum XOR (A, B), B_{in})
   = DNA XOR (Quantum XOR (0, 0), 1)
   = DNA XOR (0, 1)

= DNA XOR (TGGATC, ACCTAG) [Using NMR relaxation]
= ACCTAG
Carry = DNA OR (Quantum AND (B_{in}, NOT (XOR (A, B))), Quantum AND (NOT (A), B))
= DNA OR (Quantum AND (1, 1), Quantum AND (1,0))
= DNA OR (1, 0)
= DNA OR (ACCTAG, TGGATC) [Using NMR relaxation]
= ACCTAG

3. For inputs A, B, B_{in} = 0, 1, 0 **Sum** = DNA XOR (Quantum XOR (A, B), B_{in})
= DNA XOR (Quantum XOR (0, 1), 0)
= DNA XOR (1, 0)
= DNA XOR (ACCTAG, TGGATC) [Using NMR relaxation]
= ACCTAG
Carry = DNA OR (Quantum AND (B_{in}, NOT (XOR (A, B))), Quantum AND (NOT (A), B))
= DNA OR (Quantum AND (0, 1), Quantum AND (1, 1))
= DNA OR (Quantum AND (0, 1), 1)
= DNA OR (0, 1)
= DNA OR (TGGATC, ACCTAG) [Using NMR relaxation]
= ACCTAG
4. For inputs A, B, B_{in} = 0, 1, 1 **Sum** = DNA XOR (Quantum XOR (A, B), B_{in})
= DNA XOR (Quantum XOR (0, 1), 1)
= DNA XOR (1, 1)
= DNA XOR (ACCTAG, ACCTAG) [Using NMR relaxation]
= TGGATC
Carry = DNA OR (Quantum AND (B_{in}, NOT (XOR (A,B))), Quantum AND (NOT (A), B))
= DNA OR (Quantum AND (1, 0), Quantum AND (1, 1))
= DNA OR (Quantum AND (1, 0), 1)
= DNA OR (0, 1)
= DNA OR (ACCTAG, TGGATC) [Using NMR relaxation]
= ACCTAG

14.2.1.9 Quantum-DNA 2-to-1 Multiplexer Circuit at Room Temperature

A multiplexer (MUX) is a device that can receive multiple input signals and synthesize a single output signal in a recoverable manner for each input signal. It is also an integrated system that usually contains a certain number of data inputs and a single output. To create a Multiplexer, one NOT, two AND, and an OR gates are required. Figure 14.12 describes the Quantum-DNA circuit of the Multiplexer. A Multiplexer receives three inputs and produces one output containing "Y".

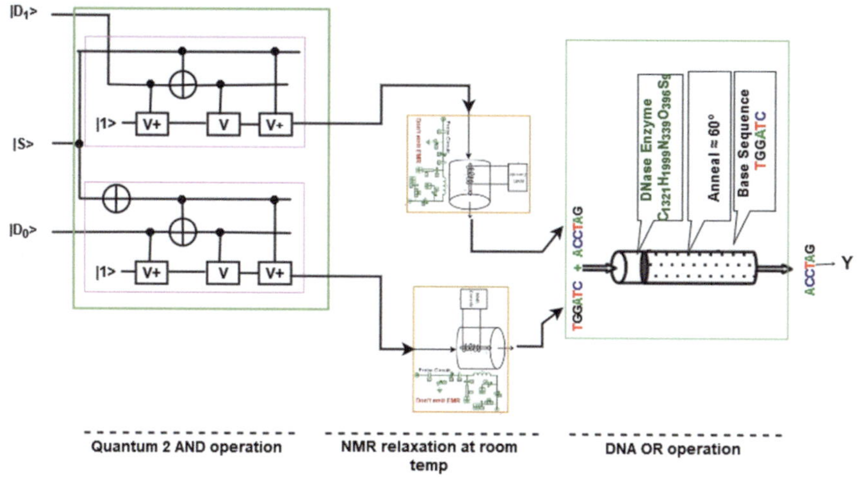

Fig. 14.12 Quantum-DNA 2-to-1 multiplexer

Table 14.6 Outputs of quantum-DNA full subtractor operation

|A>	|B>	|B$_{in}$>	SUM	Carry
|0>	|0>	|0>	TGGATC	TGGATC
|0>	|0>	|1>	ACCTAG	ACCTAG
|0>	|1>	|0>	ACCTAG	ACCTAG
|0>	|1>	|1>	TGGATC	ACCTAG
|1>	|0>	|0>	ACCTAG	TGGATC
|1>	|0>	|1>	TGGATC	TGGATC
|1>	|1>	|0>	TGGATC	TGGATC
|1>	|1>	|1>	ACCTAG	ACCTAG

1. **Design Procedure**
   To design a quantum-DNA 2-to-1 multiplexer, quantum operations and DNA operations are used to operate the input qubit for their corresponding outputs. The quantum operations will be used for receiving the input qubits and the DNA operations will be used to produce the final output against the corresponding set of inputs. Each time, the quantum-DNA multiplexer will receive three qubits as input. After operating in a certain number of quantum operations, the qubit will be turned into a corresponding DNA sequence by using an NMR relaxation room temperature probe. By using a room temperature probe and corresponding components of NMR relaxation, the excited qubit turns into the ground state and produces a DNA sequence. Then the DNA sequence is processed through DNA operations and outputs are received. Here, Fig. 14.12 describes quantum-DNA multiplexer using quantum and DNA operations.

Table 14.7 Outputs of quantum-DNA 2-to-1 multiplexer operation

|S>	|D1>	|D0>	Y
|0>	|0>	|0>	TGGATC
|0>	|0>	|1>	TGGATC
|0>	|1>	|0>	ACCTAG
|0>	|1>	|1>	ACCTAG
|1>	|0>	|0>	TGGATC
|1>	|0>	|1>	ACCTAG
|1>	|1>	|0>	TGGATC
|1>	|1>	|1>	ACCTAG

From the Fig. 14.12, it is obvious that the Quantum-DNA 2-to-1 multiplexer consists of three quantum operations and one DNA operation. Here, two AND and one NOT are used as Quantum operation and further one OR DNA operations is used.

2. **Working Procedure**

The working procedure of the Quantum-DNA 2-to-1 Multiplexer is given below for each pattern of input qubits. Here, DNA sequence ACCTAG = TRUE for qubit |1 > and DNA sequence TGGATC = FALSE for qubit |0 >.

The working procedure of the Quantum-DNA 2-to-1 Multiplexer is given below for 4 patterns of input qubits and the outputs for different combinations of inputs are given in Table 14.7.

1. For inputs S, D_1, D_0 = 0, **Y** = DNA OR (Quantum AND (D_1, S), AND (NOT (S), D_0))
   = DNA OR (Quantum AND (0, 0), AND (NOT (0), 0)
   = DNA OR (0, 0)
   = DNA OR (TGGATC, TGGATC) [Using NMR relaxation]
   = TGGATC
2. For inputs S, D_0, D_1 = 0, 0, 1 **Y** = DNA OR (Quantum AND (D_1, S), AND (NOT (S), D_0))
   = DNA OR (Quantum AND (1, 0), AND (NOT (0), 0)
   = DNA OR (0, 0)
   = DNA OR (TGGATC, TGGATC) [Using NMR relaxation]
   = TGGATC
3. For inputs S, D_0, D_1 = 0, 1, 0 **Y** = DNA OR (Quantum AND (D_1, S), AND (NOT (S), D_0))
   = DNA OR (Quantum AND (0, 0), AND (NOT (0), 1))
   = DNA OR (0, AND (1, 1)
   = DNA OR (0, 1)
   = DNA OR (TGGATC, ACCTAG) [Using NMR relaxation]
   = ACCTAG

4. For inputs S, $D_0, D_1 = 0, 1, 1$ **Y** = DNA OR (Quantum AND (D_1, S), AND (NOT (S), D_0))
   = DNA OR (Quantum AND (1, 0), AND (NOT (0), 1))
   = DNA OR (0, AND (1, 1))
   = DNA OR (0, 1)
   = DNA XOR (TGGATC, ACCTAG) [Using NMR relaxation]
   = ACCTAG

14.2.2 NMR Relaxation at 0 K

NMR room temperature relaxation process is not fully suitable. In NMR relaxation at room temperature probes, decoherence problems, polarization, and scaling problems can occur. Because, it is known that noise is proportional to temperature. If noise is high, then the temperature will be high. In-room temperature NMR relaxation processes have high temperatures that's why noise can be reached at high and that does not give us a full recovery sequence. And this problem solves the Cryogenic probe which operates at 0 K temperature. Here, the provided temperature is small and the noise is tiny. So, it is possible to get more recovery sequences from the superposition state.

14.2.2.1 Quantum-DNA AND Operation at 0 K Temperature

Figure 14.13 shows the conversion of the qubit to a DNA sequence from a quantum AND operation at 0 K temperature. The qubit output of AND gate passes through the NMR relaxation process to create a corresponding DNA sequence. The design and working procedure of the conversion of the qubit into DNA sequence from an AND operation are explained in this section.

1. **Design Procedure**
   Quantum AND operation circuit is prepared using two V+ and two NOT and one V gate. Here three qubits |X> and |Y> and a constant qubit (ancilla) |0>. From |A0> and |A1> inputs, two lines to constant output and target output line (|Q> = XY). After getting qubit from quantum AND operation, NMR relaxation is applied. In NMR relaxation, a room temperature probe is used. By NMR relaxation qubit of superposition state turns into a normal ground state and produce the DNA sequence. Figure 14.13 shows Quantum-DNA AND operation using cryogenic probe at 0 K.
2. **Working Procedure**
   At first, in this Quantum AND operational circuit, there are three inputs as qubits, two qubits are inputs in which the other is the ancilla qubit. Ancilla qubit is used here to reverse the gate. Ancilla qubit is also used for error correction and it is

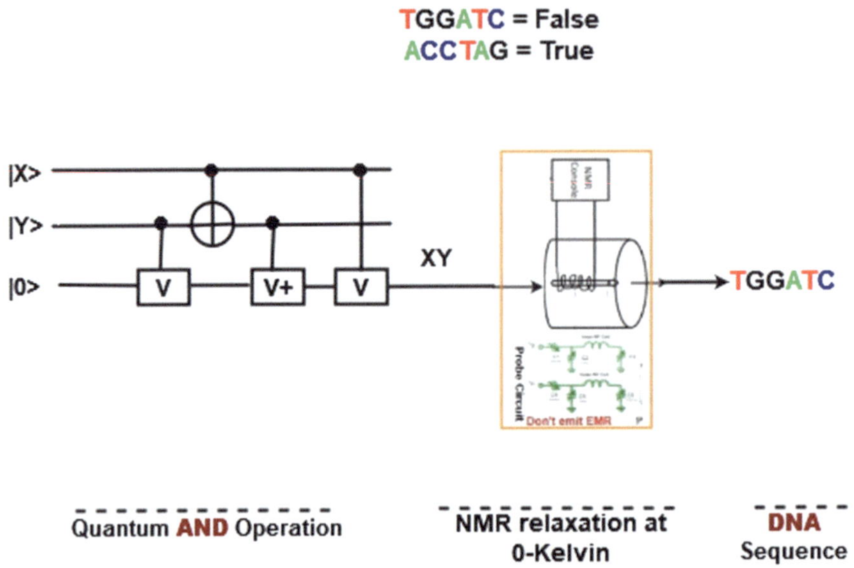

Fig. 14.13 Quantum-DNA AND operation using cryogenic probe at 0 K

Table 14.8 Outputs of data conversion for quantum AND operation

|X>	|Y>	|Q> = XY	DNA sequence
|0>	|0>	|0>	TGGATC
|0>	|1>	|0>	TGGATC
|1>	|0>	|0>	TGGATC
|1>	|1>	|1>	ACCTAG

called check qubit. Here in quantum AND gate, 0 is used because it helps to reverse the gate otherwise, it will be an error in output without an ancilla qubit. After processing input going through the quantum AND operation, it provides some output as a qubit. The provided output from quantum AND operation will be used as an input in NMR relaxation and the ground state is found. When any molecule comes to the ground state then the real sequence is produced. The DNA sequence is used for DNA logic gate operation. Here doing NMR relaxation is applied by using a cryogenic probe at 0 K temperature and in relaxation, it doesn't need to emit EMR.

Here, DNA sequence ACCTAG = TRUE for quantum qubit |1 > and DNA sequence TGGATC = FALSE for quantum qubit |0 >. The outputs for different combinations of inputs are given in Table 14.8.

14.2.2.2 Quantum-DNA OR Operation at 0 K

Figure 14.14 shows the conversion of the qubit to a DNA sequence from a quantum OR operation at 0 K temperature. The qubit output of OR operation passes through the NMR relaxation process to create a corresponding DNA sequence. The design and the working procedure of the conversion of the qubit into DNA sequence from a quantum OR operation are explained in this section.

1. **Design Procedure**
   Quantum OR operation circuit is prepared using two V+ and two V gates, three qubits |X> and |Y> and a constant qubit (ancilla) |0 >. From |X > and |Y > inputs, two lines to constant output and the target output line (|Q> = X + Y) can be expressed as X + Y. After getting qubit from quantum OR operation, NMR relaxation is applied. In NMR relaxation, a room temperature probe is used. By NMR relaxation qubit of superposition state turns into a normal ground state and the DNA sequence has produced.
2. **Working Procedure**
   At first, in this quantum OR operational circuit, there are three inputs as qubits, two qubits are input and the other is the ancilla qubit. Here in quantum OR gate, |0> is used as an ancilla qubit because without this qubit it is not possible to reverse the gate and an error will come in output without an ancilla qubit.
   After processing input going through the quantum OR operation, it's getting some outputs as qubits. The provided output from quantum OR operation will be used as an input in NMR relaxation and the ground state is found. When any molecule

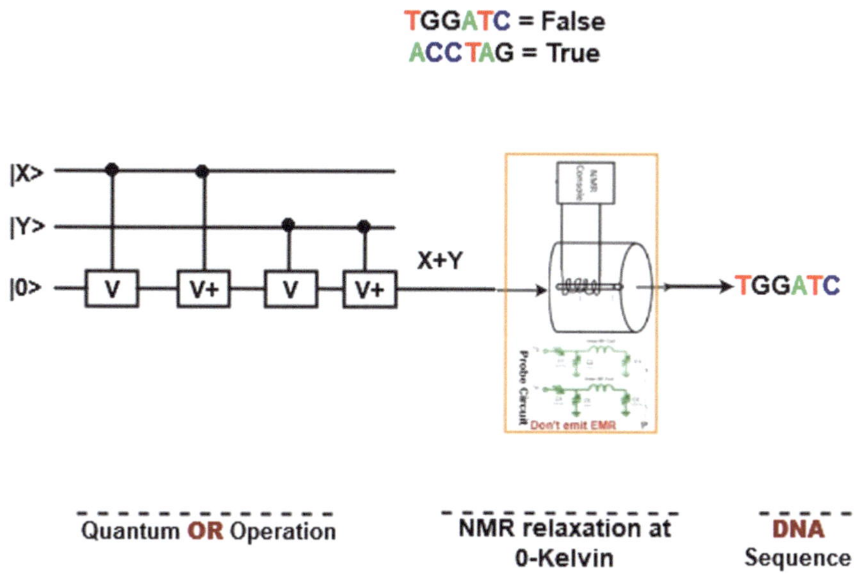

Fig. 14.14 Quantum-DNA OR operation using cryogenic probe at 0 K

comes to the ground state then the real sequence is found. The DNA sequence is used for DNA operation. Here doing NMR relaxation using a cryogenic probe at 0 K and in relaxation, it doesn't need to emit EMR.
Here, DNA sequence ACCTAG = TRUE for qubit |1 > and DNA sequence TGGATC = FALSE for qubit |0 >. The outputs for different combinations of inputs are given in Table 14.2.

14.2.2.3 Quantum-DNA XOR Operation at 0 K

Figure 14.15 shows the conversion of the qubit to a DNA sequence from a quantum XOR operation at 0 K temperature. The qubit output of XOR operation passes through the NMR relaxation process to create a corresponding DNA sequence. The design and working procedure of the conversion of the qubit into DNA sequence from an XOR operation are explained in this sections.

1. **Design Procedure**
   Quantum XOR operation circuit is prepared using one Controlled NOT or CNOT gate. Here, there are two qubits |X> and |Y>. From |Y > inputs, one line goes through the output and the target output is X XOR Y. After getting qubit from quantum XOR operation, NMR relaxation is needed. In NMR relaxation, a room temperature probe is used. By NMR relaxation qubit of superposition state turns into a normal ground state and DNA sequence is obtained.
2. **Working Procedure**
   At first, in this Quantum XOR operational circuit, there aretwo inputs work as qubits and no need of any ancilla qubits. After processing input going through the Quantum XOR operational gate, it's getting some output as a qubit. The provided

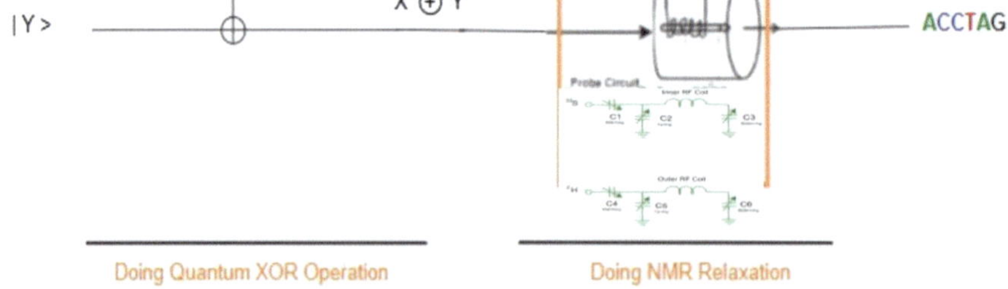

Fig. 14.15 Quantum-DNA XOR operation using cryogenic probe at 0 K

output from Quantum OR operation will be used as an input in NMR relaxation and find the ground state. When any molecule comes to the ground state then the real sequence is found. The DNA sequence is used for DNA operation. Here doing NMR relaxation using a cryogenic probe at 0 K and in relaxation, it doesn't need to emit EMR.

Here, DNA sequence ACCTAG = TRUE for qubit $|1>$ and DNA sequence TGGATC = FALSE for qubit $|0>$. The outputs for different combinations of inputs are given in Table 14.4.

14.2.2.4 Quantum-DNA Full Adder at 0 K

In Sect. 14.2.1, Quantum-DNA Full Adder and its working procedure are described already but here the design procedure of Quantum-DNA Full Adder by using a Cryogenic probe will be described. The output of different combinations is also shown in Table 14.5.

1. **Design Procedure**
   To design a Quantum-DNA Full Adder, Quantum and DNA operations are used to operate the input qubit for their corresponding outputs. The Quantum operations will be used for receiving the input qubits and the DNA operations will be used to produce the final output against the corresponding set of inputs. Each time, the Quantum-DNA Full Subtractor will receive three qubits as input. After operating in a certain number of Quantum operations, the qubit will be turned into a corresponding DNA sequence by NMR relaxation using a cryogenic probe at 0 K. The process of producing a DNA sequence against a qubit state is explained in Sect. 14.2.1. By using a cryogenic probe at 0 K and corresponding components of NMR relaxation, the excited qubit turns into a ground state and produces a DNA sequence. Then the DNA sequence is processed through DNA operations and outputs are received. Here, Fig. 14.16 describes Quantum-DNA Full Adder using Quantum and DNA operations.
   From the Fig. 14.16, it is found that the Quantum-DNA Full Adder consists of three Quantum operations and two DNA operations. Here, two AND and one XOR are used as Quantum operations and further one XOR and one OR DNA operations are used.

14.2.2.5 Quantum-DNA Full Subtractor at 0 K

In Sect. 14.2.1, Quantum-DNA Full Subtractor and its working procedure are described already but here the design procedure of Quantum-DNA Full Subtractor using Cryogenic probe will be discused. The output of different combinations is also shown in Table 14.6.

To design a Quantum-DNA Full Subtractor, Quantum and DNA operations are used to operate the input qubit for their corresponding outputs. The Quantum oper-

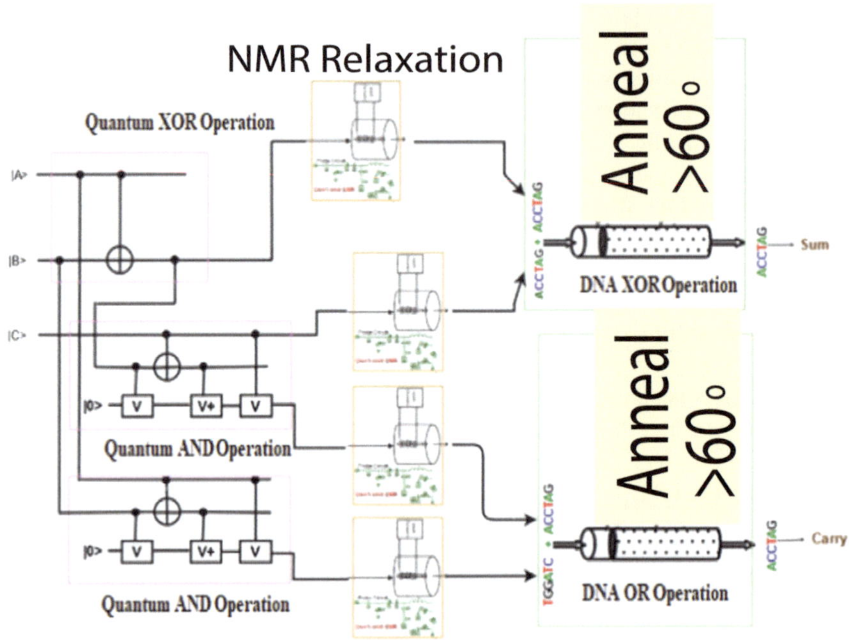

Fig. 14.16 Quantum-DNA full adder using cryogenic probe at 0 K

ations will be used for receiving the input qubits and the DNA operations will be used to produce the final output against the corresponding set of inputs. Each time, the Quantum-DNA Full Subtractor will receive three qubits as input. Here, Fig. 14.17 describes Quantum-DNA Full Subtractor using Quantum and DNA operations.

After operating in a certain number of Quantum operations, the qubit will be turned into a corresponding DNA sequence by using NMR relaxation using a Cryogenic probe at 0 K. The process of producing a DNA sequence against a qubit state is explained in Sect. 8.2.1. By using a Cryogenic probe at 0 K and corresponding components of NMR relaxation, the excited qubit turns into ground state and produces a DNA sequence. Then the DNA sequence is processed through DNA operations and outputs are received.

From the Figure, it is obvious that the Quantum-DNA Full Subtractor consists of three Quantum operations and two DNA operations. Here, two AND and one XOR are used as Quantum operations and further one XOR and one OR DNA operations are used.

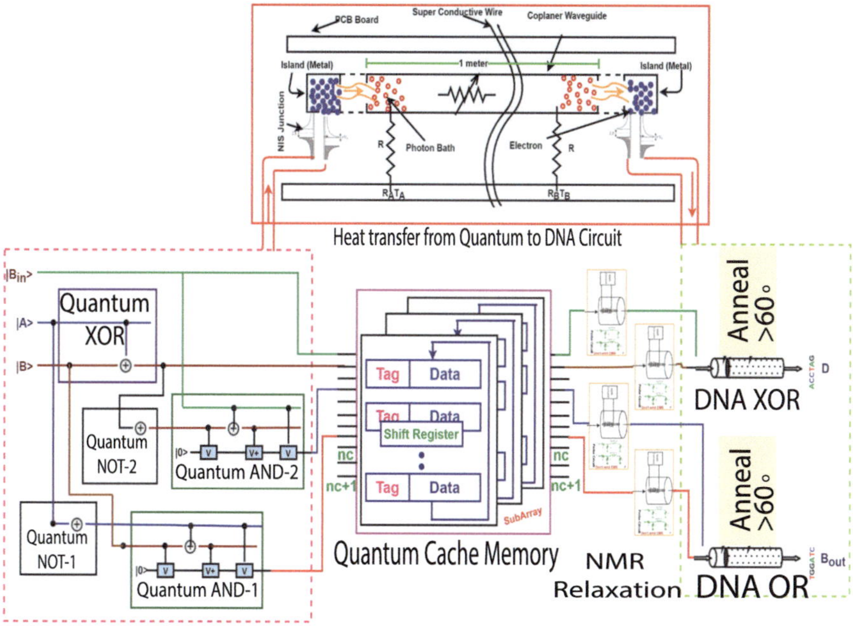

Fig. 14.17 Quantum-DNA full subtractor using cryogenic probe at 0 K

14.2.2.6 Quantum-DNA 2-to-1 Multiplexer Circuit at 0 K

In Sect. 14.2.1.9, Quantum-DNA 2-to-1 Multiplexer and its working procedure are already highlighted. Overall, the design procedure of Quantum-DNA Multiplexer using Cryogenic probe is illustrated in Fig. 14.18.

To design a Quantum-DNA 2-to-1 Multiplexer, Quantum and DNA operations are used to operate the input qubit for their corresponding outputs. The Quantum operations will be used for receiving the input qubits and the DNA operations will be used to produce the final output against the corresponding set of inputs. Each time, the Quantum-DNA Multiplexer will receive three qubits as input. After operating in a certain number of Quantum operations, the qubit will be turned into a corresponding DNA sequence by NMR relaxation using a Cryogenic probe at 0 K. The process of producing a DNA sequence against a qubit state is explained in Sect. 14.2.1. By using a Cryogenic probe at 0 K and corresponding components of NMR relaxation, the excited qubit turns into ground state and produces a DNA sequence. Then the DNA sequence is processed through DNA operations and outputs are received. Here, Fig. 14.18 describes Quantum-DNA Multiplexer using Quantum and DNA operations.

From the figure, the Quantum-DNA 2-to-1 Multiplexer consists of three quantum operations and one DNA operations. Here, two AND, one NOT are used as Quantum operations and further one OR DNA operations are used.

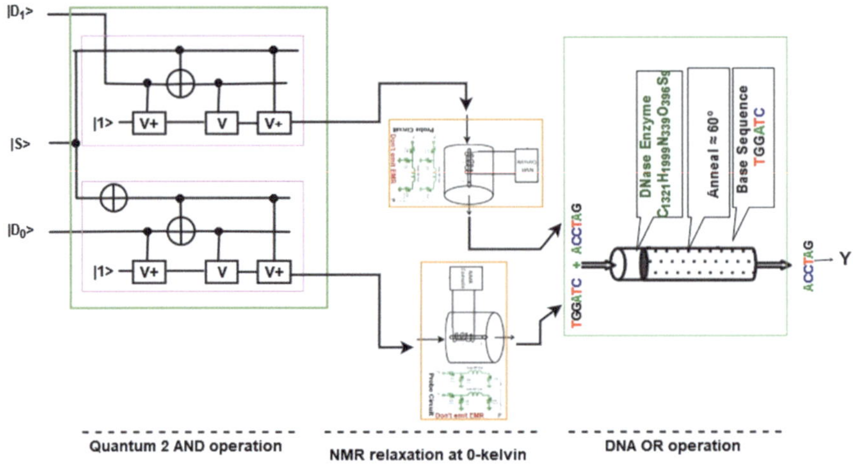

Fig. 14.18 Quantum-DNA full 2-to-1 multiplexer using cryogenic probe at 0 K

14.2.2.7 Quantum-DNA Multiplier Circuit at 0 K

A binary multiplier is a combinational logic circuit or digital device used for multiplying two binary numbers. The two numbers are more specifically known as multiplicand and multiplier and the result is known as a product. The multiplicand and multiplier can be of various bit sizes. The product's bit size depends on the bit size of the multiplicand & multiplier. The bit size of the product is equal to the sum of the bit size of the multiplier multiplicand. To create a quantum-DNA multiplier circuit, four AND operations in quantum, two XOR and, two AND DNA operational gates are required. Figure 14.19 describes the quantum-DNA circuit of the 2-qubit multiplication.

1. **Design Procedure**
   To design a Quantum-DNA Multiplier, it needs to use Quantum and DNA operations to operate the input qubit for their corresponding outputs. The Quantum operations will be used for receiving the input qubits and the DNA operations will be used to produce the final output against the corresponding set of inputs. Each time, the Quantum-DNA Multiplier will receive four qubits as input. After operating in a certain number of Quantum operations, the qubit will be turned into a corresponding DNA sequence by NMR relaxation using a Cryogenic probe at 0 K. The process of producing a DNA sequence against a qubit state is explained in Sect. 14.2.1. By using a Cryogenic probe at 0 K and corresponding components of NMR relaxation, the excited qubit turns into ground state and produces a DNA sequence. Then the DNA sequence is processed through DNA operations and outputs are received. Here, Fig. 14.19 describes Quantum-DNA Multiplier using Quantum and DNA operations.

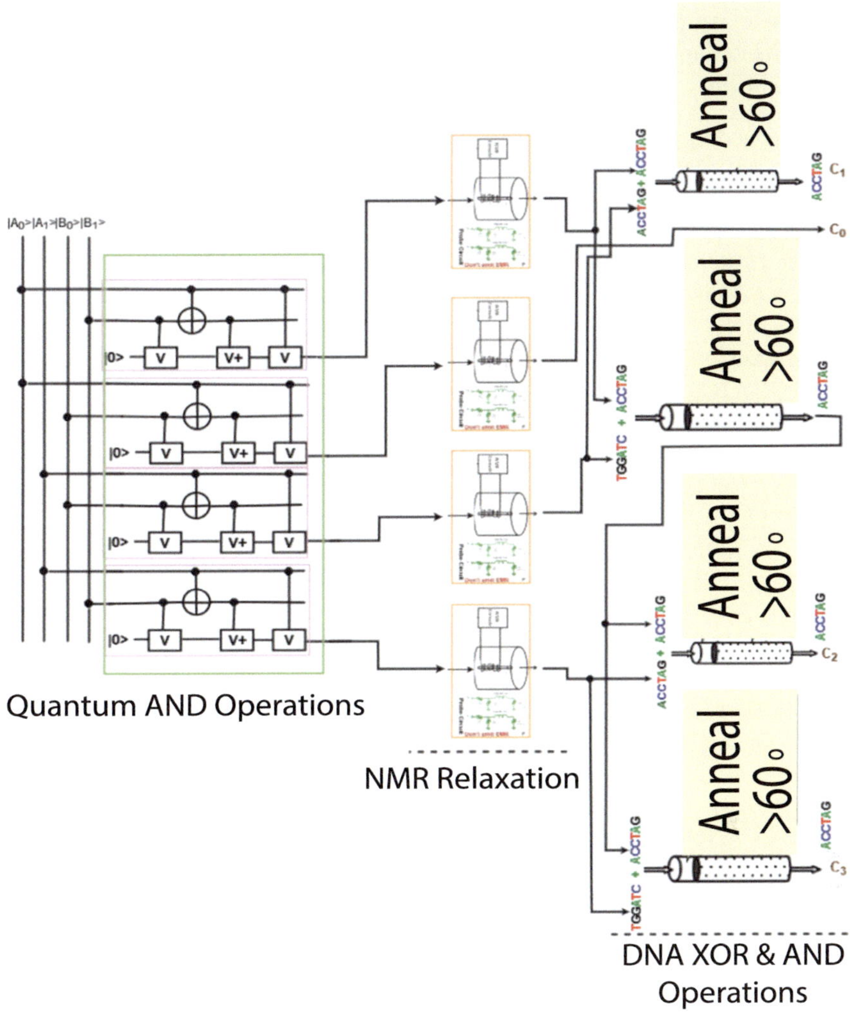

Fig. 14.19 Quantum-DNA multiplier operational circuit

From the quantum-DNA circuit in Fig. 14.19, four AND quantum operation circuits perform with qubits. The output of all quantum operation goes through NMR relaxation at 0 K. It produces DNA sequence as an output and it can be used as an input in all DNA operational circuits. In the DNA operational circuit, it uses two DNA AND and two DNA XOR operations.

2. **Working Procedure**

   The working procedure of the quantum-DNA multiplier is given below for each pattern of input qubits. Here, DNA sequence ACCTAG = TRUE for qubit $|1>$ and DNA sequence TGGATC = FALSE for qubit $|0>$.

The working procedure of the quantum-DNA multiplier is given below for 4 patterns of input qubits and the outputs for different combinations of inputs are given in Table 14.9.

1. For inputs A_0, A_1, B_0, B_1 = 0, $\mathbf{C_0}$ = DNA (Quantum AND (A_0, B_0))
   = DNA (Quantum AND (0, 0))
   = DNA (0)
   = TGGATC [Using NMR relaxation]
   $\mathbf{C_1}$ = DNA XOR (Quantum AND (A_0, B_1), Quantum AND (A_1, B_0))
   = DNA XOR (Quantum AND (0, 0), AND (0, 0)
   = DNA XOR (0, 0)
   = DNA XOR (TGGATC, TGGATC) [Using NMR relaxation]
   = TGGATC
   $\mathbf{C_2}$ = DNA XOR (DNA AND (Quantum AND (A_0, B_1), Quantum AND (A_1, B_0)), Quantum AND (A_1, B_1))
   = DNA XOR (DNA AND (Quantum AND (0, 0), Quantum AND (0, 0)), Quantum AND (0, 0))
   = DNA XOR (DNA AND (0, 0), 0)
   = DNA XOR (DNA AND (TGGATC, TGGATC), TGGATC) [Using NMR relaxation]
   = DNA XOR (TGGATC, TGGATC)
   = TGGATC
   $\mathbf{C_3}$ = DNA AND (DNA AND (Quantum AND (A_0, B_1), Quantum AND (A_0, B_1)), Quantum AND (A_1, B_1))
   = DNA AND (DNA AND (Quantum AND (0, 0), Quantum AND (0, 0)), Quantum AND (0, 0))
   = DNA AND (DNA AND (0, 0), 0)
   = DNA AND (DNA AND (TGGATC, TGGATC), TGGATC) [Using NMR relaxation]
   = DNA AND (TGGATC, TGGATC)
   = TGGATC
2. For inputs A_0, A_1, B_0, B_1 = 1, 1, 1, 0 $\mathbf{C_0}$ = DNA (Quantum AND (A_0, B_0))
   = DNA (Quantum AND (1, 1))
   = DNA
   = ACCTAG [Using NMR relaxation]
   $\mathbf{C_1}$ = DNA XOR (Quantum AND (A_0, B_1), Quantum AND (A_1, B_0))
   = DNA XOR (Quantum AND (1, 0), AND (1, 1))
   = DNA XOR (0, 1)
   = DNA XOR (TGGATC, ACCTAG) [Using NMR relaxation]
   = ACCTAG
   $\mathbf{C_2}$ = DNA XOR (DNA AND (Quantum AND (A_0, B_1), Quantum AND (A_1, B_0)), Quantum AND (A_1, B_1))
   = DNA XOR (DNA AND (Quantum AND (1, 0), Quantum AND (1, 1)), Quantum AND (1, 0))
   = DNA XOR (DNA AND (0, 1), 0)

= DNA XOR (DNA AND (TGGATC, ACCTAG), TGGATC) [Using NMR relaxation]
= DNA XOR (TGGATC, TGGATC)
= TGGATC
C$_3$ = DNA AND (DNA AND (Quantum AND (A_0, B_1), Quantum AND (A_1, B_0)), Quantum AND (A_1, B_1))
= DNA AND (DNA AND (Quantum AND (1, 0), Quantum AND (1, 1)), Quantum AND (1, 0))
= DNA AND (DNA AND (0, 1), 0)
= DNA AND (DNA AND (TGGATC, ACCTAG), TGGATC) [Using NMR relaxation]
= DNA AND (TGGATC, TGGATC)
= TGGATC

3. For inputs A_0, A_1, B_0, B_1 = 1, 1, 1, 1 **C**$_0$ = DNA (Quantum AND (A_0, B_0))
= DNA (Quantum AND (1, 1))
= DNA
= ACCTAG [Using NMR relaxation]
C$_1$ = DNA XOR (Quantum AND (A_0, B_1), Quantum AND (A_1, B_0))
= DNA XOR (Quantum AND (1, 1), AND (1, 1))
= DNA XOR (1, 1)
= DNA XOR (ACCTAG, ACCTAG) [Using NMR relaxation]
= TGGATC
C$_2$ = DNA XOR (DNA AND (Quantum AND (A_0, B_1), Quantum AND (A_1, B_0)), Quantum AND (A_1, B_1))
= DNA XOR (DNA AND (Quantum AND (1, 1), Quantum AND (1, 1)), Quantum AND (1, 1))
= DNA XOR (DNA AND (1, 1) ,1)
= DNA XOR (DNA AND (ACCTAG, ACCTAG), ACCTAG) [Using NMR relaxation]
= DNA XOR (ACCTAG, TGGATC)
= TGGATC
C$_3$ = DNA AND (DNA AND (Quantum AND (A_0, B_1), Quantum AND (A_1, B_0)), Quantum AND (A_1, B_1))
= DNA AND (DNA AND (Quantum AND (1, 1), Quantum AND (1, 1)), Quantum AND (1, 1))
= DNA AND (DNA AND (1, 1) ,1)
= DNA AND (DNA AND (ACCTAG, ACCTAG), ACCTAG) [Using NMR relaxation]
= DNA AND (ACCTAG, ACCTAG)
= ACCTAG

4. For inputs A_0, A_1, B_0, B_1 = 1, 1, 0, 1 **C**$_0$ = DNA (Quantum AND (A_0, B_0))
= DNA (Quantum AND (1, 0))
= DNA (0)
= TGGATC [Using NMR relaxation]
C$_1$ = DNA XOR (Quantum AND (A_0, B_1), Quantum AND (A_1, B_0))

= DNA XOR (Quantum AND (1, 1), AND (1, 0))
= DNA XOR (1, 1)
= DNA XOR (ACCTAG, TGGATC) [Using NMR relaxation]
= ACCTAG
C_2 = DNA XOR (DNA AND (Quantum AND (A_0, B_1), Quantum AND (A_1, B_0)), Quantum AND (A_1, B_1))
= DNA XOR (DNA AND (Quantum AND (1, 1), Quantum AND (1, 0)), Quantum AND (1, 1))
= DNA XOR (DNA AND (1, 0), 1)
= DNA XOR (DNA AND (ACCTAG, TGGATC), ACCTAG) [Using NMR relaxation]
= DNA XOR (TGGATC, ACCTAG)
= ACCTAG
C_3 = DNA AND (DNA AND (Quantum AND (A_0, B_1), Quantum AND (A_1, B_0)), Quantum AND (A_1, B_1))
= DNA AND (DNA AND (Quantum AND (1, 1), Quantum AND (1, 0)), Quantum AND (1, 1))
= DNA AND (DNA AND (1, 0), 1)
= DNA AND (DNA AND (ACCTAG, TGGATC), ACCTAG) [Using NMR relaxation]
= DNA AND (TGGATC, ACCTAG)
= TGGATC (Table 14.9)

Table 14.9 Outputs of quantum-DNA multiplier operation

$\lvert A_0\rangle$	$\lvert A_1\rangle$	$\lvert B_0\rangle$	$\lvert B_1\rangle$	C_3	C_2	C_1	C_0
$\lvert 0\rangle$	$\lvert 0\rangle$	$\lvert 0\rangle$	$\lvert 0\rangle$	TGGATC	TGGATC	TGGATC	TGGATC
$\lvert 0\rangle$	$\lvert 0\rangle$	$\lvert 0\rangle$	$\lvert 1\rangle$	TGGATC	TGGATC	TGGATC	TGGATC
$\lvert 0\rangle$	$\lvert 0\rangle$	$\lvert 1\rangle$	$\lvert 0\rangle$	TGGATC	TGGATC	TGGATC	TGGATC
$\lvert 0\rangle$	$\lvert 0\rangle$	$\lvert 1\rangle$	$\lvert 1\rangle$	TGGATC	TGGATC	TGGATC	TGGATC
$\lvert 0\rangle$	$\lvert 1\rangle$	$\lvert 0\rangle$	$\lvert 0\rangle$	TGGATC	TGGATC	TGGATC	TGGATC
$\lvert 0\rangle$	$\lvert 1\rangle$	$\lvert 0\rangle$	$\lvert 1\rangle$	TGGATC	TGGATC	TGGATC	ACCTAG
$\lvert 0\rangle$	$\lvert 1\rangle$	$\lvert 1\rangle$	$\lvert 0\rangle$	TGGATC	TGGATC	ACCTAG	TGGATC
$\lvert 0\rangle$	$\lvert 1\rangle$	$\lvert 1\rangle$	$\lvert 1\rangle$	TGGATC	TGGATC	ACCTAG	ACCTAG
$\lvert 1\rangle$	$\lvert 0\rangle$	$\lvert 0\rangle$	$\lvert 0\rangle$	TGGATC	TGGATC	TGGATC	TGGATC
$\lvert 1\rangle$	$\lvert 0\rangle$	$\lvert 0\rangle$	$\lvert 1\rangle$	TGGATC	TGGATC	ACCTAG	TGGATC
$\lvert 1\rangle$	$\lvert 0\rangle$	$\lvert 1\rangle$	$\lvert 0\rangle$	TGGATC	ACCTAG	TGGATC	TGGATC
$\lvert 1\rangle$	$\lvert 0\rangle$	$\lvert 1\rangle$	$\lvert 1\rangle$	TGGATC	ACCTAG	ACCTAG	TGGATC
$\lvert 1\rangle$	$\lvert 1\rangle$	$\lvert 0\rangle$	$\lvert 0\rangle$	TGGATC	TGGATC	TGGATC	TGGATC
$\lvert 1\rangle$	$\lvert 1\rangle$	$\lvert 0\rangle$	$\lvert 1\rangle$	TGGATC	TGGATC	ACCTAG	ACCTAG
$\lvert 1\rangle$	$\lvert 1\rangle$	$\lvert 1\rangle$	$\lvert 0\rangle$	TGGATC	ACCTAG	ACCTAG	TGGATC
$\lvert 1\rangle$	$\lvert 1\rangle$	$\lvert 1\rangle$	$\lvert 1\rangle$	ACCTAG	TGGATC	TGGATC	ACCTAG

14.2.3 Trapped Ion

An ion trap is actually an 'electric-field-test-tube' that contains gaseous ions. These gaseous ions can be either positively charged or negatively charged. The two most common ion traps are the Penning trap that is used as a combinational via of electric and magnetic fields, and the other one is Paul trap which is used as a combinational via of static and oscillating electric fields. In this section, the Paul trap will be used for the above purpose. Paul traps are commonly used as components of a mass spectrometer. But here this is used as a little bit different way to reach the desired destination. The Ion trap procedure is shown in Fig. 14.20.

14.2.3.1 Components of Trapped Ions

Trap Ion circuit is shown in Fig. 14.21. For trapping an ion, the following given components are needed:

1. **A Bunch of Atoms**
   As an atom, calcium can be considered. If calcium is taken in a test tube and removed all air from that tube it (calcium) looks shiny because it's taken out in the air which cannot react with oxygen.

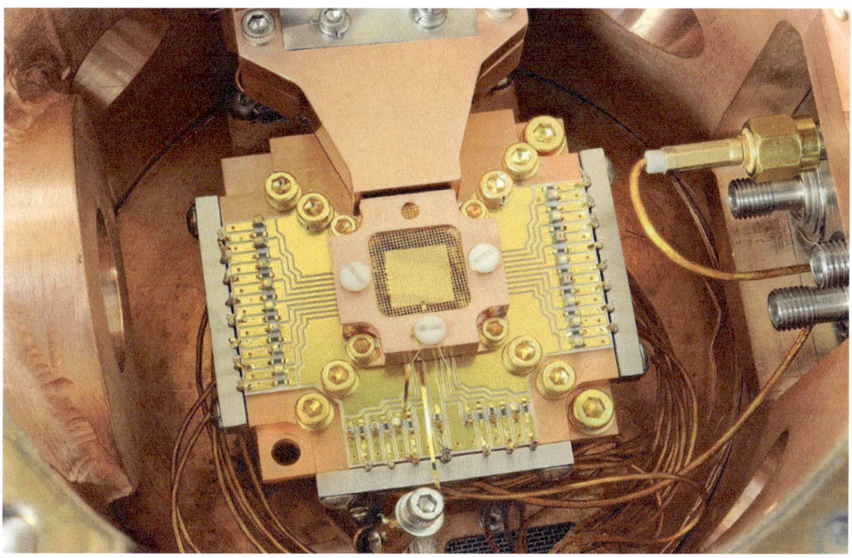

Fig. 14.20 Trapped Ion, used for experiments towards realizing a quantum computer

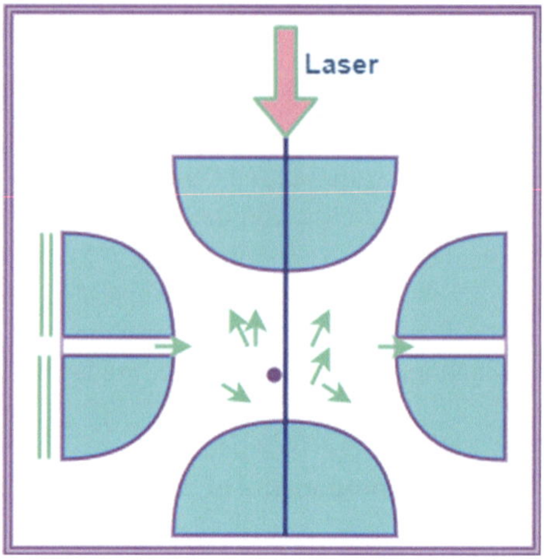

Fig. 14.21 Trapped ion circuit

2. **An Ultra-High Vacuum System**
   In a vacuum, ions are trapped. It is used so that any oxygen cannot bump with calcium. If oxygen plops with calcium, trap ions will not work properly. That is why this vacuum system is used to suck all air from surrounding the trap ion.
3. **Calcium Oven**
   A metallic tube is used to run electricity outside the vacuum so that calcium can be heated and it can be converted into vapor gas that spreads the whole trap is actually called a calcium oven.
4. **Laser**
   When a laser is applied to the electron that rotates around the nucleus, it flips the one outer electron that is actually an ion.
5. **End-Cap Electrodes**
   Leading the trap, ion accumulation basically allows efficient ions to hold the injected ions with low kinetic energies. In this case, end-cap electrodes carry small DC voltages. However, these DC potentials are used to trap ions in the axial dimension.
6. **Blade Electrodes**
   The trap consists of four blade-shaped electrodes of which two opposing ones are connected to an RF voltage while the other two are connected to the ground. It also includes two end-cap electrodes that are connected to a positive DC voltage.

14.2.3.2 Working Principle of Trapped Ions

In Paul trap, an oscillating electric potential usually combined with a static component, $U_0 + V_0 \cos\Omega t$, is applied between the ring and the pair of end-cap electrodes. It creates a potential of the form

$\Phi = U_0 + V_0 \cos\Omega t \, 2d^2 \, (r^2 - 2z^2)$

Since the trapping field is inhomogeneous, the average force acting on the particle, taken over many oscillations of the field, is not zero. Depending on the amplitude and frequency of the field, the net force may be convergent toward the center of the trap leading to confinement or divergent leading to the loss of the particle. Thus, although the electric force alternately causes convergent and divergent motion of the particle in any given direction, it is possible by appropriate choice of field amplitude and frequency to have a time-averaged restoring force in all three dimensions toward the center of the trap as required for confinement. The conditions for stable confinement of an ion with mass M and charge Q in the Paul field may be derived by solving the equation of motion:

$\partial 2u \; \partial t^2 = Q \; Md^2 \; (U_0 + V_0 \cos\Omega t) \; u$

$u = x, y, z.$

The three-dimensional limitation of charged particles is the minimum possible force at any stage of space so that the corresponding force is directed to three dimensions. In general, there may be a voluntary form of dependence of the magnitude of this arm on coordinates; however, it is advantageous to have a binding force with a condition, as it simplifies the analytical description of particle motion. Thus, it is assumed that

$$F \propto -r \tag{14.1}$$

It follows from

$$F = -gradU \tag{14.2}$$

Where $U = Q\Phi$ is the potential energy, that in general the required function Φ is a quadratic form in the Cartesian coordinates x, y, and z:

$$\Phi = \Phi_0/d^2(Ax^2 + By^2 + Cz^2) \tag{14.3}$$

where A, B, and C are constants, d is a normalizing factor, and Φ_0 can be a time-dependent function. If an attempt is made to achieve such a constraint by using an electric field acting on an ion of charge Q, it is found that to satisfy Laplace's equation $\Delta\Phi = 0$, the coefficients must satisfy $A + B + C = 0$. For the interesting case of rotational symmetry around the z-axis, this leads to $A = B = 1$ and $C = -2$, giving us the quadrupolar form

$$\Phi = \Phi_0 d^2(x^2 + y^2 - 2z^2) = \Phi_0 d^2(\rho^2 - 2z^2) \tag{14.4}$$

with $\rho^2 = x^2 + y^2$. If the radial distance from the center ($\rho = z = 0$) of a hyperbolic trap to the ring electrode is called r_0, and the axial distance to an end-cap is z_0, the equations for the hyperbolic electrode surfaces are

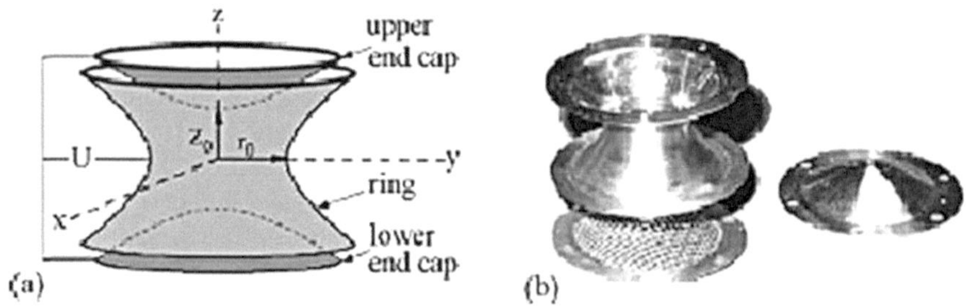

Fig. 14.22 Basic arrangement for paul and penning traps (**a**). For static trapping, cylindrical penning traps are used (**b**)

$$\rho^2 - 2z^2 = r_0{}^2 \tag{14.5}$$

$$\rho^2 - 2z^2 = -2z_0{}^2 \tag{14.6}$$

If the potential difference between the ring and end-caps is taken to be φ_0, then

$$d^2 = r_0{}^2 + 2z_0{}^2 \tag{14.7}$$

From the difference in the symptoms between the radial and axial terms, it is seen that the source of probability has a saddle point, with one coordinate being minimum but the other being maximum. Earnshaw's theorem states that it is not possible to create minimum electronic possibilities in empty space. Nevertheless, it is possible to propagate Earnshaw's theorem by superimposing a magnetic field along the z-axis to create what is called a Penning trap or trap Paul using a time-dependent electric field.

The electrons that make up a quadrilateral possibility contain three hyperbolic sheets of revolution: a ring electrode and two end-caps (Fig. 14.22(a)) which share the same asymptotic cone. The size of the device, in a variety of applications, ranges from a few centimeters for a fraction of the characteristic dimension d to one millimeter. The trapped charged particles are confined to a very area of the trap, the location of which can be centered using an additional dc field.

In recent years, various trap geometries have become commonplace that are easy to produce and align and allow optical access to the trapped particles without further modification (Fig. 14.22(b)): linear Paul traps that use four parallel rods electronically. An AC voltage applied between adjacent electrodes leads to similar dynamic prison in the three-dimensional case. Axial captures are supplied by a static voltage to the end electrodes.

The internal electrode surfaces are hyperboloids. The dynamic stabilization in the Paul trap is given by an AC voltage V_0 cost. The static stabilization in the Penning trap is given by a DC voltage $U = U_0$ and an axial magnetic field. A photograph of a trap with $\rho = 1$ cm is given. One of the endcaps is formed as mesh to allow optical access to trapped ions (Fig. 14.23).

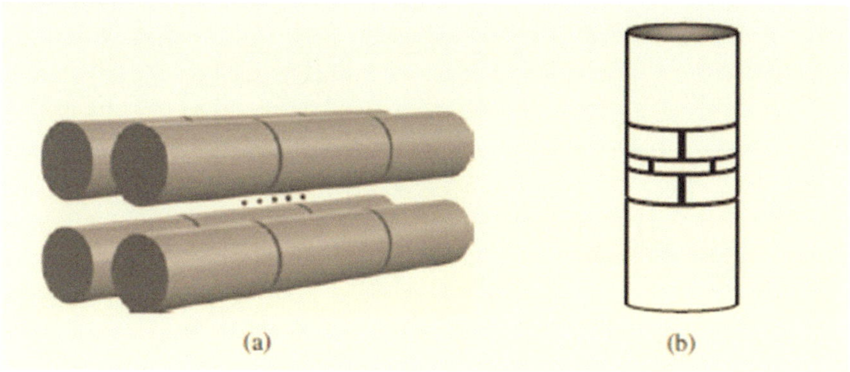

Fig. 14.23 Linear paul trap (**a**) and Open end-cap cylindrical penning trap (**b**)

A radio-frequency field applied to the rods of the linear Paul trap confines charged particles in the radial direction, DC voltage at the end segments serves for radial trapping. The Penning trap has guard electrodes between the central ring and the endcaps to compensate partly for deviations from the ideal quadrupole potential near the trap center. Ion excitation can be performed by RF fields applied to segments of the electrodes. These electrode geometries produce a harmonious binding force near the exact center of the classical form. Further away from the source, potentially higher order will become significant. For cylindrical Penning traps, this can be partially reduced by additional compensating electrodes placed between the ring and end-caps as shown in Fig. 14.23.

14.2.3.3 Quantum-DNA AND Operation

Figure 14.24 shows the quantum-DNA AND operation using trap ion. The qubit output of the quantum NAND operation passes through the trap ion to get the corresponding DNA sequence. The obtained DNA sequence then passes through the DNA NOT operation to get the output of the quantum-DNA NAND gate operation. The design and the working procedures of the quantum-DNA AND operation are explained in the this sections.

1. **Design Procedure**
   In this section, a method is established to make an AND gate by combining a quantum gate a and DNA gate. This new gate is called the quantum-DNA AND gate. In this circuit, the following terms are used:
   Quantum NAND Gate Operation
   The AND gate is easily formed from the NAND gate. After that, doing NOT gate of NAND gate, AND gate is possible to achieve. In this quantum gate, constant 1 which acts as the target bit has been assumed. If 0 is considered as a target bit then output will always be the reverse of the desired output. For example, in

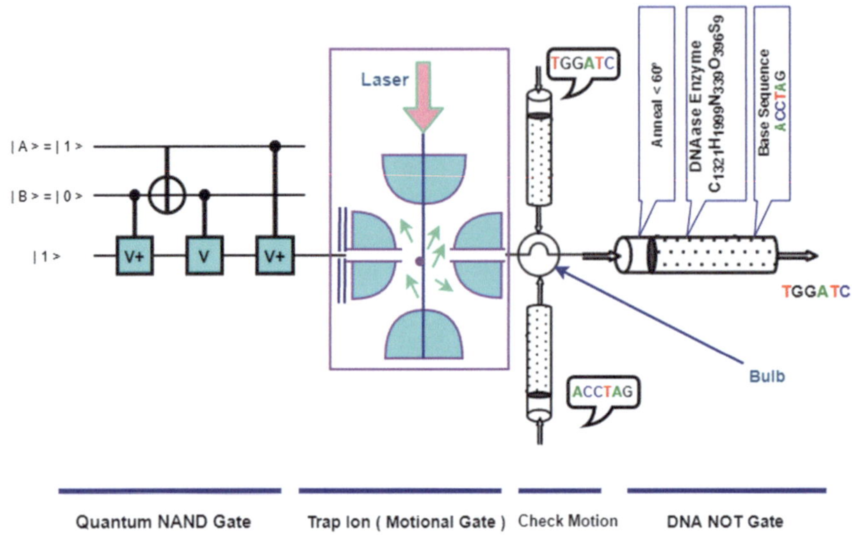

Fig. 14.24 Quantum-DNA AND operation

quantum NAND gate, the target bit is assumed as 0 then for control bit |1> and |1> combination the 1st v+ gate will work which is connected with |B>(|1>). For being target bit 0 the v+ gate will be small w. The 2nd v gate will work when the |A> value is |1> and the CNOT from of |B> is 1. In the |1> and |1> combination the |A> is |1> but CNOT form of |B> is 0. So, the middle gate will not change. So, in this gate, the previous qubit (small v) is achieved where the 1st v+ gate is placed. Now, the last v+ gate is connected with |A> (|1>). As a result, this gate value will change. For small w the last v+ gate will be 1 which violates the NAND gate truth table. Because for |1> and |1> combination the output should always be 0. But in quantum gates, if the target bit is assumed 0, the output is the reverse of real output. Not only for the NAND gate but also all gates it will show reverse output of their real output. That is why, in a quantum circuit, the target bit is always assumed 1.

Ion trap

After quantum NAND gate operation, ion has been trapped for doing motional gate.

DNA NOT Gate Operation

Based on motional gate a DNA NOT operation has been performed to get AND gate.

2. **Working Principle**

   To the execute quantum-DNA AND gate, the following steps are performed.

   (a) First of all, Quantum NAND gate is performed here where |A> and |B> act as control bit and |1> acts as the a target bit that is constant here. Target bit will work when the control bit |A> or |B> is 1, otherwise it (target bit) will remain the same. For example, if both control qubits (|A> and |B>) is 0, there will be no change in V or V+ gate. That is the reason of target bit will also remain same and that will be the NAND gate output (|1>). But if the control qubits |A> and |B> correspond to |0 > and |1> respectively, then 1st V+ gate will work as it is connected with |A1>. However, from V and V+ truth table, it is seen that for control bit <A1|(<1|), V+ gate will be W. The middle gate is V. This gate will work when the CNOT gate is active. On the other hand, CNOT gate will work when |A> is 1. If CNOT gate is active, the qubit of |B> will change and the middle V gate will work. But in this circuit, for being |A> is 0, the CNOT gate will never active and control bit |B> will not be able to change its qubit. As a result, the middle V gate will not work. But the last V+ gate is connected with control bit |A>. In this combination, for |A>=1, gates V for V+ and for V the last V+ gate will be 1 that is shown in Table 14.10. This 1 is actually quantum NAND gate output for control qubit |A> (|0>) and |B> (|1>). Similarly, all combinations of NAND gates will work.

(b) For each combination, the NAND gate will produce one qubit which will be injected in the trap ion and in the trap, ion there already will be kept Ca_+ ion. Comparing Quantum qubit and Ca_+ ion qubit, they make either a motional gate or no motional gate.

Depending on the motional gate a bulb is opened. If it is a motional gate then ACCTAG tube will open where ACCTAG DNA sequence represents true or 1. On the other hand, if it is not a motional gate TGGATC DNA sequence tube will open which represents 0 or false. Now, its turn to do DNA NOT gate. In DNA NOT gate, a base DNA sequence (ACCTAG) is kept. This sequence will react with that sequence and will come from the bulb. If they make a bond, it will give output 1. Otherwise, it will give output 0.
Truth table of V and V+ gate is shown in Table 14.10 and truth table of quantum-DNA AND operation is given in Table 14.11.

Table 14.10 Truth table of V and V+ gates

A(control)	B(target)	Q1(V gate)	Q1(V+ gate)
|0>	X	X	X
|1>	0	v	w
|1>	1	V	W
|1>	v	1	0
|1>	V	0	1
|1>	w	0	1
|1>	W	1	0

Table 14.11 Truth table of quantum-DNA AND operation

Control bit |A>	Control bit |B>	Target bit(const) |1>	Output
|0>	|0>	|1>	TGGATC
|0>	|1>	|1>	TGGATC
|1>	|0>	|1>	TGGATC
|1>	|1>	|1>	ACCTAG

14.2.3.4 Quantum-DNA OR Operation

Figure 14.25 shows the quantum-DNA OR operation using trap ion. The qubit output of the quantum NOR operation passes through the trap ion to get the corresponding DNA sequence. The obtained DNA sequence then passes through the DNA NOT operation to get the output of the quantum-DNA OR gate operation. The design and working procedures of the quantum-DNA OR gate operation are explained in this sections.

1. **Design Procedure**
   In this portion, an OR gate is formed by combining quantum gate and DNA gate. And this new gate is called the quantum-DNA gate. In this circuit, the following terms are made.
   Quantum NOR Gate Operation
   The OR gate is easily formed from the NOR gate. After that, doing NOT gate using NOR gate, OR gate is also possible to achieve. In this quantum gate, Constant

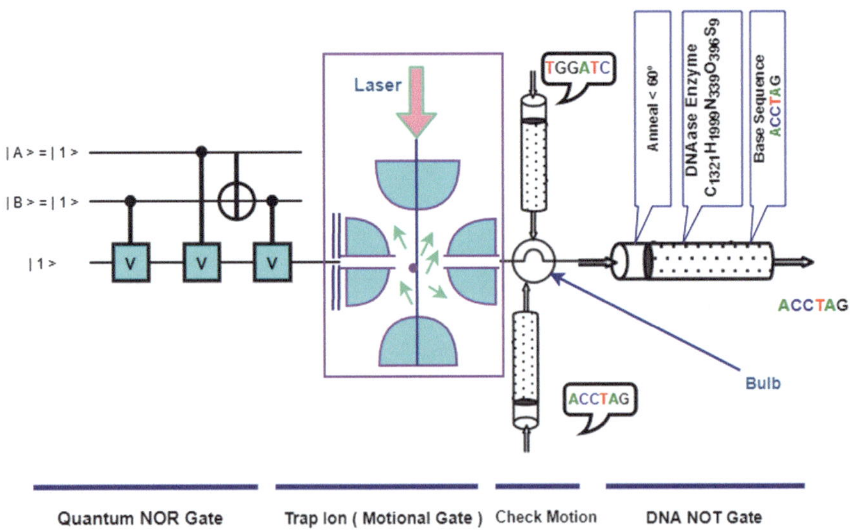

Fig. 14.25 Quantum-DNA OR operation

|1> which acts as the target bit has been assumed. If 0 is considered as a target bit then output will always be the reverse of real output.

Trapped Ion

After quantum NOR gate operation, Ion has been trapped for doing motional gate.

DNA NOT Gate Operation

Based on motional gate a DNA NOT gate operation has been performed to get OR gate.

2. **Working Principle**

   In the quantum-DNA OR gate, the following steps are performed:

   (a) First of all, Quantum NOR gate is performed here where |A> and |B> act as the control qubit and |1> acts as a target bit that is constant here. Target bit will work when the control qubit |A> or |B> is 1, otherwise it (target bit) will remain the same. For example, if both control qubit (|A> and |B>) is 0, Then output will be exactly that qubit which is considered as a target qubit. That is why for |0> and |0> combination, NOR gate output will be (|1>). But if control qubit |A> and |B> are respectively |1 > and |1> then 1st V gate will work as it is connected with |B>. For being target bit 1, the v gate will be now big V that is shown in V and V+ truth table and for big V the 2nd v gate which is connected with |A> (|1>) will be 0. In this circuit last v gate will work when the CNOT form of |B> and |A> is |1>. But for |1> and |1> combinations |A> is |1> but CNOT form of |B> is 0 as control bit of |B> is |1>. That is why, ultimately 0 will be got as the output that basically had been achieved from the middle v gate. Not only for this gate but also all gates will work like this.

   (b) For each combination, the NOR gate will produce one qubit which will be injected in the trap ion and in the trap ion there already will be kept Ca+ ion. Comparing Quantum qubit and Ca+ ion qubit, they make either a motional gate or non-motional gate.

   (c) Depending on the motional gate a bulb is opened. If it is a motional gate then ACCTAG tube will open where ACCTAG represents true or 1. On the other hand, if it is not a motional gate TGGATC will open which represents 0 or false. Now, its turn to do DNA NOT gate. In DNA NOT gate, a base DNA sequence (ACCTAG) is kept. This sequence will react with that sequence and it will come from the bulb. If they make a bond, it will give output 1. Otherwise, it will give output 0.

The truth table of V and V+ gates is shown in Table 14.12 and truth table of quantum-DNA OR operation is given in Table 14.13.

Table 14.12 Truth table of V and V+ gates

A(control)	B(target)	Q0(V gate)	Q1(V+ gate)
$\vert 0>$	X	X	X
$\vert 1>$	0	v	w
$\vert 1>$	1	V	W
$\vert 1>$	v	1	0
$\vert 1>$	V	0	1
$\vert 1>$	w	0	1
$\vert 1>$	W	1	0

Table 14.13 Truth table of quantum-DNA OR operation

Control bit $\vert A>$	Control bit $\vert B>$	Target bit(const) $\vert 1>$	Output
$\vert 0>$	$\vert 0>$	$\vert 1>$	TGGATC
$\vert 0>$	$\vert 1>$	$\vert 1>$	ACCTAG
$\vert 1>$	$\vert 0>$	$\vert 1>$	ACCTAG
$\vert 1>$	$\vert 1>$	$\vert 1>$	ACCTAG

14.3 Data Conversion in DNA-Quantum Circuits

In place of standard silicon-based computer technology, DNA computing uses biological components such as DNA, biochemistry, and molecular biology. When applied to issues that can be separated into independent, non-sequential tasks, the DNA computer has demonstrable benefits over conventional computers. The reason for this is because DNA strands can store a lot of data and do numerous operations at the same time, allowing them to solve decomposable issues considerably faster. On the other hand, the fastest computation system can be defined as a Quantum computation system, which works with qubits $|1>$, $|0>$, and both $|1>$ and $|0>$ works at the same time. In addition, Quantum computer calculations are especially promising for analyzing or simulating extremely complicated processes involving large volumes of data.

So, to find a super fast computation system with huge memory, a DNA-Quantum computation system can be developed. This DNA-Quantum computation system can merge all advantages of quantum computing and DNA computing. It will be able to compute parallel operations at super-fast speed. In a DNA-Quantum circuit, operate inputs by DNA operations and provide output in qubits. In this case, it needs to use NMR for converting DNA sequences to quantum qubits. Section 14.3.1 will describe the procedure for converting DNA sequence to the qubit.

14.3.1 Nuclear Magnetic Resonance

The acronym Nuclear Magnetic Resonance (NMR) stands for Nuclear Magnetic Resonance, which is an analytical chemical technique that allows us to examine comprehensive information about molecules. NMR is a technique for observing how molecules behave and interact in a variety of materials. NMR is used in quality control and research to determine the content and purity of a sample as well as its molecular structure. NMR is a physical phenomenon in which electromagnetic radiation is absorbed and emitted by the nucleus under a magnetic field. NMR generates a strong magnetic field, which excites the nucleus of molecules, causing them to exist in superposition. This energy has a specific resonance frequency that is determined by the strength of the magnetic field as well as the magnetic characteristics of the atom's isotopes. The detection of certain quantum-mechanical magnetic characteristics of the atomic nucleus is possible with NMR. Different components of NMR relaxation are as follows.

1. ***Magnet***: In magnet, superconducting magnets as shim coils, liquid helium, and nitrogen containers can be used.
2. ***Probe***: One important part of the probe is the RF coil. It controls temperature and molecules become superpositioned by the impact of this component. Figure 2 describes the circuit of the probe. There are two types of NMR probes which are going to be described in the following subsection.
3. ***Console***: Electronics for generating RF pulse, power and gradient amplifiers, lock system, temperature control with almost all the components of NMR controlled by the console.
4. ***Computer***: When getting all the data from the console, the computer is seen as a spectrum. Computers are used for data storage, processing, analysis components, and communication between other components.

14.3.2 Structure of NMR

The basic schematic setup of NMR is discussed in this section. Figure 14.26 shows the schematic figure, which holds superconducting coil generations or magnets. In this NMR, the samples are provided into the tube where RF coils exist. RF coil works for transmitting a signal into a sample. After collecting the signal from the components, the console digitalized the data and the computer visualized this data as a spectrum.

Figure 14.27 is the outfit and inner structure of NMR. In NMR, the probe contains the radiofrequency (RF) coils, tuned at specific frequencies for specific nuclei in a given magnetic field.

There are two types of probes for NMR such as Room temperature Probe and Cryogenic Probe. The room temperature probe of NMR is shown in Fig. 14.28.

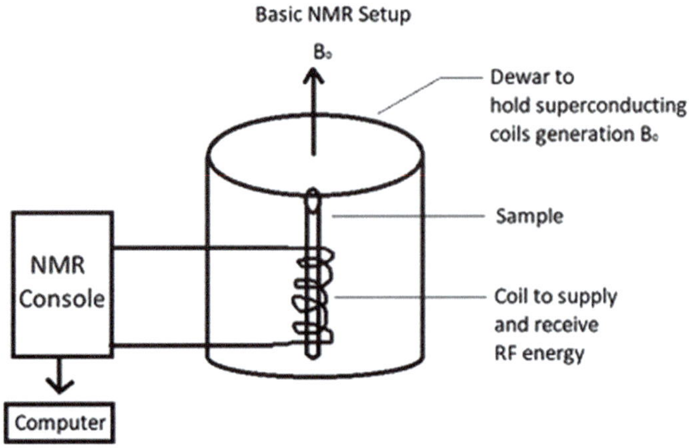

Fig. 14.26 Schematic figure of NMR

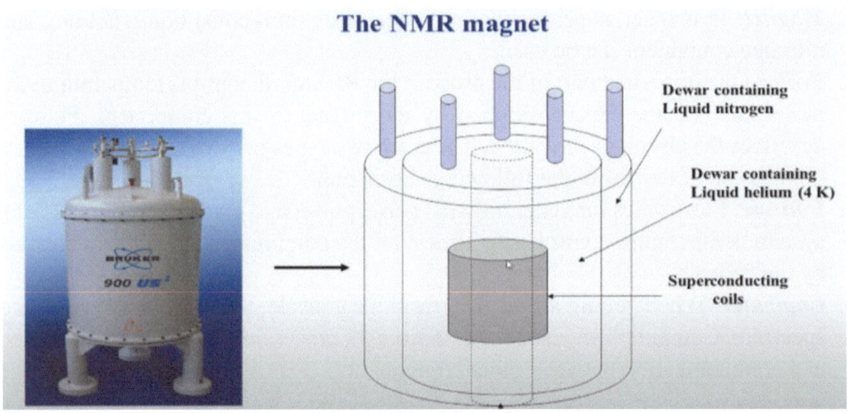

Fig. 14.27 Outfit and inner structure of NMR

Cryogenic probe's main characteristic is that its working temperature is very low. The cryogenic probes have two different coils compared to room temperature probes, the inner coil and the outer coil, which is described in Fig. 14.29. Figure 14.28 describes the room temperature probe of NMR. Same as like room temperature probe, the cryogenic probes need EMR in the NMR process. In NMR, it needs strong magnetic fields to give more energy to the NMR solvent and push them into an excited state that's why EMR is emitted.

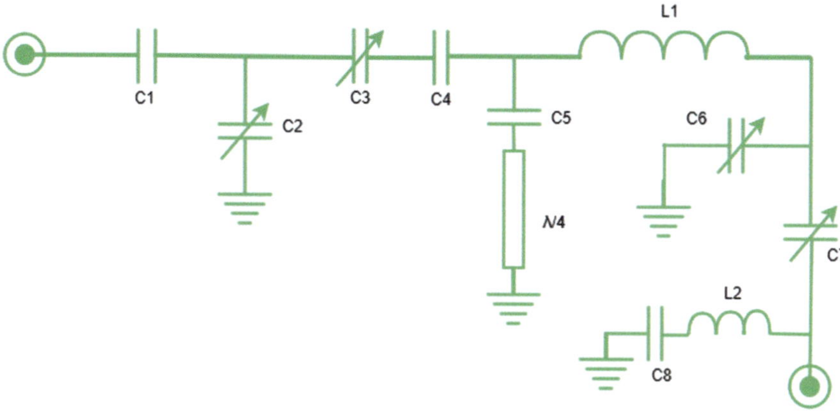

Fig. 14.28 Room temperature probe of NMR

14.3.3 Working Procedure of NMR

NMR is an analytical chemistry technique that creates a strong magnetic field and makes molecules nucleus excited, when the molecules exist in superposition.

Figure 14.30 describes a molecule that does not have any spin at the beginning of the NMR process. When molecules are provided into the NMR Probe as a sample then molecules are bound by a magnetic field. Molecules' nuclear spin started spinning but they exist in the ground state still because the magnetic field doesn't create a strong magnetic field.

Fig. 14.29 Circuit of **a** Inner RF coil **b** Outer RF coil in cryogenic probe

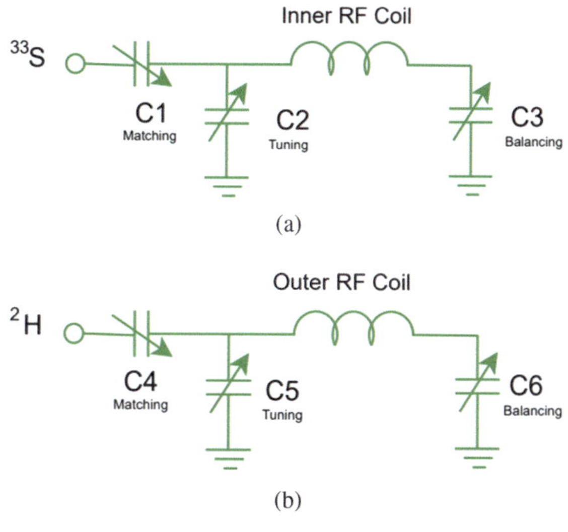

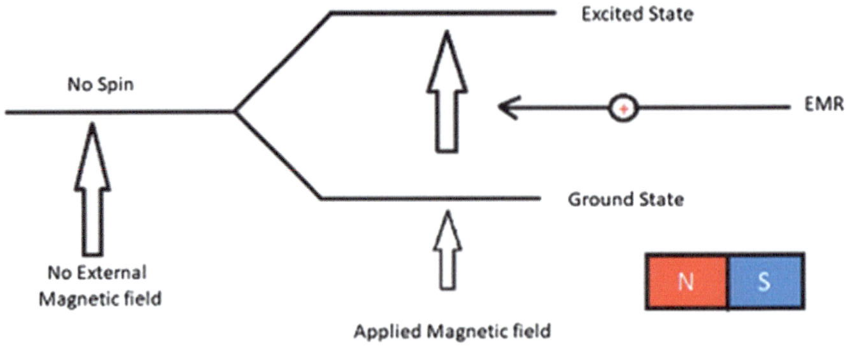

Fig. 14.30 Spins states realization of a qubit

To activate the probe and to make the magnetic field stronger it is needed to emit Electron Magnetic Resonance, EMR. After selecting the molecule to pass through the NMR process, it needs to use this molecule as a sample in NMR to produce a qubit. An NMR probe is a part of an NMR spectrometer, which works in terms of exciting the nuclear spins and detecting the NMR signal. The probe goes into the center of the magnetic field, and the sample is inserted into the probe to perform the NMR experiment. The probe contains the radiofrequency (RF) coils, tuned at specific frequencies for specific nuclei in a given magnetic field. The probe also contains the necessary hardware to control the sample temperature. From the probe, molecules make themselves in superposition.

For going from the ground state to the excited state molecules needs more energy. Measurement of energy works from the formula,

$E = hv$

By emitting the Electron Magnetic Resonance (EMR), the magnetic field becomes strong and molecules become excited and jump into excited and superposition states. By using NMR only two states of the molecule can go to the superposition state. So, it can be said that only molecules can have the superposition state which molecules $I = \frac{1}{2}$.

It is known that,
Nucleus = proton + electron

So, if protons and neutrons both numbers are odd then $I = 1$ and if both numbers are even then $I = 0$ but if one is odd and one is even then $I = 1/2, 3/2, 5/2$. But $I = \frac{1}{2}$ can be only NMR solvent.
Because, the formula, $m = 2l + 1$, defines the state of a molecule. Where, m is the magnetic quantum number.
So, for $l = 1$
$= 2.1 + 1$
$= 3$

So, if I = 1 then there will be three states.
If, I = 0
= 2.0 + 1
= 1
So, if I = 0 then there will be only 1 state.
But when I = $\frac{1}{2}$, the states are plus $\frac{1}{2}$ and minus $\frac{1}{2}$. That's why when I = $\frac{1}{2}$, the superposition can occur in the molecule. As a result, the DNA sequence molecules can be used because in DNA molecules, the hydrogen is found.

14.3.4 DNA Sequence to Qubits Using NMR

DNA computing is an emerging field for the researcher and it contains biology and molecular biology. Researchers right now find another interesting and useful topic for the future called quantum biology. Quantum biology is mainly a study between quantum computing and DNA computing.

DNA has the characteristics of enabling classical logical operation using DNA sequence. DNA prefers to be in double-stranded form, while single-stranded DNA naturally migrates towards complementary sequences to form double-stranded complexes. Complementary sequences pair the bases adenine (A) with thymine (T) and cytosine (C) with guanine (G). DNA sequences pair in an antiparallel manner, with the 5' end of one sequence pairing with the 3' end of the complementary sequence.

DNA is used for performing different operations such as NOT, AND, OR, NAND, XOR, NOR, and XNOR, which are called DNA operations. Each DNA operation input will be the single standard sequence, if it is assumed true then the complementary DNA sequence will be false. Suppose ACTCGT is the input sequence then the complementary sequence TGAGCA. In DNA computing, while designing the DNA logic gate, a predetermined single strand sequence can be supplied to induce an appropriate chemical reaction. This sequence also helps to evaluate output value whether it is true or false.

Fluorescent labels can be used to detect the presence or absence of the double-stranded sequence. The presence of a fluorescently labeled double-stranded sequence will only work if the single-stranded labeled sequences are removed. This can be accomplished using deoxyribonuclease (DNase) enzymes.

14.3.4.1 DNA Operations

DNA-based computing can do billions of operations simultaneously which is more than digital computer's capability. In addition, DNA computing can provide huge memory in small spaces. The base of DNA computing is the DNA operations. In this section, different types of DNA operations will be described.

1. **DNA Circuit for NOT Operation**

Fig. 14.31 DNA-based implementation of the NOT operation

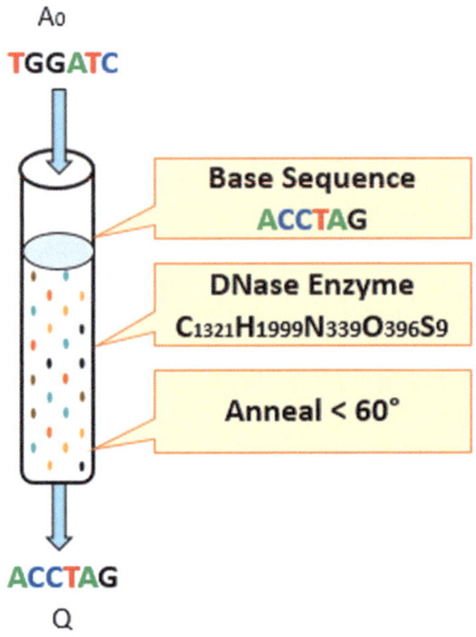

Table 14.14 Inputs and outputs of DNA NOT operation

A0	Q
TGGATC	ACCTAG
ACCTAG	TGGATC

DNA-Based Implementation of the NOT operation is shown in Fig. 14.31. Only one input is supplied, and the output is the corresponding complementary sequence. Firstly, need to choose the base mixture carefully. If the base mixture contains the sequence TGGATC it will work as a buffer. If the base mixture would have the sequence ACCTAG it will work as NOT operation. Table 14.14 shows the inputs and outputs of DNA NOT operation. It is seen previously where the sequence ACCTAG represents a "true" input and the sequence TGGATC represents a "false" input. The NOT gate, often referred to as an inverter, is one of the simplest DNA-based operations. If the input sequence is "false," then TGGATC will bind with the provided ACCTAG sequence to form a double-stranded sequence. The DNase doesn't affect the sequences, and the double-stranded sequence will be observed, which gives a "true" evaluation. Conversely, if the input sequence is "true", then ACCTAG will not bind with the provided ACCTAG sequence. The DNase will destroy both sequences, and no double-stranded sequences will be observed, which gives a "false" evaluation.

2. **DNA Circuit for OR Operation**

   First, it needs to choose the base mixture carefully. Then check each combination of input with the base mixture to match the DNA OR output. If the base mixture would have the sequence ACCTAG it would work as a DNA NAND operation. If

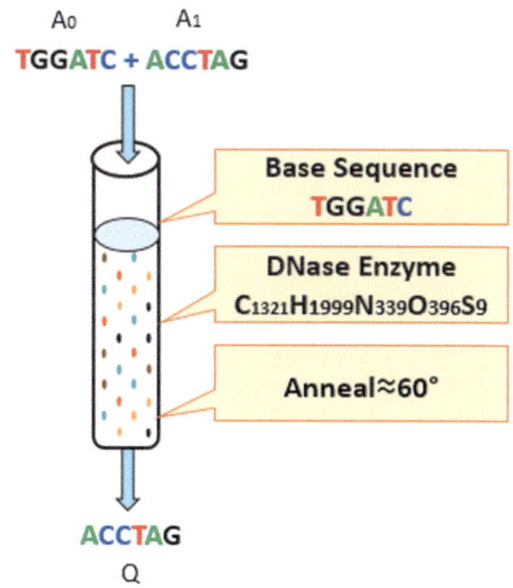

Fig. 14.32 DNA-based implementation of the OR operation

the base mixture contains the sequence TGGATC, for each combination of input sequences it works as DNA OR Operation. The DNA-based implementation of the OR operation is shown in Fig. 14.32 and Table 14.15 shows the inputs and outputs of DNA OR operation.

Consider the example above where the sequence ACCTAG represents a "true" input and the sequence TGGATC represents a "false" input. The OR gate evaluates "true" if one or both of the gate inputs are "true." If a double-stranded sequence is observed, then the result is "true"; otherwise, the result is "false." If both of the input sequences are "true" ACCTAG sequences, then one sequence will combine with the supplied "false" TGGATC sequence to produce a double stranded sequence. The DNase will destroy the remaining input sequences and the double stranded sequence will generate a "true" evaluation. If one input sequence is "false" and the other is "true," then the "true" ACCTAG sequence will combine with either of the "false" TGGATC sequences to produce a double stranded sequence. The DNase will destroy the remaining "false" sequences and the gate will result a "true" evaluation. If both input sequences are "false" TGGATC sequences, then none of the sequences will combine with the supplied "false" sequence. The DNase will destroy all sequences in the mixture, and it gives a "false" evaluation of the gate.

3. **DNA Circuit for NOR Operation**

NOR gate is the combination of OR and NOT gate. The output of the OR gate will go through the NOT gate to generate NOR gate output. At first, it needs to choose an OR gate (Fig. 14.32) and then choose a NOT gate (Fig. 14.31). Finally, it is necessary to check each combination of input with the base mixture of OR gate and the output of the OR gate check with the base mixture of NOT gate to match the NOR gate output. The DNA-based implementation of the NOR operation is

Table 14.15 Inputs and outputs of DNA OR operation

A0	A1	Q
TGGATC	TGGATC	TGGATC
TGGATC	ACCTAG	ACCTAG
ACCTAG	TGGATC	ACCTAG
ACCTAG	ACCTAG	ACCTAG

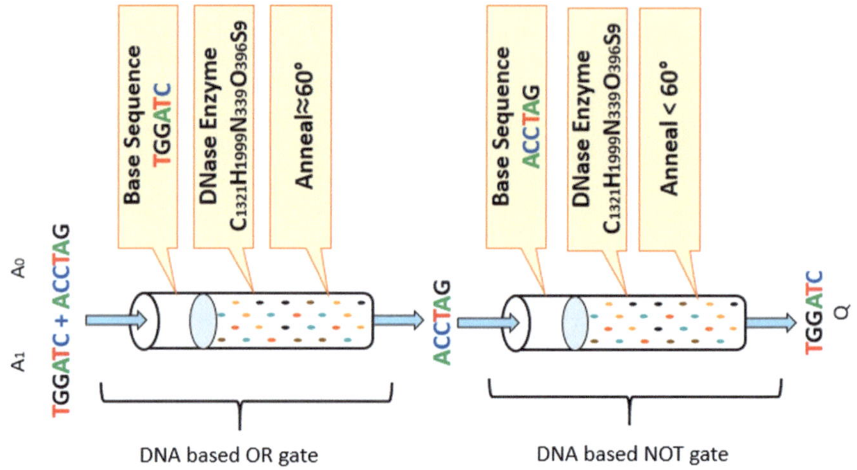

Fig. 14.33 DNA-based implementation of the NOR operation

Table 14.16 Inputs and outputs of DNA NOR operation

A0	A1	Q
TGGATC	TGGATC	ACCTAG
TGGATC	ACCTAG	TGGATC
ACCTAG	TGGATC	TGGATC
ACCTAG	ACCTAG	TGGATC

shown in Fig. 14.33 and Table 14.16 shows the inputs and outputs of DNA NOR operation.

The NOR gate, which evaluates "true" only when both inputs are "false", which is created by applying the NOT gate to the output of the OR gates. Continuing with the example above, if both of the input sequences are "false" TGGATC sequences, then one will combine with the supplied "false" TGGATC sequence in OR gate to produce a double-stranded molecule. The DNase will destroy the remaining input sequences and the double-stranded sequence will generate a "false" TGGATC sequence. This "false" sequence will go through NOT gate and result in a "true" evaluation. If one input sequence is "false" and the other is "true", then the "false" TGGATC input sequence will combine with either of the "true" ACCTAG sequences in OR gate to produce the necessary double-stranded sequence.

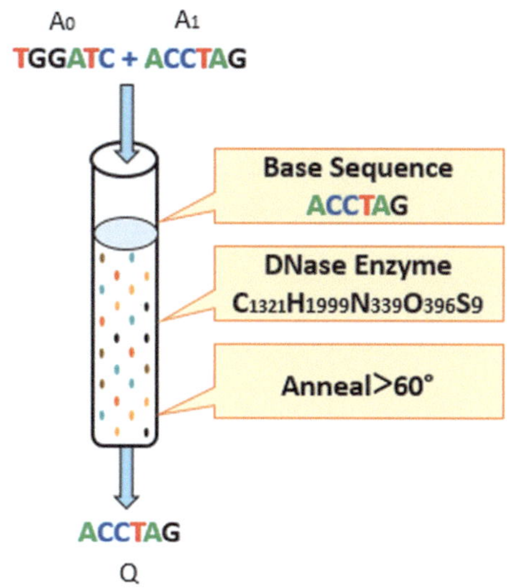

Fig. 14.34 DNA-based implementation of the NAND gate

Table 14.17 Inputs and outputs of DNA NAND operation

A0	A1	Q
TGGATC	TGGATC	ACCTAG
TGGATC	ACCTAG	ACCTAG
ACCTAG	TGGATC	ACCTAG
ACCTAG	ACCTAG	TGGATC

DNAse will then destroy the remaining "false" sequences and the double-stranded sequence will generate a "true" ACCTAG sequence. This "true" sequence will go through NOT gate and it generates a "false" evaluation. Finally, if both input sequences are "true" ACCTAG sequences, then neither will combine with the supplied "false" sequence in OR gate. DNase will destroy all sequences in the mixture, where the result is in a "true" sequence. This "true" sequence will go through NOT gate and it generates a "false" evaluation.

4. **DNA Circuit for NAND Operation**

Firstly, it needs to choose the base mixture carefully. Then check each combination of input base mixture to match the NAND gate output. If the base mixture contains the sequence TGGATC, then it will work as OR gate. If the base mixture would have the sequence ACCTAG, for each combination of input sequences it works as a NAND gate operation. The DNA-based implementation of the NAND gate operation is shown in Fig. 14.34, and Table 14.17 shows the inputs and outputs of DNA NAND operation.

The NAND gate evaluates "true" if inputs are not both "true." Thus, introducing the "true" sequence in the base mixture will require at least one of the inputs be "false" to form a double-stranded sequence. The DNase will destroy any single-stranded sequence in the mixture. Continuing with the example above, if

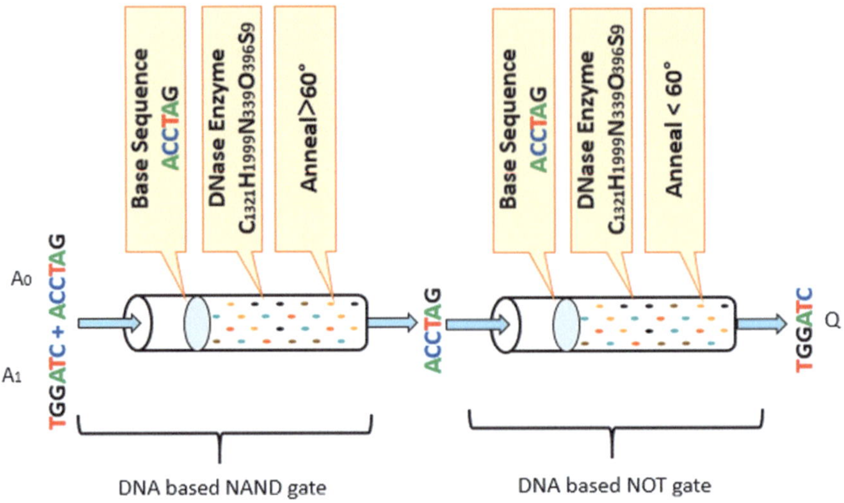

Fig. 14.35 DNA-based AND operation

both of the input sequences are "false" TGGATC sequences, then one will combine with the supplied "true" ACCTAG sequence to produce a double-stranded molecule. The DNase will destroy the remaining input sequences and the double-stranded sequence will result in a "true" evaluation. If one input sequence is "false" and the other is "true", then the "false" TGGATC input sequence will combine with either of the "true" ACCTAG sequences to produce the necessary double-stranded sequence. The DNase will then destroy the remaining "false" sequence and the gate will result in a "true" evaluation. Finally, if both input sequences are "true" ACCTAG sequences, where no one will combine with the supplied "true" sequence. The DNase will destroy all sequences in the mixture, and result in a "false" evaluation.

5. **DNA Circuit for AND Operation**
   AND gate is the combination of NAND and NOT gates. The output of the NAND gate will go through the NOT gate to generate AND gate output. At first, it needs to choose a NAND gate (Fig. 14.34) and then choose a NOT gate (Fig. 14.31). Finally, check each combination of input with the base mixture of NAND gate and the output of the NAND gate check with the base mixture of NOT gate to match the AND gate output. The DNA-based implementation of the AND operation is shown in Fig. 14.35, and Table 14.18 shows the inputs and outputs of DNA AND operation.
   The AND gate, which evaluates "true" only when both inputs are "true", and it is created by applying the NOT gate to the output of the NAND gate. Continuing with the example above, if both of the input sequences are "false" TGGATC sequences, then one will combine with the supplied "true" ACCTAG sequence in NAND gate to produce a double stranded molecule. The DNase will destroy

Table 14.18 Inputs and outputs of DNA AND operation

A0	A1	Q
TGGATC	TGGATC	TGGATC
TGGATC	ACCTAG	TGGATC
ACCTAG	TGGATC	TGGATC
ACCTAG	ACCTAG	ACCTAG

the remaining input sequence, and the double stranded sequence will generate a "true" ACCTAG sequence. This "true" sequence will go through NOT gate and result in a "false" evaluation. If one input sequence is "false" and the other is "true", then the "false" TGGATC input sequence will combine with either of the "true" ACCTAG sequences in NAND gate to produce the necessary double-stranded sequence. The DNase will then destroy the remaining "false" sequences and the double-stranded sequence will generate a "true" ACCTAG sequence. This "true" sequence will go through NOT gate and result in a "false" evaluation. Finally, if both input sequences are "true" ACCTAG sequences, then no sequence will combine with the supplied "true" sequence in NAND gate. The DNase will destroy all sequences in the mixture, which results in a "false" sequence. This "false" sequence will go through NOT gate and result in a "true" evaluation.

6. **DNA Circuit for XOR Operation**
   For XOR gate design, there is no need for any base mixture as none of the sequences produce the XOR output. Actually, it needs to check the input sequence. For sequences to have opposite values, they are complementary and will bind together to form a double-stranded sequence. For each combination of input sequences, the output will be matched. The DNA-based implementation of the

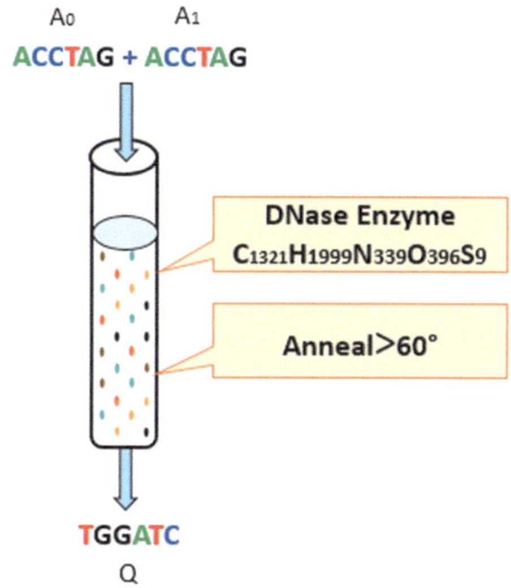

Fig. 14.36 DNA-based implementation of the XOR operation

Table 14.19 Inputs and outputs of DNA XOR operation

A0	A1	Q
TGGATC	TGGATC	TGGATC
TGGATC	ACCTAG	ACCTAG
ACCTAG	TGGATC	ACCTAG
ACCTAG	ACCTAG	TGGATC

XOR gate operation is shown in Fig. 14.36, and Table 14.19 shows the inputs and outputs of DNA XOR operation.

The XOR gate evaluates "true" only if exactly one of the input sequences evaluates "true." In DNA-based gates, the XOR gate is the most simplistic design in that no external sequences need to be supplied to the gate. For the sequences to have opposite values TGGATC and ACCTAG, they are complementary, and they will bind together to form a double-stranded sequence and the output is "true". If the inputs are not complementary (TGGATC, TGGATC or ACCTAG, ACCTAG), sequences will not bind to one another and the DNase will destroy both input sequences when the output is "false".

7. **DNA Circuit for XNOR Operation**

   XNOR gate is the combination of XOR and NOT gates. The output of the XOR gate will go through the NOT gate to generate XNOR gate output. Firstly, it needs to choose a XOR gate and then choose a NOT gate. Finally, check the input sequence of XOR. For the sequences to have opposite values, they are complementary which will bind together to form a double-stranded sequence. Finally, the output of the XOR gate checks with the base mixture of NOT gate to match the XNOR gate output. The DNA-based implementation of the XNOR gate operation is shown in Fig. 14.37, Table 14.20 shows the inputs and outputs of DNA XNOR operation.

   The XNOR gate, which evaluates "false" only when any inputs are "false", which is created by applying the NOT gate to the output of the XOR gate. The XOR gate evaluates "true" only if exactly one of the input sequences evaluates "true." In DNA-based logic gates, the XOR gate is the most simplistic design where no external sequences need to be supplied to the gate. For sequences to have opposite values TGGATC and ACCTAG, they are complementary, which will bind together to form a double-stranded sequence and the output is "true". This "true" sequence will go through DNA NOT gate and result in a "false" evaluation. If the inputs are not complementary (TGGATC, TGGATC or ACCTAG, ACCTAG), the sequences will not bind to one another and the DNase will destroy both input sequences and the output is "false". This "false" sequence will go through DNA NOT gate and result a "true" evaluation.

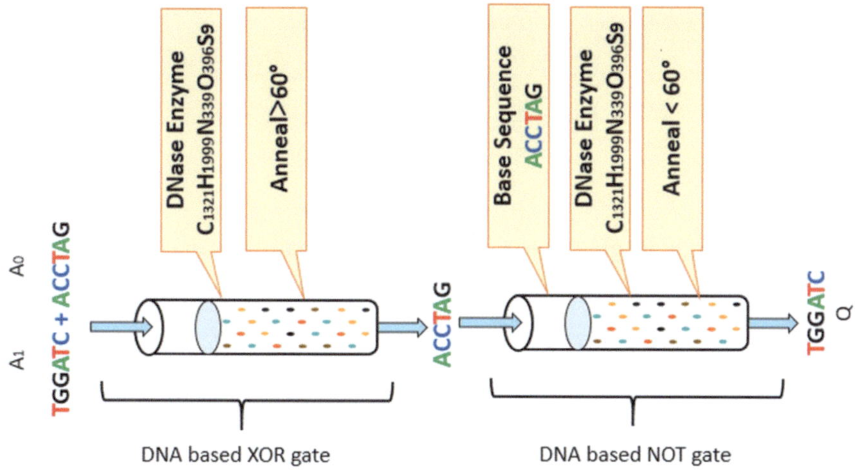

Fig. 14.37 DNA-based implementation of the XNOR operation

Table 14.20 Inputs and outputs of DNA XNOR operation

A0	A1	Q
TGGATC	TGGATC	ACCTAG
TGGATC	ACCTAG	TGGATC
ACCTAG	TGGATC	TGGATC
ACCTAG	ACCTAG	ACCTAG

14.3.4.2 DNA-Quantum Operations Using NMR at Room Temperature

DNA-Quantum computing merges all the advantages of individual DNA computing and Quantum computing. The DNA-based computing is mostly chosen for its simultaneous operation capability. On the other hand, quantum computing is an emergent process of speedy computing which is more capable than a digital computer. In addition, Quantum computing can perform calculations in a few seconds for which today's supercomputer would take ages or decades. DNA computing can provide huge memory in small spaces. So, the large memory space and fast computation can be possible by using the DNA-Quantum computing process. The base of DNA-Quantum computing is nothing, but it is DNA and Quantum operational gates. In this section, different types of DNA-quantum operational gates will be described.

1. **DNA-Quantum NOT Operation**
   The DNA NOT operation is prepared to produce the output as a complementary sequence of input. To produce the output, it needs a single input that contains single strands of DNA sequence. Also it needs a base sequence and anneal temperature for the DNA operation. After getting the output DNA sequence from DNA NOT operation, it needs to go through the NMR process. In NMR, a room temperature probe has been used. By performing the NMR operation on the DNA sequence,

Table 14.21 Inputs and outputs of DNA-quantum NOT operation

Input	Output	Quantum qubit, $\lvert X\rangle$
TGGATC	ACCTAG	$\lvert 1\rangle$
ACCTAG	TGGATC	$\lvert 0\rangle$

the qubit of normal ground state turns into a superposition state or ground state according to the input sequence in NMR. The output of the NMR process will be qubit.

The DNA-Quantum NOT operation is the inverter operation, which is one of the simplest DNA-Quantum operations. If the input DNA sequence is "false", the TGGATC will bind with the provided ACCTAG sequence to form a double-stranded sequence. The DNase will not affect the sequences, and the double-stranded sequence will be observed, representing a "true" evaluation. Conversely, if the input sequence is "true", then ACCTAG will not bind with the provided ACCTAG sequence, where the DNase will destroy both sequences, and no double-stranded sequences will be observed, representing a "false" evaluation.

When getting the DNA sequence from DNA NOT operation, this sequence will go through NMR as a sample and dive into the probe. Using the NMR process at room temperature and by emitting EMR, this biomolecule will get a superposition state and produce a qubit.

Here, the DNA sequence ACCTAG = TRUE for qubit $|1\rangle$ and the DNA sequence TGGATC = FALSE for qubit $|0\rangle$. The outputs for different combinations of inputs are given in Table 14.21.

Figure 14.38 shows the process of converting DNA sequences to qubits of DNA-Quantum NOT operation.

2. **DNA-Quantum OR Operation**

   To produce the output, it needs two inputs which has single strands of DNA sequence. Also, it needs a base sequence and anneal temperature for the DNA operation. To produce the expected output for OR operation the base mixture should be chosen carefully. Then, it is better to check each combination of input with base mixture to match the DNA OR gate output. If the base mixture would have the sequence ACCTAG then it works as a DNA NAND gate. If the base mixture has the sequence TGGATC, then for each combination of input sequences it works as OR gate operation.

   After getting the output DNA sequence from DNA OR operation, it needs to go through the NMR process. In NMR, a room temperature probe is used. By doing NMR for the DNA sequence, the qubit of normal ground state turns into superposition state or ground state according to the input sequence in NMR. The output of the NMR process is a qubit. Figure 14.39 shows the process of converting DNA sequences to qubits of DNA-Quantum OR operation.

   The DNA-quantum OR operation is the most important DNA-quantum operation. Let us assume two sequences ACCTAG to represents a "true" input and the

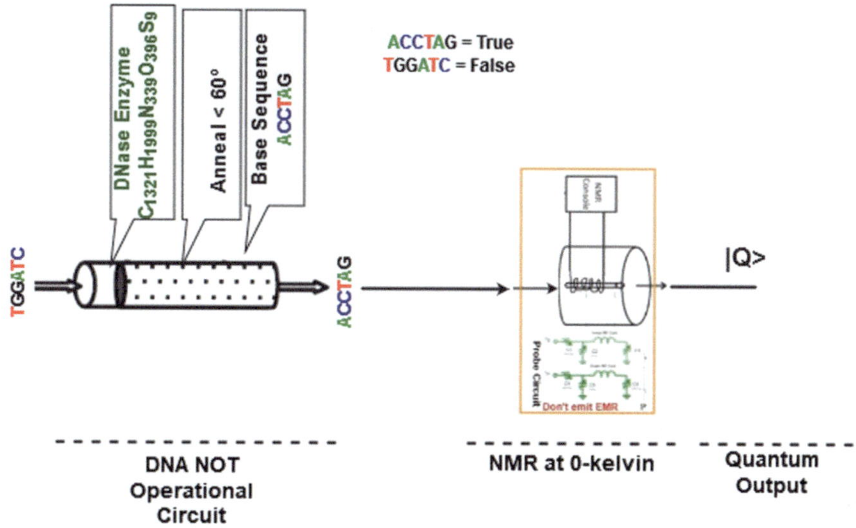

Fig. 14.38 DNA-quantum NOT operation

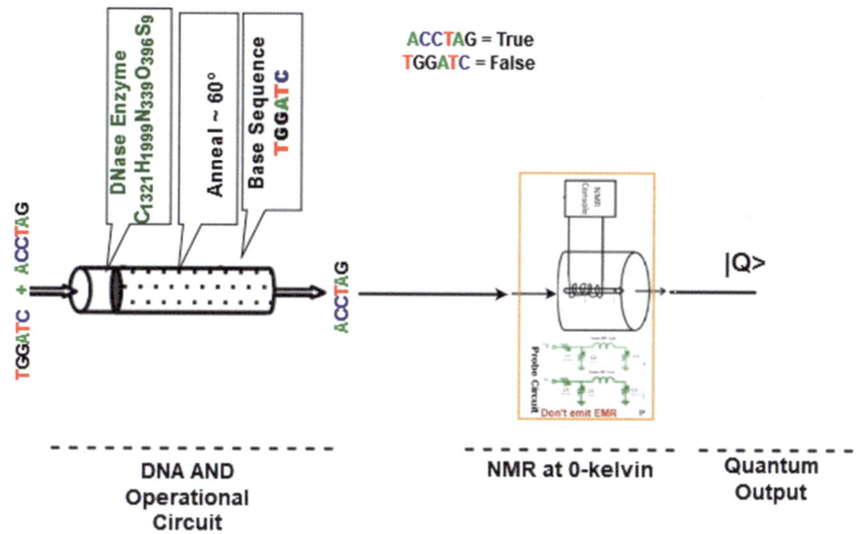

Fig. 14.39 DNA-quantum OR operation

sequence TGGATC represents a "false" input. The gate evaluates "true" if one or both of the gate inputs are "true." If a double-stranded sequence is observed, then the result is "true"; otherwise, the result is "false." If both of the input sequences are "true" ACCTAG sequences, then one sequence will combine with the supplied "false" TGGATC sequence to produce a double-stranded sequence. The DNase

Table 14.22 Inputs and outputs of DNA-quantum OR operation

$Input_1$	$Input_2$	Output	Quantum qubit, |X>
TGGATC	TGGATC	TGGATC	|0>
TGGATC	ACCTAG	ACCTAG	|1>
ACCTAG	TGGATC	ACCTAG	|1>
ACCTAG	ACCTAG	ACCTAG	|1>

will destroy the remaining input sequences and the double-stranded sequence will result in a "true" evaluation. If one input sequence is "false" and the other is "true," then the "true" ACCTAG sequence will combine with either of the "false" TGGATC sequences to produce a double-stranded sequence. The DNase will destroy the remaining "false" sequences and the gate will generate a "true" evaluation. If both input sequences are "false" TGGATC sequences, then combined with the supplied "false" sequence. The DNase will destroy all sequences in the mixture, and it produces a "false" evaluation of the gate.

When getting the DNA sequence from DNA OR operation, this sequence will go through NMR as a sample and dive into the probe. Using the NMR process at room temperature and by emitting EMR, this biomolecule will get a superposition state and produce a qubit.

Here, DNA sequence ACCTAG = TRUE for qubit |1 > and DNA sequence TGGATC = FALSE for qubit |0 >. The outputs for different combinations of input are given in Table 14.22.

3. **DNA-Quantum XOR Operation**

To produce output, it needs two inputs which has single strands of DNA sequence and no need for a base sequence. But it needs anneal temperature for the DNA XOR operation. It is required to check the input sequence. For sequences to have opposite values, they are complementary which will bind together to form a double-stranded sequence. For each combination of input sequences, the output needs to be checked.

After getting the output DNA sequence from DNA XOR operation, it is needed to provide it through the NMR process. In NMR, a room temperature probe is used. By performing NMR for the DNA sequence, the qubit of normal ground state turns into superposition state or ground state according to the input sequence in NMR. The output of the NMR operation is a qubit. Figure 14.40 shows the process of converting DNA sequences to qubits of DNA-Quantum XOR operation.

The DNA-Quantum XOR operation is another most important DNA-Quantum operation. Consider two sequences, where ACCTAG represents a "true" input and the sequences TGGATC represents a "false" input. The XOR gate evaluates "true" if both sequence of the gate inputs are complementary to each other. They make a double stranded DNA sequence. If a double stranded sequence is observed, then the result is "true"; otherwise, the result is "false." If both of the input sequences are different to each other ACCTAG and TGGATC sequences, then these sequences

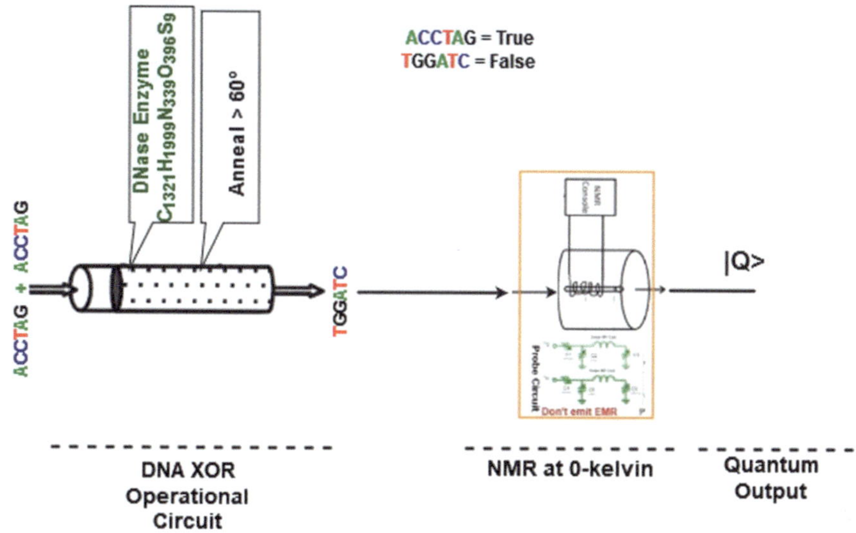

Fig. 14.40 DNA-quantum XOR operation

will combine and produce a double-stranded sequence. The DNase will destroy the remaining input sequence and the double-stranded sequence will result in a "true" evaluation. If both input sequences are "false" or "true," then the sequence will not combine with each other. The DNase will destroy all sequences in the mixture, and it gives a "false" evaluation of the operation.

When getting the DNA sequence from DNA XOR operation, this sequence will go through NMR as a sample and dive into the probe. Using the NMR process at room temperature and by emitting EMR, this biomolecule will get a superposition state and become a qubit.

Here, DNA sequence ACCTAG = TRUE for qubit $|1>$ and DNA sequence TGGATC = FALSE for qubit $|0>$. The outputs for different combinations of input are given in Table 14.23.

4. **DNA-Quantum AND Operation**

To produce the output, it needs two inputs which consist of single strands of DNA sequence. Also it needs a base sequence and anneal temperature for the DNA operation. To produce the expected output for AND operation, it is needed to choose the base mixture carefully. Then it is better to check each combination of input with the base mixture to match the AND gate output. If the base mixture has the sequence TGGATC it will work as DNA NOR gate. If the base mixture would have the sequence ACCTAG, for each combination of inputs sequences it works as DNA AND operation.

After getting the output DNA sequence from DNA AND operation, there is a need to provide it through the NMR process. In NMR, a room temperature probe is used. By doing NMR for the DNA sequence, the qubit of normal ground state

Table 14.23 Inputs and outputs of DNA-quantum XOR operation

$Input_1$	$Input_2$	Output	Quantum qubit, |Q>
TGGATC	TGGATC	TGGATC	|0>
TGGATC	ACCTAG	ACCTAG	|1>
ACCTAG	TGGATC	ACCTAG	|1>
ACCTAG	ACCTAG	TGGATC	|0>

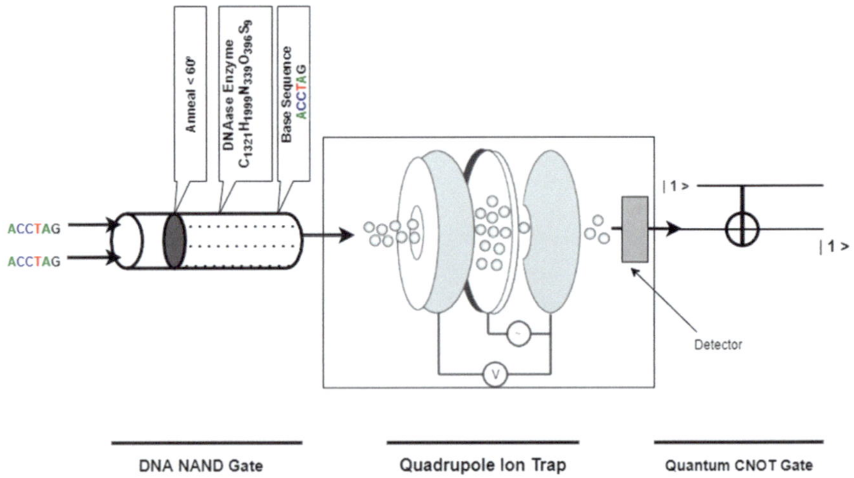

Fig. 14.41 DNA-quantum AND operation

turns into superposition state or ground state according to the input sequence in NMR (Fig. 14.41). The output of the NMR process is quantum qubit.

The DNA-Quantum AND operation is the most important logic operation in DNA-Quantum Computing. Consider the two sequences, where ACCTAG represents a “true” input and the sequence TGGATC represents a “false” input. The AND gate evaluates “true” if both of the gate inputs are “true.” If one of the sequences is “false” then the output will be “false”. The DNase will destroy the remaining input sequence and the double-stranded sequence will result a “true” evaluation. If one input sequence is “false” and the other is “true,” then the “true” ACCTAG sequence will combine with either of the “false” TGGATC sequences to produce a double-stranded sequence. The DNase will destroy the remaining “false” sequences and the gate will result a “true” evaluation. If both input sequences are “false” TGGATC sequences, then neither will combine with the supplied “false” sequence. The DNase will destroy all sequences in the mixture, which results a “false” evaluation of the gate.

When getting the DNA sequence from DNA AND operation, this sequence will go through NMR as a sample and dive into the probe. Using the NMR process at

Table 14.24 Input and output of DNA-quantum AND operation

$Input_1$	$Input_2$	Output	Quantum qubit, |A>
TGGATC	TGGATC	TGGATC	|0>
TGGATC	ACCTAG	TGGATC	|0>
ACCTAG	TGGATC	TGGATC	|0>
ACCTAG	ACCTAG	ACCTAG	|1>

room temperature and by emitting EMR, this biomolecule will get a superposition state and become a qubit.
Here, DNA sequence ACCTAG = TRUE for qubit |1 > and DNA sequence TGGATC = FALSE for qubit |0 >. The outputs for different combinations of inputs are given in Table 14.24.

5. **DNA-Quantum Full Adder at Room Temperature**
A full adder is an arithmetic circuit which adds three inputs and produces two outputs. The first two inputs are A and B and the third input is an input carry as C_{in}. The output carry is designated as C_{out} and the normal output is designated as S which is Sum. A Full Adder logic is designed in such a manner that can take eight inputs together to create a qubit adder and cascade the carry qubit from one adder to another. To create a Full Adder, one OR, two AND and two XOR operations are required. Figure 14.42 describes the DNA-Quantum circuit of the Full Adder.
To design a DNA-Quantum Full Adder, DNA and Quantum operations are used to operate the input DNA sequence for their corresponding output qubits. The DNA operations will be used for receiving the input sequence and the Quantum operations will be used to produce the final output against the corresponding set of inputs. Each time, the DNA-Quantum Full adder will receive three DNA sequences as input. After performing a certain number of DNA operations, the qubit will be turned into corresponding qubits by using NMR at room temperature probe. By using a room temperature probe and corresponding components of the NMR process, the stable or grounded state qubits get excited by using EMR to provide an output as a qubit. Then, the qubits are processed through quantum operations; and outputs are generated.
Here, Fig. 14.42 describes DNA-Quantum Full adder using DNA and Quantum operations.
From the Fig. 14.42, it is obvious that the DNA- Quantum Full Adder consists of three DNA operations and two Quantum operations. Here, two AND and one XOR are used as DNA operations and further one XOR and one OR Quantum operations are used.
The working procedure of the DNA-Quantum Full adder is given below for each pattern of the input DNA sequence. Here, DNA sequence ACCTAG = TRUE for qubit |1 > and DNA sequence TGGATC = FALSE for qubit |0 >.

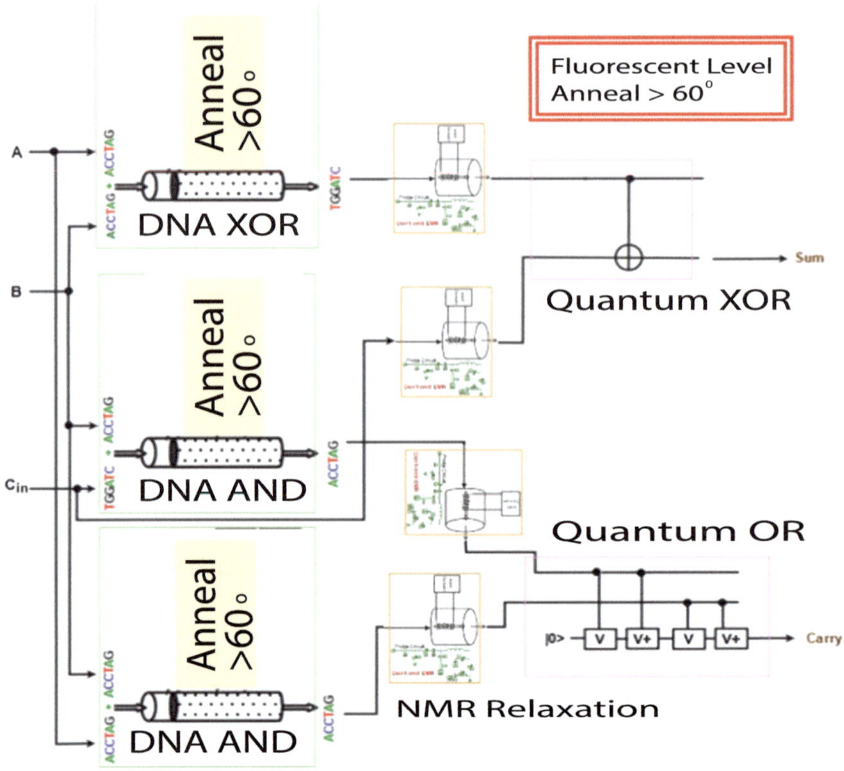

Fig. 14.42 DNA-quantum full adder at room temperature

Table 14.25 Outputs of DNA-quantum full adder operation

A	B	C_{in}	SUM	Carry
TGGATC	TGGATC	TGGATC	|0>	|0>
TGGATC	TGGATC	ACCTAG	|1>	|0>
TGGATC	ACCTAG	TGGATC	|1>	|0>
TGGATC	ACCTAG	ACCTAG	|0>	|1>
ACCTAG	TGGATC	TGGATC	|1>	|0>
ACCTAG	TGGATC	ACCTAG	|0>	|1>
ACCTAG	ACCTAG	TGGATC	|0>	|1>
ACCTAG	ACCTAG	ACCTAG	|1>	|1>

The working procedure of the DNA-quantum Full Adder is given below for 4 out of 8 patterns of input DNA sequences. In addition, all inputs and outputs combinations are tabulated in Table 14.25.

(a) For inputs A, B, C_{in} = TGGATC, **Sum** = Quantum XOR (DNA XOR (A, B), C)
= Quantum XOR (DNA XOR (TGGATC, TGGATC), TGGATC)
= Quantum XOR (TGGATC, TGGATC)
= Quantum XOR (0, 0) [Using NMR]
= 0
Carry = Quantum OR (DNA AND (C, XOR (A, B)), DNA AND (A,B))
= Quantum OR (DNA AND (TGGATC, XOR (TGGATC, TGGATC)), DNA AND (TGGATC, TGGATC))
= Quantum OR (DNA AND (TGGATC, XOR (TGGATC, TGGATC)), DNA AND (TGGATC, TGGATC))
= Quantum OR (TGGATC, TGGATC)
= Quantum OR (0, 0) [Using NMR]
= 0

(b) For inputs A, B, C_{in} = TGGATC, TGGATC, ACCTAG **Sum** = Quantum XOR (DNA XOR (A, B), C)
= Quantum XOR (DNA XOR (TGGATC, TGGATC), ACCTAG)
= Quantum XOR (TGGATC, ACCTAG)
= Quantum XOR (0, 1) [Using NMR]
= 1
Carry = Quantum OR (DNA AND (C, XOR (A, B)), DNA AND (A, B))
= Quantum OR (DNA AND (ACCTAG, XOR (TGGATC, TGGATC)), DNA AND (TGGATC, TGGATC))
= Quantum OR (DNA AND (ACCTAG, TGGATC), TGGATC)
= Quantum OR (TGGATC, TGGATC)
= Quantum OR (0, 0) [Using NMR]
= 0

(c) For inputs A, B, C_{in} = TGGATC, ACCTAG, TGGATC **Sum** = Quantum XOR (DNA XOR (A, B), C)
= Quantum XOR (DNA XOR (TGGATC, ACCTAG), TGGATC)
= Quantum XOR (ACCTAG, TGGATC)
= Quantum XOR (1, 0) [Using NMR]
= 1
Carry = Quantum OR (DNA AND (C, XOR (A, B)), DNA AND (A, B))
= Quantum OR (DNA AND (TGGATC, XOR (TGGATC, ACCTAG)), DNA AND (TGGATC, ACCTAG))
= Quantum OR (DNA AND (TGGATC, ACCTAG), TGGATC)
= Quantum OR (TGGATC, TGGATC)
= Quantum OR (0, 0) [Using NMR]
= 0
For inputs A, B, C_{in} = TGGATC, ACCTAG, ACCTAG

Sum = Quantum XOR (DNA XOR (A, B), C)
= Quantum XOR (DNA XOR (TGGATC, ACCTAG), ACCTAG)
= Quantum XOR (ACCTAG, ACCTAG)
= Quantum XOR (1, 1) [Using NMR]
= 0
Carry = Quantum OR (DNA AND (C, XOR (A, B)), DNA AND (A, B))
= Quantum OR (DNA AND (ACCTAG, XOR (TGGATC, ACCTAG)), DNA AND (TGGATC, ACCTAG))
= Quantum OR (DNA AND (ACCTAG, ACCTAG), TGGATC)
= Quantum OR (ACCTAG, TGGATC)
= Quantum OR (1, 0) [Using NMR]
= 1

6. **DNA-Quantum Full Subtractor at Room Temperature**
A full subtractor is a combinational circuit that performs subtraction of two qubits, one is minuend and the other is subtrahend, taking into account the borrow of the previous adjacent lower minuend qubit. This circuit has three inputs and two outputs. The three inputs A, B, and B_{in}, denote the minuend, subtrahend, and previous borrow respectively. The two outputs, D and B_{out} represent the difference and output borrows, respectively. To create a full subtractor, one OR, two AND, two OR, and two XOR gates are required.
To design a DNA-Quantum Full Subtractor, DNA and Quantum operations are used to operate the input qubit for their corresponding outputs. The DNA operation will be used for receiving the input DNA sequence and the Quantum operations will be used to produce the final output against the corresponding set of inputs. Each time, the DNA-Quantum Full Subtractor will receive three DNA sequences as input. After performing a certain number of DNA operations, the DNA sequence will be turned into corresponding qubits by using NMR at a room temperature probe. The process of producing a qubit against a DNA sequence is explained before. By using a room temperature probe and corresponding components of NMR, the ground qubit turns on/off excited state and produces a qubit for emitting EMR. Then the qubits are processed through Quantum operation and outputs are received. Here, Fig. 14.43 describes DNA-Quantum Full Subtractor using DNA and Quantum operation.
From Fig. 14.43, it is seen that the DNA-Quantum Full Subtractor consists of three DNA operations and two Quantum operations. Here, two AND and one XOR are used as DNA operation and further one XOR and one OR Quantum operation are used.
The working procedure of the DNA-Quantum Full Subtractor is given below for each pattern of input qubits. Here, DNA sequence ACCTAG = TRUE for qubit |1 > and DNA sequence TGGATC = FALSE for qubit |0 >.
The working procedure of the DNA-Quantum Full Subtractor is given below for 4 out of 8 patterns of input DNA sequencs. In addition, inputs and outputs of different combinations are tabulated in Table 14.26.

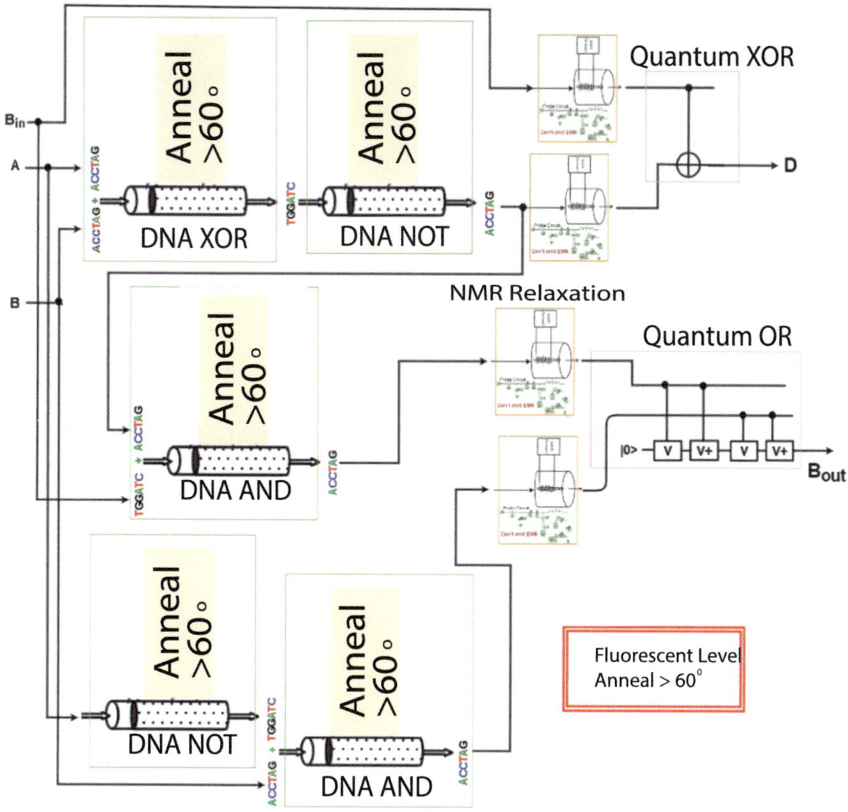

Fig. 14.43 DNA-quantum full subtractor at room temperature

Table 14.26 Outputs of DNA-quantum full subtractor operation

A	B	B_{in}	D	B_{out}
TGGATC	TGGATC	TGGATC	|0>	|0>
TGGATC	TGGATC	ACCTAG	|1>	|1>
TGGATC	ACCTAG	TGGATC	|1>	|1>
TGGATC	ACCTAG	ACCTAG	|0>	|1>
ACCTAG	TGGATC	TGGATC	|1>	|0>
ACCTAG	TGGATC	ACCTAG	|0>	|0>
ACCTAG	ACCTAG	TGGATC	|0>	|0>
ACCTAG	ACCTAG	ACCTAG	|1>	|1>

(a) For inputs A, B, B_{in} = TGGATC, **D** = Quantum XOR (DNA XOR (A, B), B_{in})
= Quantum XOR (DNA XOR (TGGATC, TGGATC), TGGATC)
= Quantum XOR (TGGATC, TGGATC)
= Quantum XOR (0, 0) [Using NMR]
= 0
$\mathbf{B}_{out}$= Quantum OR (DNA AND (B_{in}, NOT (XOR (A, B))), DNA AND (NOT (A), B))
= Quantum OR (DNA AND (TGGATC, ACCTAG), DNA AND (ACCTAG, TGGATC))
= Quantum OR (TGGATC, TGGATC)
= Quantum OR (0, 0) [Using NMR]
= 0
(b) For inputs A, B, B_{in} = TGGATC, TGGATC, ACCTAG **D** = Quantum XOR (DNA XOR (A, B), B_{in})
= Quantum XOR (DNA XOR (TGGATC, TGGATC), ACCTAG)
= Quantum XOR (TGGATC, ACCTAG)
= Quantum XOR (0, 1) [Using NMR]
= 1
$\mathbf{B}_{out}$ = Quantum OR (DNA AND (B_{in}, NOT (XOR (A, B))), DNA AND (NOT (A), B))
= Quantum OR (DNA AND (ACCTAG, ACCTAG), DNA AND (ACCTAG, TGGATC))
= Quantum OR (ACCTAG, 0)
= Quantum OR (1, 0) [Using NMR]
= 1
(c) For inputs A, B, B_{in} = TGGATC, ACCTAG, TGGATC **D** = Quantum XOR (DNA XOR (A, B), B_{in})
= Quantum XOR (DNA XOR (0, ACCTAG), 0)
= Quantum XOR (ACCTAG, 0)
= Quantum XOR (1, 0) [Using NMR]
= 1
$\mathbf{B}_{out}$ = Quantum OR (DNA AND (B_{in}, NOT (XOR (A, B))), DNA AND (NOT (A), B))
= Quantum OR (DNA AND (TGGATC, ACCTAG), DNA AND (ACCTAG, ACCTAG))
= Quantum OR (DNA AND (TGGATC, ACCTAG), ACCTAG)
= Quantum OR (TGGATC, ACCTAG)
= Quantum OR (0, 1) [Using NMR]
= 1
(d) For inputs A, B, B_{in} = TGGATC, ACCTAG, ACCTAG **D** = Quantum XOR (DNA XOR (A, B), B_{in})
= Quantum XOR (DNA XOR (TGGATC, ACCTAG), ACCTAG)
= Quantum XOR (ACCTAG, ACCTAG)
= Quantum XOR (1, 1) [Using NMR]

= 0
B_{out} = Quantum OR (DNA AND (B_{in}, NOT (XOR (A, B))), DNA AND (NOT (A), B))
= Quantum OR (DNA AND (ACCTAG, TGGATC), DNA AND (ACCTAG, ACCTAG))
= Quantum OR (DNA AND (ACCTAG, TGGATC), ACCTAG)
= Quantum OR (TGGATC, ACCTAG)
= Quantum OR (1, 0) [Using NMR]
= 1

7. **DNA-Quantum 2-to-1 Multiplexer at Room Temperature**
A multiplexer (MUX) is a device that can receive multiple input signals and synthesize a single output signal in a recoverable manner for each input signal. It is also an integrated system that usually has a certain number of data inputs and a single output. To create a Multiplexer, one NOT, two AND, and an OR gates are required. Figure 14.44 describes the DNA-Quantum circuit of the Multiplexer. A Multiplexer receives three inputs and produces one output having "Y".
To design a DNA-Quantum 2-to-1 Multiplexer, it is required to use DNA and Quantum operational gates so that it can operate the input DNA sequences for their corresponding outputs. The DNA operation will be used for receiving the input DNA sequence and the Quantum operation will be used to produce the final output against the corresponding set of inputs. Each time, the DNA-Quantum Multiplexer will receive three qubits as input. After performing a certain number of DNA operations, the sequence will be turned into corresponding qubits by using NMR at a room temperature probe. By using a room temperature probe and corresponding components of NMR, the ground qubit turns on/off excited state and produces a qubit for emitting EMR. Then the qubits are processed through Quantum operations and outputs are received.
From the Fig. 14.44, it is found that the Quantum-DNA 2-to-1 multiplexer consists of three DNA operations and one Quantum operation. Here, two AND and one NOT are used as DNA operations and further one OR Quantum operation is also used.
The working procedure of the Quantum-DNA 2-to-1 multiplexer is given below for each pattern of input DNA sequence. Here, DNA sequence ACCTAG = TRUE for qubit |1 > and DNA sequence TGGATC = FALSE for qubit |0 > are used.
The working procedure of the DNA-Quantum 2-to-1 multiplexer is given below for 4 out of 8 patterns of input DNA sequences. In addition, inputs and outputs of different combinations are tabulated in Table 14.27.

(a) For inputs S, D_1, D_0 = TGGATC, **Y** = Quantum OR (DNA AND (D_1, S), AND (NOT (S), D_0))
= Quantum OR (DNA AND (TGGATC, TGGATC), AND (NOT (TGGATC), TGGATC)
= Quantum OR (TGGATC, TGGATC)
= Quantum OR (0, 0) [Using NMR]
= 0

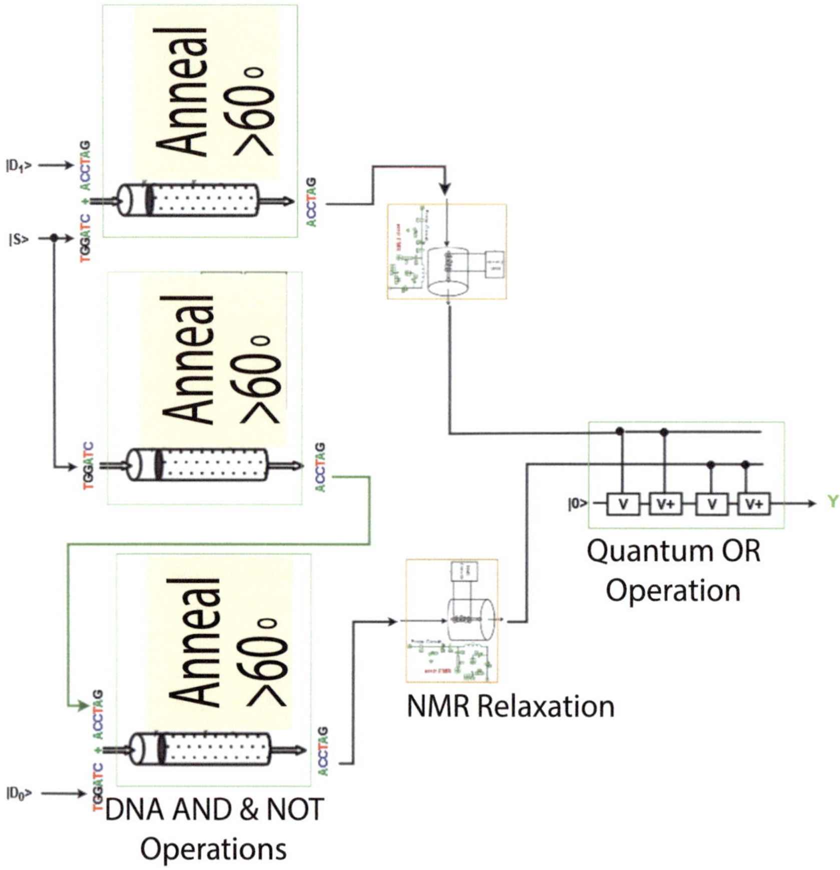

Fig. 14.44 DNA-quantum 2-to-1 multiplexer at room temperature

Here, Fig. 14.44 describes DNA-quantum multiplexer using DNA and quantum operational gates.

(b) For inputs S, D_0, D_1 = TGGATC, TGGATC, ACCTAG **Y** = quantum OR (DNA AND (D_1, S), AND (NOT (S), D_0))
= Quantum OR (DNA AND (ACCTAG, TGGATC), AND (NOT (TGGATC), TGGATC)
= Quantum OR (TGGATC, TGGATC)
= Quantum OR (0, 0) [Using NMR]
= 1

(c) For inputs S, D_0, D_1 = TGGATC, ACCTAG, TGGATC **Y** = Quantum OR (DNA AND (D_1, S), AND (NOT (S), D_0))
= Quantum OR (DNA AND (TGGATC, TGGATC), AND (NOT (TGGATC), ACCTAG))
= Quantum OR (TGGATC, AND (ACCTAG, ACCTAG)

Table 14.27 Outputs of a DNA-quantum 2-to-1 multiplexer operation

S	D1	D0	Y
TGGATC	TGGATC	TGGATC	|0>
TGGATC	TGGATC	ACCTAG	|0>
TGGATC	ACCTAG	TGGATC	|1>
TGGATC	ACCTAG	ACCTAG	|1>
ACCTAG	TGGATC	TGGATC	|0>
ACCTAG	TGGATC	ACCTAG	|1>
ACCTAG	ACCTAG	TGGATC	|0>
ACCTAG	ACCTAG	ACCTAG	|1>

= Quantum OR (TGGATC, ACCTAG)
= Quantum OR (0, 1) [Using NMR]
= 1

(d) For inputs S, D_0, D_1 = TGGATC, ACCTAG, ACCTAG **Y** = Quantum OR (DNA AND (D_1, S), AND (NOT (S), D_0))
= Quantum OR (DNA AND (ACCTAG, TGGATC), AND (NOT (TGGATC), ACCTAG))
= Quantum OR (TGGATC, AND (ACCTAG, ACCTAG))
= Quantum OR (TGGATC, ACCTAG)
= Quantum XOR (0, 1) [Using NMR]
= 1

8. **DNA-Quantum Multiplier at Room Temperature**
A quantum multiplier is a combinational logic circuit or quantum device used for multiplying two quantum numbers. The two numbers are more specifically known as multiplicand and multiplier and the result is known as a product. The multiplicand and multiplier can be of various qubit sizes. The product's qubit size depends on the qubit size of the multiplicand and multiplier. The qubit size of the product is equal to the sum of the qubit size of the multiplier multiplicand. To create a DNA-Quantum Multiplier circuit, four DNA AND operations, two XOR, and two AND Quantum operations are required. Figure 14.45 describes the DNA-Quantum circuit of the 2-bit multiplication.
To design a DNA-Quantum multiplier, Quantum and DNA operational gates are used to operate the input DNA sequence for their corresponding outputs. The DNA operation will be used for receiving the input DNA sequence and the Quantum operational gates will be used to produce the final output against the corresponding set of inputs. Each time, the DNA-Quantum multiplier will receive four qubits as input. After performing a certain number of DNA operations, the DNA sequence will be turned into corresponding qubits by NMR at room temperature probe. By using a room temperature probe and corresponding components of the NMR process, the DNA sequence will be used to produce a state of the quantum qubit. Then the quantum qubit is processed through Quantum operations and outputs are

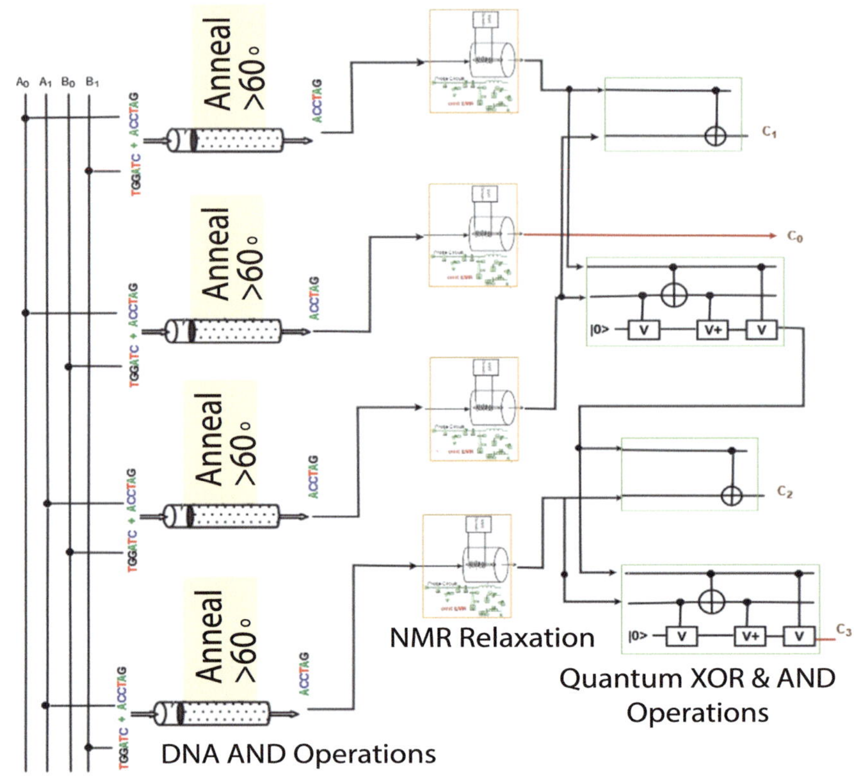

Fig. 14.45 DNA-Quantum multiplier operational circuit

received. Here, Fig. 14.45 describes DNA-Quantum Multiplier using DNA and Quantum operations.

From the DNA-Quantum circuit in Fig. 14.45, it is found that four (4) AND DNA circuits perform with DNA sequences. and it can be used as an input in all Quantum circuits. In Quantum circuit it uses two Quantum AND and two Quantum XOR operations.

The working procedure of the DNA-Quantum Multiplier is given below for each pattern of input DNA sequences. Here, DNA sequence ACCTAG = TRUE for qubit I1> and DNA sequence TGGATC = FALSE for qubit |0 >.

The working procedure of the DNA-Quantum Multiplier is given below for 4 patterns of input DNA sequences and the outputs for different combinations of inputs are given in Table 14.28.

(a) For inputs A_0, A_1, B_0, B_1 = TGGATC, $\mathbf{C}_0$ = Quantum (DNA AND (A_0, B_0))
= Quantum (DNA AND (TGGATC, TGGATC))
= Quantum (TGGATC)

Table 14.28 Outputs of DNA-quantum multiplier operation

A_0	A_1	B_0	B_1	$\|C_3>$	$\|C_2>$	$\|C_1>$	$\|C_0>$
TGGATC	TGGATC	TGGATC	TGGATC	$\|0>$	$\|0>$	$\|0>$	$\|0>$
TGGATC	TGGATC	TGGATC	ACCTAG	$\|0>$	$\|0>$	$\|0>$	$\|0>$
TGGATC	TGGATC	ACCTAG	TGGATC	$\|0>$	$\|0>$	$\|0>$	$\|0>$
TGGATC	TGGATC	ACCTAG	ACCTAG	$\|0>$	$\|0>$	$\|0>$	$\|0>$
TGGATC	ACCTAG	TGGATC	TGGATC	$\|0>$	$\|0>$	$\|0>$	$\|0>$
TGGATC	ACCTAG	TGGATC	ACCTAG	$\|0>$	$\|0>$	$\|0>$	$\|1>$
TGGATC	ACCTAG	ACCTAG	TGGATC	$\|0>$	$\|0>$	$\|1>$	$\|0>$
TGGATC	ACCTAG	ACCTAG	ACCTAG	$\|0>$	$\|0>$	$\|1>$	$\|1>$
ACCTAG	TGGATC	TGGATC	TGGATC	$\|0>$	$\|0>$	$\|0>$	$\|0>$
ACCTAG	TGGATC	TGGATC	ACCTAG	$\|0>$	$\|0>$	$\|1>$	$\|0>$
ACCTAG	TGGATC	ACCTAG	TGGATC	$\|0>$	$\|1>$	$\|0>$	$\|0>$
ACCTAG	TGGATC	ACCTAG	ACCTAG	$\|0>$	$\|1>$	$\|1>$	$\|0>$
ACCTAG	ACCTAG	TGGATC	TGGATC	$\|0>$	$\|0>$	$\|0>$	$\|0>$
ACCTAG	ACCTAG	TGGATC	ACCTAG	$\|0>$	$\|0>$	$\|1>$	$\|1>$
ACCTAG	ACCTAG	ACCTAG	TGGATC	$\|0>$	$\|1>$	$\|1>$	$\|0>$
ACCTAG	ACCTAG	ACCTAG	ACCTAG	$\|1>$	$\|0>$	$\|0>$	$\|1>$

= 0 [Using NMR]
C_2 = Quantum XOR (DNA AND (A_0, B_1), DNA AND (A_1, B_0))
= Quantum XOR (DNA AND (TGGATC, TGGATC), DNA AND (TGGATC, TGGATC))
= Quantum XOR (TGGATC, TGGATC)
= Quantum XOR (0, 0) [Using NMR]
= 0
C_2 = Quantum XOR (Quantum AND (DNA AND (A_0, B_1), DNA AND (A_1, B_0)), DNA AND (A_1, B_1))
= Quantum XOR (Quantum AND (DNA AND (TGGATC, TGGATC), DNA AND (TGGATC, TGGATC)), DNA AND (TGGATC, TGGATC))
= Quantum XOR (Quantum AND (TGGATC, TGGATC), TGGATC)
= Quantum XOR (Quantum AND (0, 0), 0) [Using NMR]
= Quantum XOR (0, 0)
= 0
C_3 = Quantum AND (Quantum AND (DNA AND (A_0, B_1), DNA AND (A_0, B_1)), DNA AND (A_1, B_1))
= Quantum AND (Quantum AND (DNA AND (TGGATC, TGGATC), DNA AND (TGGATC, TGGATC)), DNA AND (TGGATC, TGGATC))
= Quantum AND (Quantum AND (TGGATC, TGGATC), TGGATC)
= Quantum AND (Quantum AND (0, 0), 0) [Using NMR]
= Quantum AND (0, 0)
= 0

(b) For inputs A_0, A_1, B_0, B_1 = ACCTAG, ACCTAG, ACCTAG, TGGATC $\mathbf{C}_0$
= Quantum (DNA AND (A_0, B_0))
= Quantum (DNA AND (ACCTAG, ACCTAG))
= Quantum (ACCTAG)
= 1 [Using NMR]
$\mathbf{C}_1$ = Quantum XOR (DNA AND (A_0, B_1), DNA AND (A_1, B_0))
= Quantum XOR (DNA AND (ACCTAG, TGGATC), AND (ACCTAG, ACCTAG))
= Quantum XOR (TGGATC, ACCTAG)
= Quantum XOR (0, 1) [Using NMR]
= 1
$\mathbf{C}_2$ = Quantum XOR (Quantum AND (DNA AND (A_0, B_1), DNA AND (A_1, B_0)), DNA AND (A_1, B_1))
= Quantum XOR (Quantum AND (DNA AND (ACCTAG, TGGATC), DNA AND (ACCTAG, ACCTAG)), DNA AND (ACCTAG, TGGATC))
= Quantum XOR (Quantum AND (TGGATC, ACCTAG)TGGATC)
= Quantum XOR (Quantum AND (0, 1), 0) [Using NMR]
= Quantum XOR (0, 0)
= 0
$\mathbf{C}_3$ = Quantum AND (Quantum AND (DNA AND (A_0, B_1), DNA AND (A_0, B_1)), DNA AND (A_1, B_1))
= Quantum AND (Quantum AND (DNA AND (ACCTAG, TGGATC), DNA AND (ACCTAG, ACCTAG)), DNA AND (ACCTAG, TGGATC))
= Quantum AND (Quantum AND (TGGATC, ACCTAG)TGGATC)
= Quantum AND (Quantum AND (0, 1), 0) [Using NMR]
= Quantum AND (0, 1)
= 0

(c) For inputs A_0, A_1, B_0, B_1 = ACCTAG $\mathbf{C}_0$ = DNA (Quantum AND (A_0, B_0))
= DNA (Quantum AND (1, 1))
= DNA (ACCTAG)
= 1 [Using NMR]
$\mathbf{C}_1$ = Quantum XOR (DNA AND (A_0, B_1), DNA AND (A_1, B_0))
= Quantum XOR (DNA AND (ACCTAG, ACCTAG), AND (ACCTAG, ACCTAG))
= Quantum XOR (ACCTAG, ACCTAG)
= Quantum XOR (1, 1) [Using NMR]
= 0
$\mathbf{C}_2$ = Quantum XOR (Quantum AND (DNA AND (A_0, B_1), DNA AND (A_1, B_0)), DNA AND (A_1, B_1))
= Quantum XOR (Quantum AND (DNA AND (ACCTAG, ACCTAG), DNA AND (ACCTAG, ACCTAG)), DNA AND (ACCTAG, ACCTAG))
= Quantum XOR (Quantum AND (ACCTAG, ACCTAG), ACCTAG)
= Quantum XOR (Quantum AND (1, 1), 1) [Using NMR]
= Quantum XOR (1, 1)
= 0

$\mathbf{C_3}$ = Quantum AND (Quantum AND (DNA AND (A_0, B_1), DNA AND (A_1, B_0))DNA AND (A_1, B_1))
= Quantum AND (Quantum AND (DNA AND (ACCTAG, ACCTAG), DNA AND (ACCTAG, ACCTAG)), DNA AND (ACCTAG, ACCTAG))
= Quantum AND (Quantum AND (ACCTAG, ACCTAG)ACCTAG)
= Quantum AND (Quantum AND (1, 1), 1) [Using NMR]
= Quantum AND (1, 1)
= 1

(d) For inputs A_0, A_1, B_0, B_1 = ACCTAG, ACCTAG, TGGATC, ACCTAG $\mathbf{C_0}$
= Quantum (DNA AND (A_0, B_0))
= Quantum (DNA AND (ACCTAG, TGGATC))
= Quantum (TGGATC)
= 0 [Using NMR]
$\mathbf{C_1}$ = Quantum XOR (DNA AND (A_0, B_1), DNA AND (A_1, B_0))
= Quantum XOR (DNA AND (ACCTAG, ACCTAG), AND (ACCTAG, TGGATC))
= Quantum XOR (ACCTAG, TGGATC)
= Quantum XOR (1, 0) [Using NMR]
= 0
$\mathbf{C_2}$ = Quantum XOR (Quantum AND (DNA AND (A_0, B_1), DNA AND (A_1, B_0)), DNA AND (A_1, B_1))
= Quantum XOR (Quantum AND (DNA AND (ACCTAG, ACCTAG), DNA AND (ACCTAG, TGGATC)), DNA AND (ACCTAG, ACCTAG))
= Quantum XOR (Quantum AND (ACCTAG, TGGATC), ACCTAG)
= Quantum XOR (Quantum AND (1, 0), 1) [Using NMR]
= Quantum XOR (0, 1)
= 1
$\mathbf{C_3}$ = Quantum AND (Quantum AND (DNA AND (A_0, B_1), DNA AND (A_1, B_0)), DNA AND (A_1, B_1))
= Quantum AND (Quantum AND (DNA AND (ACCTAG, ACCTAG), DNA AND (ACCTAG, TGGATC)), DNA AND (ACCTAG, ACCTAG))
= Quantum AND (Quantum AND (ACCTAG, TGGATC), ACCTAG)
= Quantum AND (Quantum AND (1, 0), 1) [Using NMR]
= Quantum AND (0, 1)
= 0

14.3.5 NMR at 0 K Using Cryogenic Probe

The availability of state polarization decreases as the number of qubits rises, which is an evident challenge in building large NMR quantum computers. While the exact polarization achieved relies on experimental details, the population difference between the lowest and highest energy states can be used to approximate an upper limit.

The amount of polarization accessible diminishes exponentially as the size of the system grows. If the size of the NMR sample is kept constant, the signal strength will go off exponentially with the number of qubits; conversely, exponentially large samples would be required to maintain a constant signal size. Some other issues can occur while using a room temperature probe as decoherence and scaling problems.

For solving all mentioned problems, the cryogenic probe and pseudo liquid mixture are used. The cryogenic probe works at cool temperatures and temperature is proportional to noise. So, if the cryogenic probes are used, then the decoherence problem will be solved as per several studies. But if many qubits are used then a polarization problem will be faced. Different DNA-Quantum operations are described in this section at 0 K temperature using cryogenic probes.

14.3.5.1 DNA-Quantum NOT Operation at 0 K

DNA NOT operation is prepared to produce the output as a complementary sequence of the input. To produce output, it needs a single input that contains single strands of DNA sequence. it needs, need a base sequence and anneal temperature for the DNA operation. After getting the output DNA sequence from DNA NOT operation, it is needed to provide it through the NMR process. In NMR, a cryogenic probe at 0 K is used. By doing NMR for the DNA sequence, the qubit of normal ground state turns into superposition state or ground state according to the input sequence in NMR. The output of the NMR operation is a qubit.

The DNA-Quantum NOT operational gate is the inverter operation, which is one of the simplest DNA-Quantum operations. If the input DNA sequence is "false", TGGATC will bind with the provided ACCTAG sequence to form a double stranded sequence, where the DNase will have not affect the sequences, and the double stranded sequence will be observed, representing a "true" evaluation. Conversely, if the input sequence is "true", then ACCTAG will not bind with the provided ACCTAG sequence, where the DNase DNase will destroy both sequences, and no double stranded sequences will be observed, representing a "false" evaluation.

When getting the DNA sequence from DNA NOT operation, this sequence will go through NMR as a sample and dive into the probe. Using the NMR process using cryogenic probe at 0 K and by emitting EMR, this biomolecule will get a superposition state and become a qubit (|Q> in Fig. 14.46).

Here, DNA sequence ACCTAG = TRUE for Quantum qubit $|1>$ and DNA sequence TGGATC = FALSE for Quantum qubit $|0>$.

14.3.5.2 DNA-Quantum OR Operation at 0 K

Two inputs containing single strands of DNA sequence are required to generate the output. Also it needs a base sequence and anneal temperature for the DNA operation.

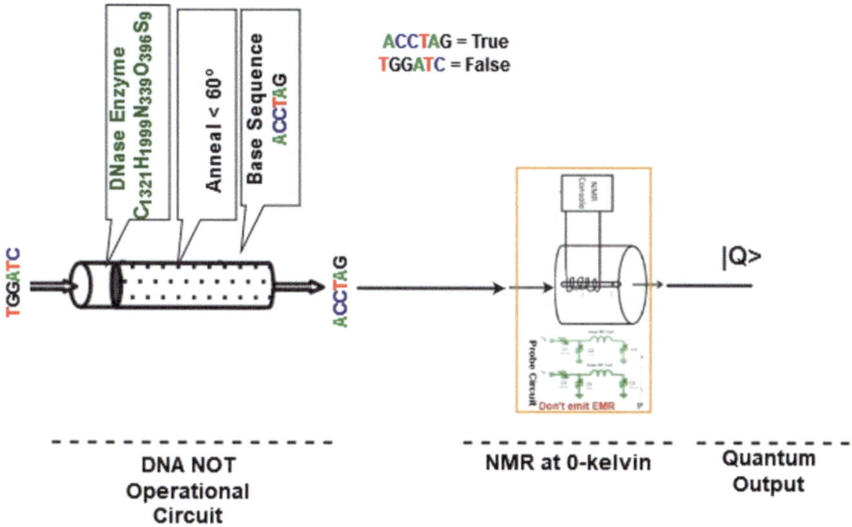

Fig. 14.46 DNA-quantum NOT operation at 0 K

To produce the expected output for OR operation, the base mixture is chosen carefully. Then check each combination of input with the base mixture to match the OR gate output. If the base mixture would have the sequence ACCTAG it would work as a NOR gate. If the base mixture has the sequence TGGATC, for each combination of input sequences it works as OR operation.

After getting the output DNA sequence from DNA OR operation, it needs to go through the NMR process. In NMR, a cryogenic probe at 0 K is used. By performing NMR for the DNA sequence, the qubit of normal ground state turns into superposition state or ground state according to the input sequence in NMR. The output of the NMR operation is a qubit. Figure 14.47 shows the DNA-Quantum OR operation at 0 K.

The DNA-Quantum OR operation is the most important DNA-Quantum operation. Consider two sequences, where sequence ACCTAG represents a "true" input and the sequence TGGATC represents a "false" input. The OR gate evaluates "true" if one or both of the gate inputs are "true." If a double-stranded sequence is observed, then the result is "true"; otherwise, the result is "false." If both of the input sequences are "true" ACCTAG sequences, then one sequence will combine with the supplied "false" TGGATC sequence to produce a double-stranded sequence. The DNase will destroy the remaining input sequence and the double-stranded sequence will result in a "true" evaluation. If one input sequence is "false" and the other is "true," then the "true" ACCTAG sequence will combine with either of the "false" TGGATC sequences to produce a double-stranded sequence. The DNase will destroy the remaining "false" sequence and the gate will result a "true" evaluation. If both input sequences are "false" TGGATC sequences, then neither will combine with the supplied "false"

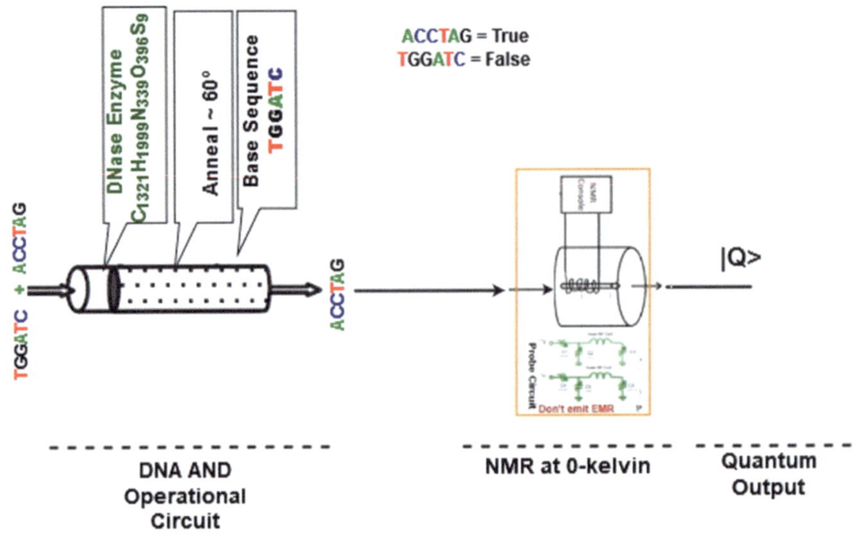

Fig. 14.47 DNA-quantum OR operation at 0 K

sequence. The DNase will destroy all sequences in the mixture, and result a "false" evaluation of the gate.

When getting the DNA sequence from DNA OR operation, this sequence will go through NMR as a sample and dive into the probe. Using the NMR process at 0 K and by emitting EMR, this biomolecule will get a superposition state and become a qubit (|Q> in Fig. 14.47).

Here, DNA sequence ACCTAG = TRUE for qubit |1 > and DNA sequence TGGATC = FALSE for qubit |0 >.

14.3.5.3 DNA-Quantum XOR Operation at 0 K

To produce the output, two inputs are needed that have single strands of DNA sequence. There is no need of any base sequence. But the anneal temperature is needed for the DNA XOR operation. It is needed to check the input sequence. For sequences to have opposite values, they are complementary which will bind together to form a double-stranded sequence. For each combination of the input sequences, the output will be matched.

After getting the output DNA sequence from the DNA XOR operation, there is a need to pass it through the NMR process. In NMR, a room temperature probe is used. By doing NMR for the DNA sequence, the qubit of normal ground state turns into superposition state or ground state according to the input sequence in NMR. The output of the NMR process is a qubit.

The DNA-Quantum XOR operational gate is the most important DNA-Quantum logic operation. Consider two sequences, where ACCTAG represents a “true” input and the sequence TGGATC represents a “false” input. The XOR gate evaluates “true” if both sequences of the gate inputs are complementary to each other. They make a double-stranded DNA sequence. If a double-stranded sequence is observed, then the result is “true”; otherwise, the result is “false.” If both of the input sequences are different to each other ACCTAG and TGGATC sequences, then these sequences will combine and produce a double-stranded sequence. The DNase will destroy the remaining input sequences and the double-stranded sequence will result a “true” evaluation. If both input sequence is “false” or “true,” then the sequence will not combine with each other. The DNase will destroy all sequences in the mixture, and results a “false” evaluation of the operation.

When getting the DNA sequence from DNA XOR operations, this sequence will go through NMR as a sample and dive into the probe. Using the NMR process at room temperature and by emitting EMR, this biomolecule will get a superposition state and become a qubit (|Q> in Fig. 14.48).

Here, DNA sequence ACCTAG = TRUE for qubit |1 > and DNA sequence TGGATC = FALSE for qubit |0 >.

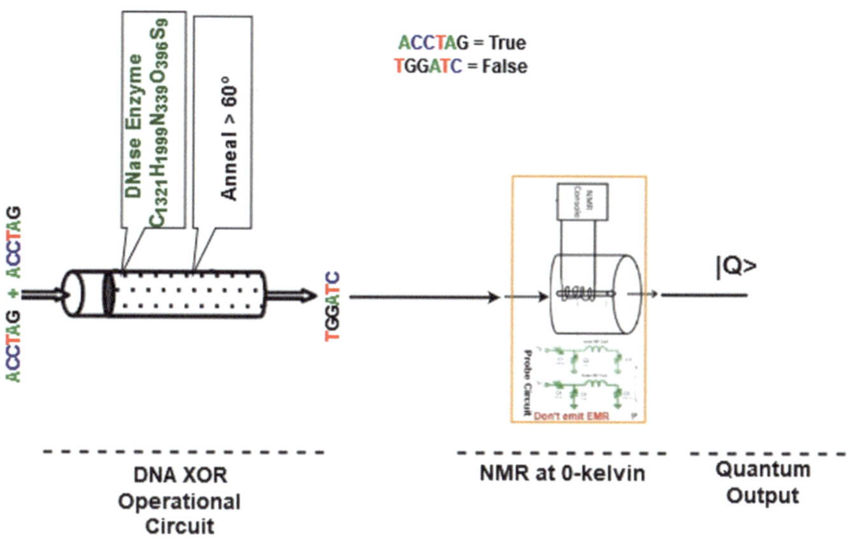

Fig. 14.48 DNA-quantum XOR operation

14.3.5.4 DNA-Quantum Full Adder at 0 K

DNA-Quantum Full Adder and its working procedure are described previously but here the design procedure of DNA-Quantum Full Adder by using a Cryogenic probe will be described. The output of different combinations is also shown in Table 14.25.

To design a DNA-Quantum Full Adder, it needs to use DNA and Quantum operation to operate the input DNA sequence for their corresponding outputs. The DNA operations will be used for receiving the input DNA sequence and the Quantum operation will be used to produce the final output against the corresponding set of inputs. Each time, the DNA-Quantum Full Subtractor will receive three DNA sequences as input. After operating a certain number of DNA operations, the DNA sequence will be turned into corresponding qubits by using the NMR process, which uses a cryogenic probe at 0 K. By using a cryogenic probe at 0 K and corresponding components of NMR, a qubit can reach the excited state or grounded state. To find the qubit in an excited state one has to emit EMR. After processing through the NMR, a DNA sequence is converted into corresponding qubits, and outputs are received. Here, Fig. 14.49 describes DNA-Quantum Full Adder by using DNA and Quantum operation.

From Fig. 14.49, it is seen that the DNA-Quantum Full Adder consists of three DNA operations and two Quantum operations. Here, two AND and one XOR are used as DNA operation and further one XOR and one OR Quantum operations are used.

14.3.5.5 DNA-Quantum Full Subtractor at 0 K

DNA-Quantum Full Subtractor and its working procedure are already described but the design procedure of a DNA-Quantum full subtractor using cryogenic probe at 0 K is described. The output of different combinations is also shown in Table 14.26. Here, Fig. 14.50 describes DNA-Quantum Full Subtractor using DNA and Quantum operations.

To design a DNA-Quantum Full Subtractor, it needs to use DNA and Quantum operations to operate the input qubit for their corresponding outputs. The DNA operations will be used for receiving the input DNA sequence and the Quantum operations will be used to produce the final output against the corresponding set of inputs. Each time, the DNA-Quantum Full Subtractor will receive three DNA sequences as inputs.

After operating in a certain number of DNA operations, the DNA sequence will be turned into corresponding qubits by NMR by using a Cryogenic probe at 0 K. By using a Cryogenic probe at 0 K and corresponding components of NMR, the neutral qubit goes to an excited state for emitting EMR. Then the qubit is processed through Quantum operations and outputs are received.

From the Fig. 14.50, it is seen that the DNA-Quantum Full Subtractor consists of three DNA operations and two Quantum operations. Here, two AND and one XOR

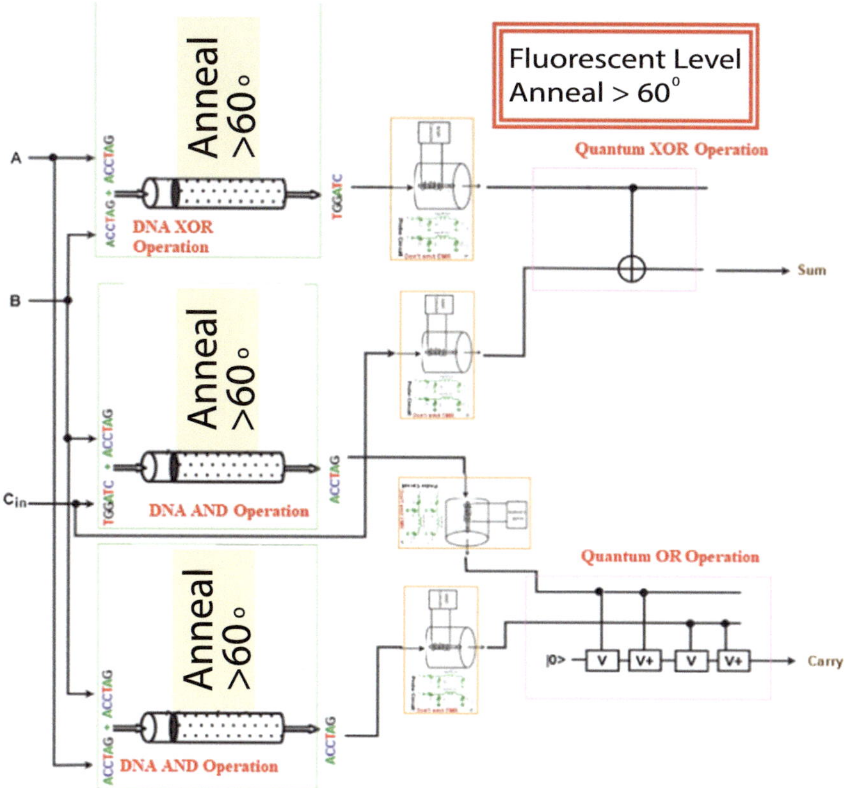

Fig. 14.49 DNA-quantum full adder at 0 K

are used as DNA operations and further one XOR and one OR Quantum operations are used.

14.3.5.6 DNA-Quantum 2-to-1 Multiplexer at 0 K

Section 14.3.4.2 has already described DNA-Quantum 2-to-1 Multiplexer and its working procedure but this section also discusses the design procedure of DNA-Quantum Multiplexer using Cryogenic probe at 0 K. The output of different combinations is also shown in Table 14.27.

To design a DNA-Quantum 2-to-1 Multiplexer, DNA and Quantum operations are used to operate the input DNA sequence for their corresponding outputs. The DNA operations will be used for receiving the input sequence and the Quantum operational gates will be used to produce the final output against the corresponding set of inputs. Each time, the DNA-Quantum multiplexer will receive three DNA sequences as inputs. After operating a certain number of DNA operations, the sequence will

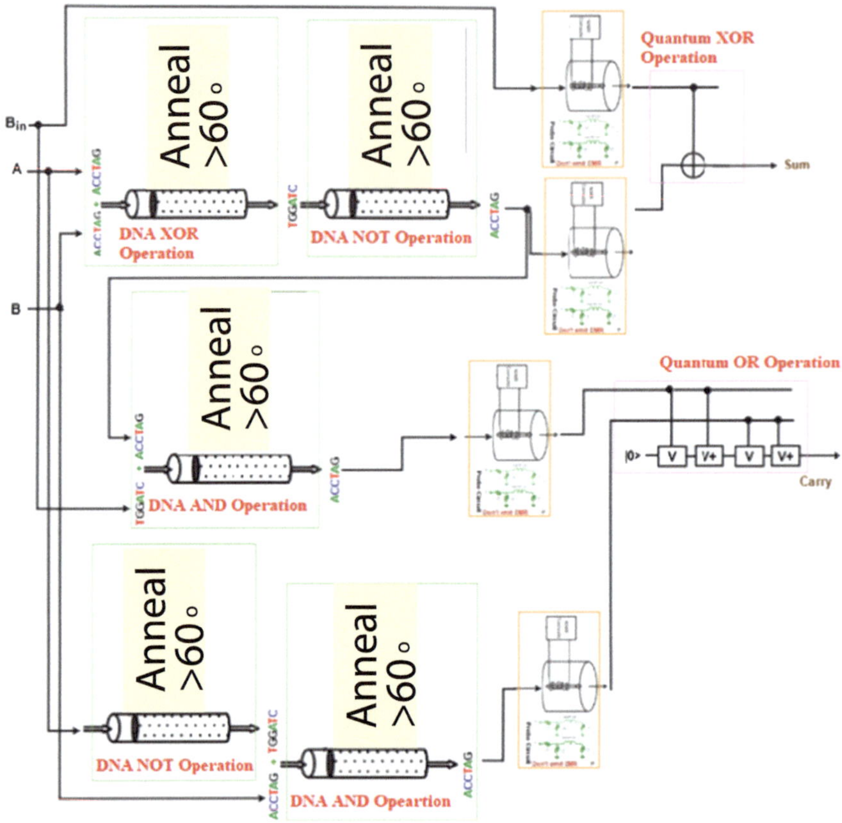

Fig. 14.50 DNA-quantum full subtractor at 0 K

be turned into corresponding qubits by using NMR. The NMR process will be performed by using a Cryogenic probe at 0 K. By using a Cryogenic probe at 0 K and corresponding components of NMR, the grounded qubit turns into super-positioned. Then the qubit is processed through Quantum operations and outputs are received. Here, Fig. 14.51 describes quantum-DNA multiplexer using DNA operations and quantum operations.

From the figure, it is found that the DNA-quantum 2-to-1 multiplexer consists of three DNA operations and one quantum operation. Here, two AND and one NOT are used as DNA operations and further one quantum OR operation is used.

14.3.6 *Quadrupole Ion Trap*

A quadrupole ion trap is such an ion trap that uses dynamic electric fields to trap any charged particles. This ion trap is also called the Paul trap or radio frequency trap

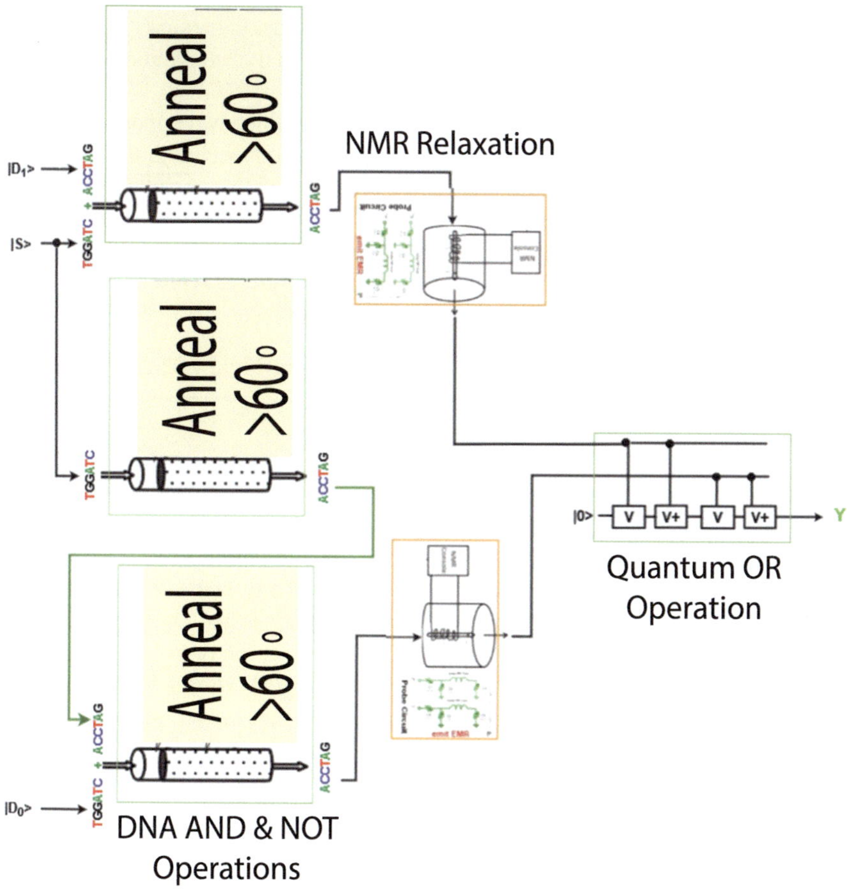

Fig. 14.51 DNA-quantum 2-to-1 multiplexer at 0 K

(RF). This trap is used as a component of a mass spectrometer and/or a trapped ion quantum computer.

14.3.6.1 Components of Quadrupole Ion Trap

The quadrupole ion trap consists of some components like ion beam, end cap, Ring electrode, RF voltage, and AC voltage. A little bit explanation of these components is given below:

1. **Ion Beam**
   An ion beam is a type of charged particle beam that can consist of ions.
2. **End-Cap Electrodes**
   Leading the trap, ion accumulation allows efficient ions to hold the injected ions

with low kinetic energies. In this case, end-cap electrodes carry small DC voltages. However, these DC potentials are used to trap ions in the axial dimension.

3. **Ring Electrodes**
   The trap consists of four blade-shaped electrodes of which two opposing ones are connected to an RF-voltage while the other two are connected to the ground. It also includes two end-cap electrodes that are connected to a positive DC voltage.
4. **RF Voltage and AC voltage**
   All Radio Frequency (RF) signals are Alternating Current (AC) signals and have an amplitude that can be measured in Volts. So an RF voltage is just an AC voltage. RF signals can be any of a wide range of frequencies.

The block diagram of quadrupole ion trap is shown in Fig. 14.52 and the circuit diagram of quadrupole ion trap is shown in Fig. 14.53.

14.3.6.2 Working Principle of Quadrupole Ion Trap

The quadrupole ion consists of two hyperbolic metal electrodes with the focuses facing each other and a hyperbolic ring electrode halfway between the other two electrodes. The injected ions are trapped by AC (oscillating) and DC (static) electric fields. The AC radio frequency voltage oscillates between the two hyperbolic metal end cap electrodes if ion excitation is desired; the driving AC voltage is applied to the

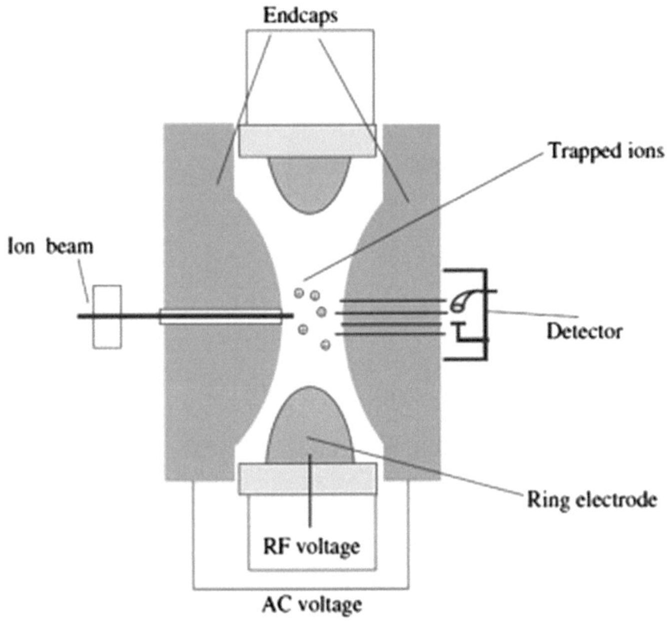

Fig. 14.52 Block diagram of quadrupole ion trap

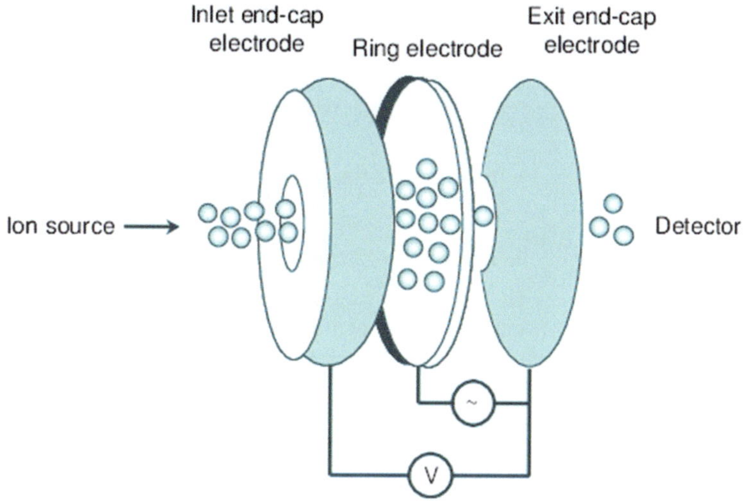

Fig. 14.53 Circuit diagram of quadrupole trap ion

ring electrode. At first, the ions are pulled up and down axially when they are pushed in radially. After that, the ions are pulled out radially and pushed in axially (from the top and bottom). In this way the ions move in a complex motion that generally involves the cloud of ions being long and narrow and then short and wide, back and forth, oscillating between the two states. The motion of these ions in this field is described by solutions to the Mathieu equation. When it is written for ion motion in a trap, the equation is

$$\frac{d^2u}{d\in^2} + [a_u - 2q_u \cos(2\in)]u = 0 \tag{14.8}$$

where μ presents the x, y, and z coordinates is a dimensionless variable given by, $\varepsilon = \Omega t/2$, and $a_{u\ and}$ q_u are dimensionless trapping parameters. The parameter Ω radial frequency of the potential is applied to the ring electrode. By using the chain rule, it can be shown that

$$\frac{d^2u}{dt^2} = \frac{\Omega^2}{4}\frac{d^2u}{d\varepsilon^2} \tag{14.9}$$

Substituting Eqs. 14.9 in 14.8 yields

$$\frac{4}{\Omega^2}\frac{d^2u}{d\epsilon^2} + [a_u - 2q_u\cos(\Omega t)]u = 0 \tag{14.10}$$

Multiplying this equation by m and rearranging terms shows us that

$$m\frac{d^2u}{d\epsilon^2} + m\frac{\Omega^2}{4}[a_u - 2q_u\cos(\Omega t)]u = 0 \tag{14.11}$$

The above equation represents the force of the ion By Newton's laws of motion formula. The forces in each dimension are not coupled, thus the force acting on an ion. For example, the X dimension is

$$F_x = ma = m\frac{d^2x}{dt^2} = -e\frac{\partial\emptyset}{\partial x} \tag{14.12}$$

Here is the quadrupole potential, given by

$$\emptyset = \frac{\emptyset_0}{r_0^2}(\lambda x^2 + \sigma y^2 + \gamma z^2) \tag{14.13}$$

Where, $\emptyset_0$ is the applied electric potential and λ, σ and γ are weighting factors, and r_0 is a size parameter constant. To satisfy Laplace equations, it can be
$\lambda + \sigma + \gamma = 0$
For an ion trap, $\lambda = \sigma = 1$, and $\gamma = -2$ and for a quadrupole mass filter,
$\lambda = -\sigma = 1$ and $\gamma = 0$
Transforming Eq. 14.13 into a cylindrical coordinate system with x=rcos θ, y = rsin θ and z = z
and applying the trigonometric formula sin2θ + cos 2θ = 1 gives

$$\emptyset_{r,z} = \frac{\emptyset_0}{r_0^2}(r^2 - 2z^2) \tag{14.14}$$

The applied electric potential is a combination of RF and DC given by

$$\emptyset_0 = U + V\cos\Omega t \tag{14.15}$$

where Ω = 2πv and v is the applied frequency in Hertz.

Substituting Eq. 14.15 into Eq. 14.13 with λ = 1 gives

$$\frac{\partial\emptyset}{\partial x} = \frac{2x}{r_0^2}(U + V\cos\Omega t \tag{14.16}$$

Substituting Eq. 14.16 into Eq. 14.12 leads to

$$m\frac{d^2x}{dt^2} = -\frac{2e}{r_0^2}\left(U + V cos \Omega t\right)x \tag{14.17}$$

Comparing terms on the right-hand side of Eqs. 14.7 and 14.17 leads to

$$a_x = \frac{8eU}{mr_0^2\Omega^2} \tag{14.18}$$

and

$$q_x = \frac{4eV}{mr_0^2\Omega^2} \tag{14.19}$$

Further $q_x = q_y$

$$a_z = \frac{16eU}{mr_0^2\Omega^2} \tag{14.20}$$

and

$$q_z = \frac{8eV}{mr_0^2\Omega^2} \tag{14.21}$$

The trapping of ions can be understood in terms of stability regions in form Eq. (14.18) and Eq. (14.19) space.

14.3.6.3 DNA-Quantum AND Operation with Quadrupole Ion Trap

In this portion, an AND gate is used by combining a quantum gate and a DNA gate. And this new gate is called the DNA-quantum AND gate. In this circuit, the following terms are used:

1. **DNA NAND operation**
   The AND gate is easily formed from the NAND gate. After that, doing CNOT operation of NAND gate, quantum AND operation is possible to achieve. In this quantum CNOT gate, Constant |1> which acts as the target qubit has been assumed. If 0 is considered as a target qubit then output will always be the reverse of real output.
2. **Quadrupole Ion trap**
   After DNA NAND gate operation, DNA sequence that would be gained as output has been trapped. During trap, the sequence will go to superposition which produces qubit.
3. **Quantum CNOT operation**
   After producing a qubit, it will be used as input into the CNOT gate that will produce the final output.

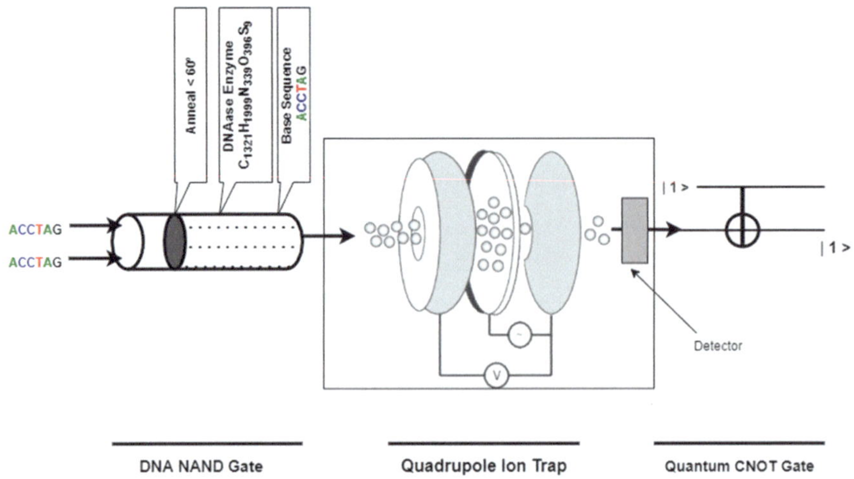

Fig. 14.54 DNA-quantum AND operation

Table 14.29 Truth table of DNA-quantum AND gate

A	B	Base	Output
TGGATC	TGGATC	ACCTAG	|0 >
TGGATC	ACCTAG	ACCTAG	|0 >
ACCTAG	TGGATC	ACCTAG	|0 >
ACCTAG	ACCTAG	ACCTAG	|1 >

Figure 14.54 shows the DNA-quantum AND Operation with quadrupole ion trap and Table 14.29 shows the truth table of DNA-quantum AND operation.

To execute DNA-quantum AND gate, the following steps are performed:

1. First of all, DNA NAND gate is performed here. The NAND gate evaluates "true" (ACCTAG) if both inputs are low or false. If one input is high and another is low or opposite of this then output will be high or true. The NAND gate provides only the false or low output when the both outputs are high or true. According to the given input the DNA sequence makes a double strand sequence. This double strand will be either "True" or "False". For example, if both inputs are high (ACCTAG), no input will make a double-strand with the base sequence that is given in a tube. As there will be no bond, the Output will be 0. Every time the DNase will destroy the single-stranded sequence in this mixture. If a double stranded sequence is observed, the result is "true" (ACCTAG); otherwise, the result is "false" (TGGATC).
2. For each combination, the NAND gate will produce one DNA sequence ("ACC-TAG" or "TRUE") or ("TGGATC" or "FALSE") which will be injected in the trap ion, and in the trap ion, they will produce qubit. This qubit will be the input

of quantum CNOT gate. In CNOT gate, control bit |1> is assumed as constant. As it is assumed 0, the output will always be reversed for the real output. After this CNOT operation, the final output will be achieved which is a combination of DNA NAND gate and quantum CNOT gate.

14.3.6.4 DNA-Quantum NOR Operation with Quadrupole Ion Trap

Quantum DNA NOR operation is shown in Fig. 14.55 using quadrupole ion trap. Here, OR operation is performed using DNA OR gate operation and NOT operation performed using quantum CNOT operation. To get the desired output of quantum NOT gate operation, an ancillary bit |1> is fixed in quantum CNOT gate operation.

1. **Design Procedure of DNA-quantum NOR Gate**
   In this portion, an OR gate is designed combining a quantum gate and a DNA gate. And this new gate is called the DNA-quantum gate. In this circuit, the following terms are used:
   (a) **Quantum NOR operation**
   The OR operation is easily formed from the NOR operation. After that, doing NOT operation on NOR operation, OR operation is possible to achieve. In this quantum CNOT gate, constant |1> acts as the target bit. If 0 is considered as a target bit then the output will always be the reverse for real output.
   (b) **Ion trap**
   After DNA NOR operation, the DNA sequence has been trapped for producing qubit.
   (c) **Quantum CNOT operation**
   After producing qubit, it will be used as input in the CNOT gate which will produce the final output.

2. **Working Principle of a DNA-quantum NOR Operation**
   To execute DNA-quantum NOR operation, the following steps are performed. The truth table of DNA-Quantum NOR operation is given in Table 14.30.

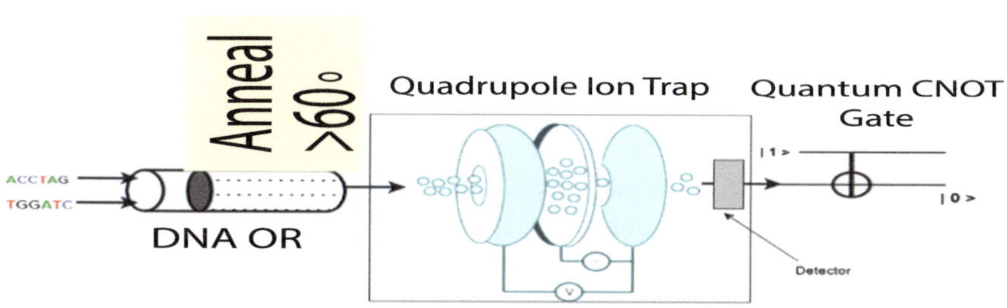

Fig. 14.55 DNA-quantum NOR operation

Table 14.30 Truth table of DNA-quantum NOR gate

A0	A1	Base	Output
TGGATC	TGGATC	TGGATC	|1 >
TGGATC	ACCTAG	TGGATC	|0 >
ACCTAG	TGGATC	TGGATC	|0 >
ACCTAG	ACCTAG	TGGATC	|0 >

(a) First of all, DNA-quantum NOR operation is performed here. The NOR operation evaluates "true" (ACCTAG) only when both inputs are low or false("TGGATC"). If one input is high and another is low or opposite of this then the output will be always low or false ("TGGATC"). Moreover, The NOR operation provides the false ("TGGATC") or low output when they're both outputs are high or true (ACCTAG). According to the given input, the DNA sequence makes a double strand sequence. This double-strand will be "True" or "False". For example, if one input is high (ACCTAG) and the other one is low ("TGGATC"), one input (TGGATC) will make a double-strand with the base sequence (ACCTAG) which is given in tube and the DNase will destroy the rest of the single stranded sequence in this mixture. If a double stranded sequence is observed, then the result is "true" (ACCTAG); otherwise, the result is "false" (TGGATC). Not only for this operation but also all operations will work like this.

(b) For each combination, the NOR operation will produce one DNA sequence ("ACCTAG" or "TRUE") or ("TGGATC" or "FALSE") which will be injected in the trap ion and in the trap ion, they will produce qubit. This qubit will be the input of quantum CNOT operation. In CNOT operation control bit |1> is assumed as constant. As it is assumed 0, the output will always be reversed for the real output. After this CNOT operation, the final output will be achieved which is a combination of DNA NAND operation and quantum CNOT operation.

14.4 Summary

Calculations of a quantum computer are especially promising for analyzing or simulating extremely complicated processes with large volumes of data. Quantum computers may help researchers to get a better and more detailed understanding of how specific particles, components, and processes interact in live cells. However, there are also possible medical applications. Most importantly, experts believe that quantum computers will go forward with artificial intelligence (AI) significantly. These could then safely and reliably take over tasks such as data evaluation or forecasting in the future.

Furthermore, DNA computing is a new technology which uses DNA molecules to create computers that are quicker than the most powerful human-built computers in the market. The DNA computers of the future will be able to work in a massively parallel fashion, completing several calculations at the same time. In practically every sector, DNA computers have made significant developments.

To advance computation, DNA-Quantum computing systems can be used, which will merge all the advantages of both DNA computing and quantum computing. In DNA-Quantum computing, input data is in DNA sequence and the output is in qubit. It uses NMR to convert DNA sequences into qubits. Again, in Quantum-DNA computing, input data is in qubit and the output is in DNA sequence. Quantum-DNA uses NMR relaxation to convert qubits into DNA sequences. Here, different types of operational circuits such as Full Adder, Half Subtractor, Full Subtractor, Multiplier, Parity checker, and multiplexer are discussed, where both DNA-Quantum and Quantum-DNA computing processes are used.

Chapter 15
Data Management Techniques

15.1 Introduction

Both quantum computing and DNA computing approaches can be used to accomplish operations at a very high speed and efficiency. There is a critical issue that occurs when merging both computing systems for a single activity. Quantum computers are a million times quicker than the world's fastest supercomputer. As a result, quantum operations produce instantaneous results. However, DNA procedures take a long time to prepare for its operations. So, the quantum computer's output qubits cannot be instantly inserted into the DNA system. Thus, when working with the Quantum-DNA system, a temporary storage device or system is required where the qubits may be kept for a very short time. So, a quantum cache memory is needed to store qubits during a Quantum-DNA computing circuit operation. The structure and working procedure of quantum and DNA cache memory are described in this chapter. In the DNA-Quantum circuit operation, a DNA cache memory is needed to store DNA sequences for temporary purpose.

As a result, when working with the quantum-DNA system and DNA-quantum system, a temporary storage device or system is required where the qubits or DNA sequences may be kept for a very short time. So, quantum cache memory is needed to operate a quantum-DNA circuit; and DNA cache memory is needed to operate a DNA-quantum circuit, which can store data for a short amount of time and can pass it to another.

15.2 Quantum Cache Memory

Cache memory is a supplementary memory system that temporarily stores frequently used instructions and data for quicker processing. It is an extremely fast memory type that holds frequently requested data and instructions so that they are immediately

H. Md. Hasan Babu, *Quantum Biocomputing in Quantum Biology Volume II*,
https://doi.org/10.1007/978-981-97-5349-9_15

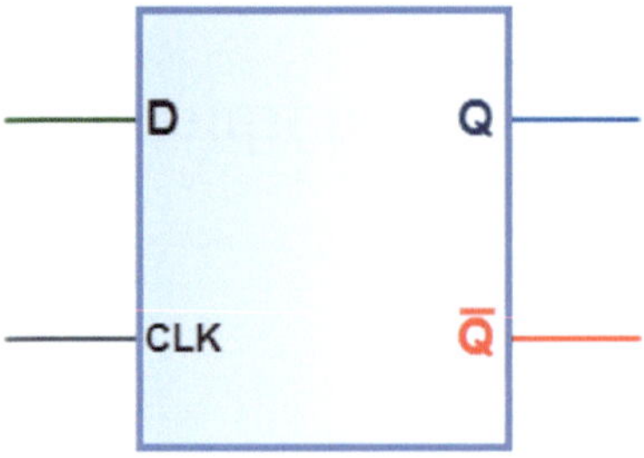

Fig. 15.1 Logic symbol of D flip-flop

available for further processing. In most microprocessors, Static Random-Access Memory (SRAM) is used as cache memory as SRAM has a very high speed. Due to this high speed, cache memory can be used to store data temporarily, which will be designed for the quantum computer.

15.2.1 D Flip-Flop

A cache memory is designed using the D flip-flops which will create a one-bit SRAM along with Read/Write and select bit as input. A Delay flip-flop or D flip-flop is an electronic circuit used to delay the change of state of its output signal (Q) until the next rising edge of a clock timing input signal occurs. Figure 15.1 shows the logic symbol for a D flip-flop. Further, Fig. 15.2 shows the circuit diagram of the Quantum D flip-flop.

D flip-flip works on two states, SET and RESET. The following truth table will show the operations of a D flip-flop. Input-output table of the D flip-flop is given in Table 15.1.

15.2.1.1 Design Procedure

Cache memory can be used in the quantum system due to its speed and reliability. It can pass and get data very frequently. The mapping and swapping techniques in the cache memory are much optimized. It is known that cache memory is nothing but SRAM which is made out of D flip-flops and other gates. Figure 15.2 shows the design of the D flip-flop, which needs two inputs |D> and |CLK> pulse.

To design the quantum D flip-flop, five quantum NAND operations including a clock pulse are used. Here, |D> is the quantum qubits that are |0> and |1>. When the CLK pulse is 1, the quantum D flip-flop transfers the |D> input to output |Q> and if the CLK pulse is 0, then the output remains unchanged.

Table 15.1 Input-output table of D flip-flop

|D>	|CLK>	|Q>	|$\underline{Q}$>
|0>	|1>	|0>	|1>
|1>	|1>	|1>	|0>
X	0	Q	$\underline{Q}$

15.2.2 *Quantum One-Qubit Cache Memory*

A one-bit quantum cache memory will be constructed using the previously designed quantum D flip-flop. The circuit diagram of the quantum one-qubit cache memory can be shown in Fig. 15.3.

The circuit of one-qubit quantum cache memory contains six quantum NAND and three quantum AND operations. Again, | R/W > = |1> indicates the READ operation and | R/W > = |0> indicates the WRITE operation. Here, |S> indicates the select qubit, where |S> = |1> means the memory which is selected.

15.2.3 *Quantum Eight-Qubit Cache Memory*

A 4-to-2 quantum cache memory is formed using 8 quantum cache memory cells (one-qubit quantum cache memory). Figure 15.4 shows a 4-to-2 quantum cache mem-

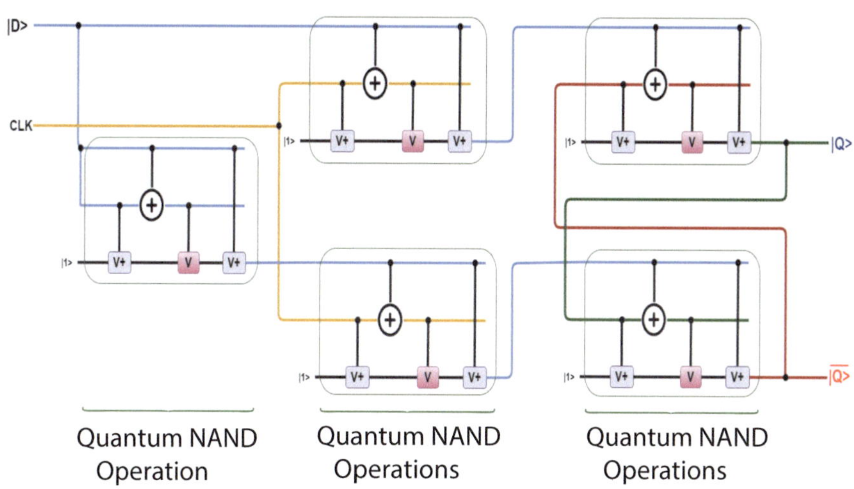

Fig. 15.2 Circuit diagram of quantum D flip-flop

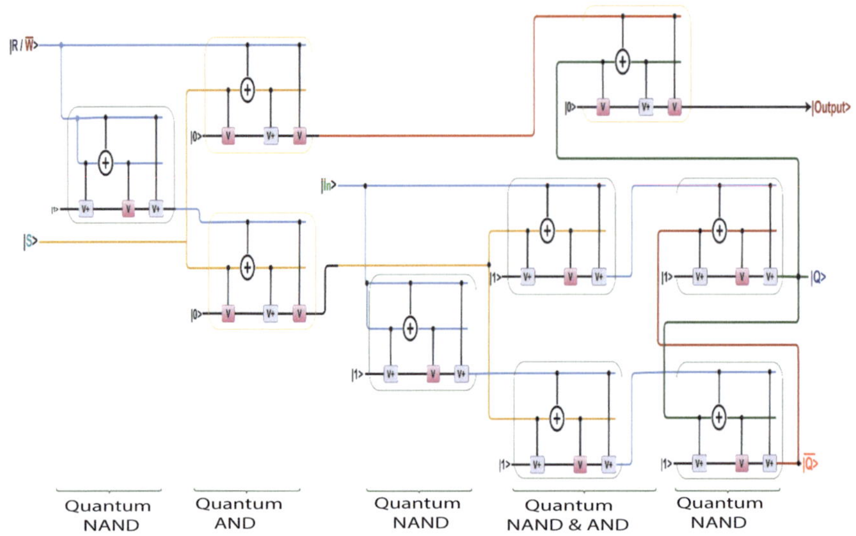

Fig. 15.3 The circuit diagram of one-bit quantum cache memory

ory that contains 8 quantum cache memory cells, providing two-qubit output and four locations (00, 01, 10, and 11).

The quantum 4-to-2 cache memory contains a quantum 2-to-4 decoder which decodes the input qubits into 4 output qubits. Here, R indicates the quantum cache memory cell. The circuit of the quantum 4-to-2 cache memory includes a 2-to-4

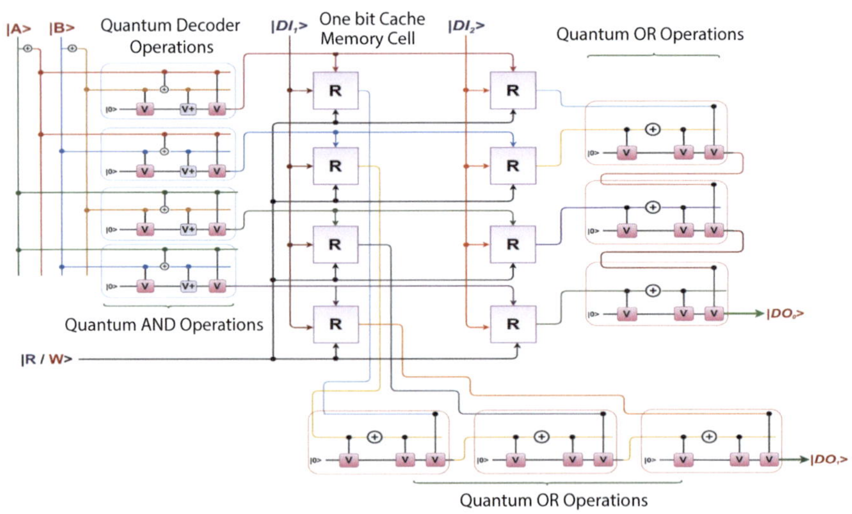

Fig. 15.4 The circuit diagram of quantum 4-to-2 cache memory

decoder, 8 quantum cache memory cells, 6 quantum OR operations, where inputs are two qubits and |R/W>| signal.

15.2.3.1 Working Procedure of the Quantum Cache Memory

Figure 15.4 shows a 4-to-2 quantum cache memory. It includes 8 quantum cache memory cells providing two-qubit output and four locations. The cache memory has a 2-to-4 quantum decoder and 8 quantum cache memory cells implemented with quantum D flip-flops and quantum gates. The four locations (00, 01, 10, and 11) in the cache memory are addressed by 2 qubits (A_1, A_0). In order to read from location 00, the address $A_1A_0 = 00$ and |R/W>=1|>. The decoder selects |0>, high. |R/W>=| 1 will apply 0 at the clock inputs of the two quantum cache memory cells of the top row and will apply 1 at the inputs of the output quantum AND operations, thus transferring the outputs of the two quantum D flip-flops to the inputs of the two quantum OR operations. The other inputs of the quantum OR operations will be 0.

Thus, the outputs of the two quantum cache memory cells of the top row will be transferred to DO_1, and DO_0, performing a READ operation.

On the other hand, consider a WRITE operation; the two-qubit data to be written is presented at $|DI_1>$ $|DI_0>$. Suppose $|A_1>$ $|A_0> = |0>$ $|0>$. The top row is selected ($0_0 = 1$). Input qubits at $|DI_1>$ and $|DI_0>$ will respectively be applied at the inputs of the D flip-flops of the top row. Because |R/W>=|>, the clock inputs of both the quantum D flip-flops of the top row are |1>. Thus, the D inputs are transferred to the outputs of the flip-flops.

Therefore, data at DI_1 DI_0 will be written into the quantum cache memory. The block diagram for a quantum cache memory is shown in Fig. 15.5.

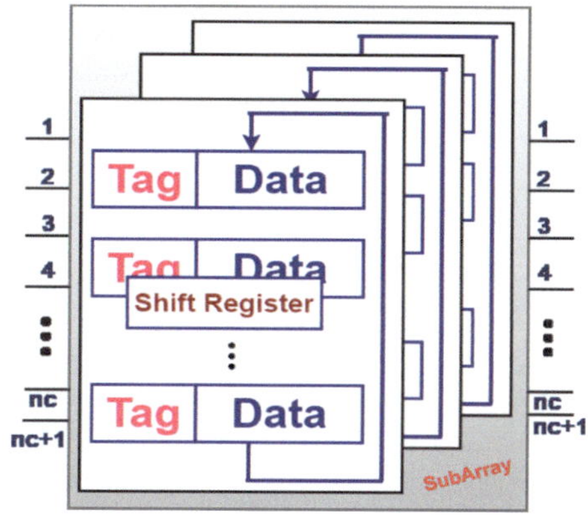

Fig. 15.5 Block diagram of a quantum cache memory

15.3 Data Management in Quantum-DNA Circuits

According to quantum computing, quantum computation is faster than classical computation systems. Quantum computers are also more powerful than supercomputers in terms of computing. They are thousands of times faster than regular computers and supercomputers at processing data. Quantum computers can execute calculations that would take a regular computer 1000 years to complete in a matter of seconds. On the other hand, the use of DNA strands to compute has led to high parallel computation that makes up for the slow processing of the chip. Memory space required by DNA is around 1 bit per cubic nanometer which is much less when compared to regular storage systems Consumption of power is almost nil as the chemical bonds in DNA produce energy to build or repair new strands. So, to find a super faster computation system with huge memory, a Quantum-DNA computation system can be developed. This Quantum-DNA computation system can merge all advantages of quantum computing and DNA computing.

In a Quantum-DNA computing system, the input will be received as a qubit and after performing a certain number of quantum operations these qubits will be turned into DNA sequences by NMR relaxation.

15.3.1 Data Management in Quantum-DNA Full Adder

A full adder is an adder that adds three inputs and produces two outputs. The first two inputs are A and B and the third input is an input carry as C_{in}. The output carry is designated as C_{out} and the normal output is designated as S which is Sum. A full adder logic is designated in such a manner that can take eight inputs together to create a qubit adder and cascade the carry qubit from one adder to another. To create a full adder, one OR, two AND, and two XOR operations are required. Figure 15.6 describes the Quantum-DNA circuit of the full adder.

15.3.1.1 Design Procedure

To design a Quantum-DNA full-adder quantum and DNA operations can be used so that it can be operated using the input qubit for their corresponding outputs. The Quantum operations will be used for receiving the input qubits and the DNA operations will be used to produce the final output against the corresponding set of inputs. Each time, the Quantum-DNA full adder will receive three qubits as input. After performing a certain number of Quantum operations, the qubit will be stored in Cache memory. To store 4 qubits, a 16-to-4 Quantum cache memory is required. After getting the qubits from cache memory, they will be turned into the corresponding DNA sequence by using an NMR relaxation room temperature probe. By using a room temperature probe and corresponding components of NMR relaxation, the

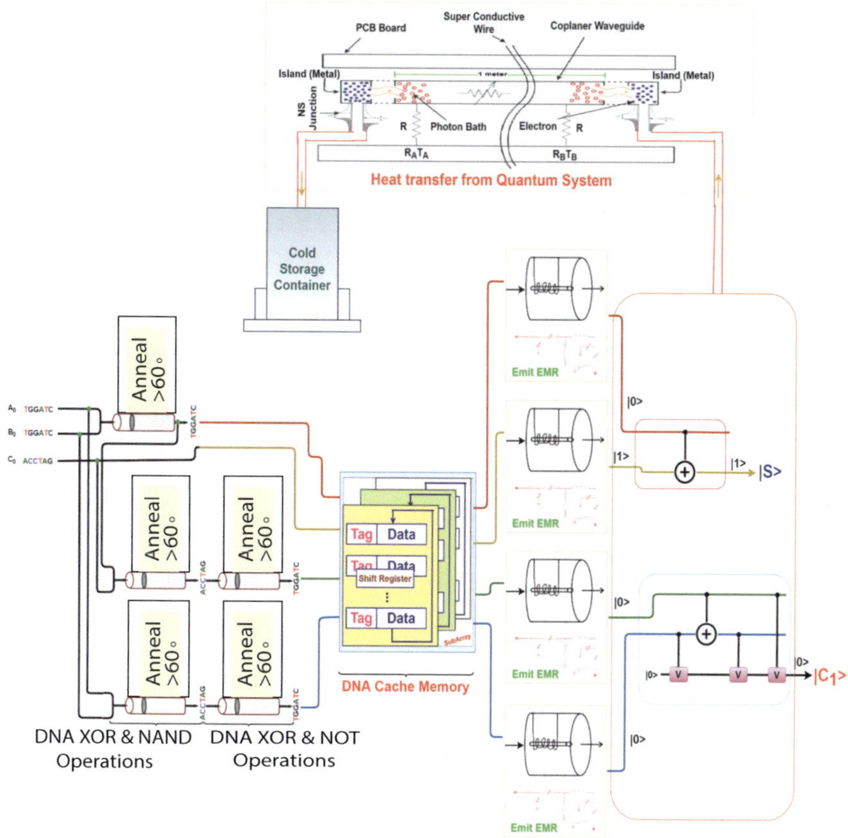

Fig. 15.6 Circuit of quantum-DNA full adder with cache memory

excited qubit turns into a ground state and produces a DNA sequence. Then the DNA sequence is processed through DNA operations and outputs are received. Here, Fig. 15.6 describes Quantum-DNA full adder using Quantum and DNA operations.

From Fig. 15.6, it is found that the Quantum-DNA full adder consists of three Quantum operational gates, two DNA operations and a 16-to-4 cache memory. Here, two AND and one XOR are used as Quantum operations and further one XOR and one OR DNA operations are used.

15.3.1.2 Working Procedure

The working procedure of the Quantum-DNA full adder is given below for each pattern of input qubits. Here, DNA sequence ACCTAG = TRUE for qubit | 1 > and DNA sequence TGGATC = FALSE for qubit | 0 > are used.

Table 15.2 Outputs of quantum-DNA full-adder operation

<table>
<tr><th>|A0></th><th>|A1></th><th>|Q></th><th>SUM</th><th>Carry</th></tr>
<tr><td>|0></td><td>|0></td><td>|0></td><td>TGGATC</td><td>TGGATC</td></tr>
<tr><td>|0></td><td>|0></td><td>|1></td><td>ACCTAG</td><td>TGGATC</td></tr>
<tr><td>|0></td><td>|1></td><td>|0></td><td>ACCTAG</td><td>TGGATC</td></tr>
<tr><td>|0></td><td>|1></td><td>|1></td><td>TGGATC</td><td>ACCTAG</td></tr>
<tr><td>|1></td><td>|0></td><td>|0></td><td>ACCTAG</td><td>TGGATC</td></tr>
<tr><td>|1></td><td>|0></td><td>|1></td><td>TGGATC</td><td>ACCTAG</td></tr>
<tr><td>|1></td><td>|1></td><td>|0></td><td>TGGATC</td><td>ACCTAG</td></tr>
<tr><td>|1></td><td>|1></td><td>|1></td><td>ACCTAG</td><td>ACCTAG</td></tr>
</table>

To store qubits in cache memory, 4 locations select line (A, B, C and D), one CLK input line, one |R/W>| input line and 4-input qubit lines in cache memory (DI_1, DI_2, DI_3, and DI_4) are needed. The output from the cache memory will be DO_0, DO_1, DO_2, and DO_3.

The working procedure of the Quantum-DNA full adder within cache memory is given below for one pattern of input qubits and the outputs for different combinations of inputs are given in Table 15.2. Here, the inputs of cache memory from the quantum gates are given as follows:

1. For inputs A, B, C = 0,
   $\mathbf{DI}_1 = A = 0$
   $\mathbf{DI}_2$ = Quantum XOR (A, B)
   = Quantum XOR (0, 0)
   = 0
   $\mathbf{DI}_3$ = Quantum XOR (XOR (A, B), C)
   = Quantum XOR (XOR (0, 0), 0)
   = Quantum XOR (0, 0)
   = 0
   $\mathbf{DI}_4$ = Quantum AND (A, B)
   = Quantum AND (0, 0)
   = 0

The output of Quantum cache memory will be converted into DNA sequence using NMR relaxation. For qubit |0> DNA sequence is TGGATC.
Sum = DNA XOR (DO_0, DO_1)
= DNA XOR (TGGATC, TGGATC)
= TGGATC
Carry = DNA OR (DO_2, DO_3)
= DNA XOR (TGGATC, TGGATC)
= TGGATC

15.3.2 Data Management in Quantum-DNA Multiplier

A binary multiplier is a combinational circuit or digital device used for multiplying two binary numbers. The two numbers are more specifically known as multiplicand and multiplier and the result is known as a product. The multiplicand and multiplier can be of various bit sizes. The product's bit size depends on the bit size of the multiplicand and multiplier. The bit size of the product is equal to the sum of the bit size of multiplier multiplicand. To create a Multiplication circuit, six AND and two XOR operations are required. Figure 15.7 describes the Quantum-DNA circuit of the two-qubit multiplication.

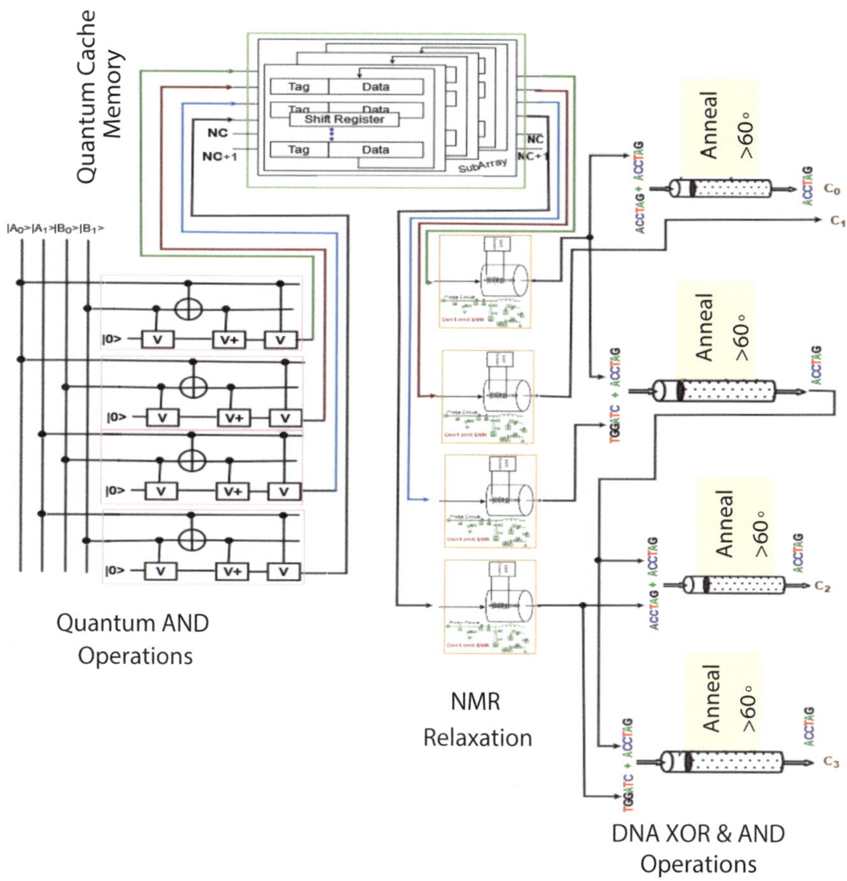

Fig. 15.7 Circuit of quantum-DNA multiplier with cache memory

15.3.2.1 Design Procedure of Quantum-DNA Multiplier

To design a quantum-DNA multiplier, quantum operations and DNA operations are used to operate the input qubit for their corresponding outputs. The quantum operations will be used for receiving the input qubits; and the DNA operations will be used to produce the final output against the corresponding set of inputs. Each time, the quantum-DNA Multiplier will receive four qubits as input. After operating in a certain number of quantum operations, the qubit will be stored in quantum cache memory. To store 4 qubits, a 16×4 quantum cache memory is needed.

After getting the qubits from cache memory, it will be turned into corresponding DNA sequence by using NMR relaxation room temperature probe. By using a room temperature probe and corresponding components of NMR relaxation, the excited qubit turns into ground state and produces a DNA sequence. Then the DNA sequence is processed through DNA operations and outputs are received. Figure 15.7 describes quantum-DNA multiplier using quantum operations and DNA operations. From Fig. 15.7, it is found that the quantum-DNA multiplier consists of four quantum operations, four DNA operations and a 16×4 cache memory. Here, four AND operations are used in quantum operation and storing qubits in cache memory and further two XOR and two DNA AND operations to provide the output.

15.3.2.2 Working Procedure of the Quantum-DNA Multiplier

The working procedure of the quantum-DNA multiplier is given below for each pattern of input qubits. Here, DNA sequence ACCTAG = TRUE for qubit $|1>$ and DNA sequence TGGATC = FALSE for qubit $|0>$ are used.

To store qubits in cache memory, 4 locations select line (A, B, C and D), one CLK input line, one |R/W>| input line and 4 input qubit lines in cache memory (DI_1, DI_2, DI_3, and DI_4). The output from the cache memory will be DO_0, DO_1, DO_2, and DO_3 are needed.

The working procedure of the quantum-DNA multiplier within cache memory is given below for one pattern of input qubits. Here, the inputs of cache memory from the quantum gates are given follows:

1. For inputs $A_0, A_1, B_0, B_1 = 0$,
   $\mathbf{DI_1}$ = Quantum AND (A_0, B_1)
   = Quantum AND (0, 0)
   = 0
   $\mathbf{DI_2}$ = Quantum AND (A_0, B_0)
   = Quantum AND (0, 0)
   = 0
   $\mathbf{DI_3}$ = Quantum AND (A_1, B_0)
   = Quantum AND (0, 0)
   = 0
   $\mathbf{DI_4}$ = Quantum AND (A_1, B_1)

= Quantum AND (0, 0)
= 0

Output of quantum cache memory will be converted into DNA sequence using NMR relaxation at room temperature. For qubit |0> DNA sequence is TGGATC.

$\mathbf{C}_0$ = DNA XOR (DO_0, DO_2)
= DNA XOR (TGGATC, TGGATC)
= TGGATC
$\mathbf{C}_1$ = TGGATC
$\mathbf{C}_2$ = DNA XOR (AND (DO_1, DO_2), DO_3)
= DNA XOR (AND (TGGATC, TGGATC), TGGATC)
= DNA XOR (TGGATC, TGGATC)
= TGGATC
$\mathbf{C}_3$ = DNA XOR (AND (DO_1, DO_2), DO_3)
= DNA XOR (AND (TGGATC, TGGATC), TGGATC)
= DNA XOR (TGGATC, TGGATC)
= TGGATC

15.3.3 Data Management in Quantum-DNA Half Subtractor

A half subtractor is a type of subtractor, an electronic circuit that performs the subtractions of numbers. The half subtractor can subtract two single digits and provide the output plus a borrow value. It has two inputs, called A and B, and two outputs D (difference) and B (borrow). Quantum-DNA half subtractor will subtract two qubits from the quantum digit |0>, and |1> and produce two outputs as DNA difference |D> and DNA borrow sequence |B>. The quantum-DNA circuit for the half subtractor is shown in Fig. 15.8.

15.3.3.1 Design Procedure

To design a quantum-DNA half subtractor, quantum and DNA operations are used to operate the input qubit for their corresponding outputs. The quantum operations will be used for receiving the input qubits and the DNA operations will be used to produce the final output against the corresponding set of inputs. Each time, the quantum-DNA half subtractor will receive two qubits as input. After performing in a certain number of quantum operations, the qubit will be stored in quantum cache memory. To store 3 qubits, an 8×3 quantum cache memory is needed.

After getting the qubits from quantum cache memory, they will be turned into corresponding DNA sequences using an NMR relaxation room temperature probe. By using a room temperature probe and corresponding components of NMR relaxation, the excited qubit turns into a ground state and produces a DNA sequence. Then the DNA sequence is processed through DNA operations and outputs are received.

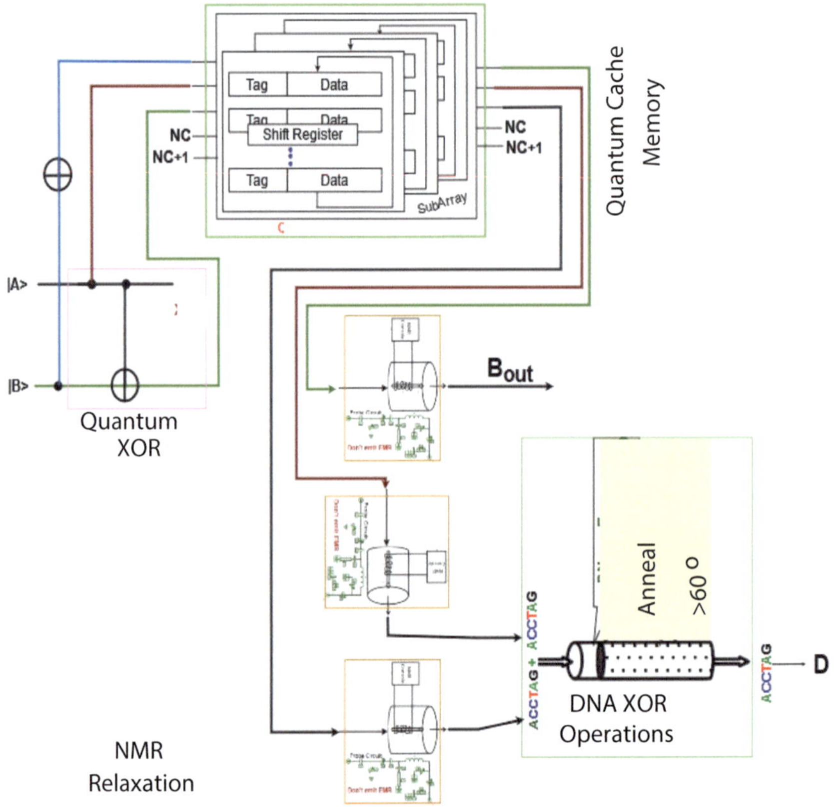

Fig. 15.8 Circuit of quantum-DNA half subtractor with cache memory

Here, Fig. 15.8 describes quantum-DNA half subtractor using quantum operations and DNA operations.

Based on the Fig. 15.8, it is found that the quantum-DNA multiplier consists of two quantum operations, one DNA operations and an 8-to-3 cache memory. Here, one XOR and one NOT operations are used in quantum operation and stored qubits in quantum cache memory and further one DNA XOR operation is needed to provide the output.

15.3.3.2 Working Procedure

The working procedure of the quantum-DNA half subtractor is given below for each pattern of input qubits. Here, DNA sequence ACCTAG = TRUE for qubit | 1 > and DNA sequence TGGATC = FALSE for qubit | 0 > are used.

Table 15.3 Outputs of quantum-DNA half subtractor operation

$\vert A\rangle$	$\vert B\rangle$	DI_1	DI_2	DI_3	Difference	Borrow
$\vert 0\rangle$	$\vert 0\rangle$	$\vert 1\rangle$	$\vert 0\rangle$	$\vert 0\rangle$	TGGATC	TGGATC
$\vert 0\rangle$	$\vert 1\rangle$	$\vert 0\rangle$	$\vert 0\rangle$	$\vert 1\rangle$	ACCTAG	ACCTAG
$\vert 1\rangle$	$\vert 0\rangle$	$\vert 1\rangle$	$\vert 1\rangle$	$\vert 1\rangle$	ACCTAG	TGGATC
$\vert 1\rangle$	$\vert 1\rangle$	$\vert 0\rangle$	$\vert 1\rangle$	$\vert 0\rangle$	TGGATC	TGGATC

To store qubits in cache memory, 3 locations select line (A, B and C), one CLK input line, one $\vert R/W\rangle\vert$ input line and 3 input qubit lines in cache memory (DI_1, DI_2 and DI_3). The output from the cache memory will be DO_0, DO_1, and DO_2 are needed.

The working procedure of the quantum-DNA half subtractor within cache memory is given below for one pattern of input qubits and the outputs for different combinations of inputs are given in Table 15.3. Here, the inputs of cache memory from the quantum gates are given as follows:

1. For inputs A = 1, B = 0,
   $\mathbf{DI}_1$ = Quantum NOT (B)
   = Quantum NOT (0)
   = 1
   $\mathbf{DI}_2$ = A = 1
   $\mathbf{DI}_3$ = Quantum XOR (A, B)
   = Quantum XOR (1, 0)
   = 1

The output of quantum cache memory will be converted into DNA sequence using NMR relaxation at room temperature. For qubit $\vert 1\rangle$ DNA sequence is ACCTAG.

$\mathbf{B}_{out}$ = DO_0 = ACCTAG.
D = DNA XOR (DO_1, DO_2)
= DNA XOR (ACCTAG, ACCTAG)
= TGGATC

15.3.4 Data Management in Quantum-DNA Full Subtractor

A full subtractor is a combinational circuit that performs subtraction of two qubits, one is minuend and other is subtrahend, taking into account the borrow of the previous adjacent lower minuend qubit. This circuit has three inputs and two outputs. The three inputs A, B and B_{in}, denote the minuend, subtrahend, and previous borrow respectively. The two outputs, D and B_{out} represent the difference and output borrows, respectively. To create a Full Subtractor, one OR, two AND, two OR and 2 XOR operational gates are required. Figure 15.9 describes the quantum-DNA circuit of full subtractor.

15.3.4.1 Design Procedure of a Quantum-DNA Full Subtractor

To design a Quantum-DNA Full Subtractor Quantum and DNA operations are used to operate the input qubit for their corresponding outputs. The Quantum operations will be used for receiving the input qubits; and the DNA operations will be used to produce the final output against the corresponding set of inputs. Each time, the Quantum-DNA Full Subtractor will receive three qubits as input. After operating in a certain number of Quantum operations, the qubit will be stored in Cache memory. To store 4 qubits, a 16-to-4 Quantum cache memory is needed.

After getting the qubits from cache memory, it will be turned into the corresponding DNA sequences using NMR relaxation room temperature probe. By using a room temperature probe and corresponding components of NMR relaxation, the excited qubit turns into ground state and produces a DNA sequence. Then the DNA sequence is processed through DNA operations and outputs are received. Here, Fig. 15.9 describes Quantum-DNA Full Subtractor using Quantum and DNA operations.

From the Fig. 15.9, it is found that the Quantum-DNA Full Subtractor consists of four Quantum operations, two DNA operations and a 16-to-4 cache memory. Here, two AND, one XOR and one NOT Quantum operations are used in Quantum operation and storing qubits in Cache memory and further one XOR and one OR DNA operations to provide the output.

15.3.4.2 Working Procedure

The working procedure of the Quantum-DNA Full Subtractor is given below for each pattern of input qubits. Here, DNA sequence ACCTAG = TRUE for qubit $|1>$ and DNA sequence TGGATC = FALSE for qubit $|0>$ are used.

To store qubits in cache memory, 4 locations select line (A, B, C and D), one CLK input line, one $|R/W>|$ input line and 4 input qubit lines in cache memory (DI_1, DI_2, DI_3, and DI_4). The output from the cache memory will be DO_0, DO_1, DO_2, and DO_3 are needed.

The working procedure of the quantum-DNA Full Subtractor within cache memory is given below for one pattern of input qubits and the outputs for different combinations of inputs are given in Table 15.4. Here, the inputs of cache memory from the quantum gates are found as follows:

1. For inputs A, B, $B_{in} = 1, 1, 0$,
   DI_1 $= B_{in} = 0$
   DI_2 = Quantum XOR (A, B)
   = Quantum AND (1, 1)
   = 1
   DI_3 = Quantum AND (NOT (XOR (A, B), B)
   = Quantum AND (NOT (XOR (1, 1), 1)
   = Quantum AND (NOT (0), 1)

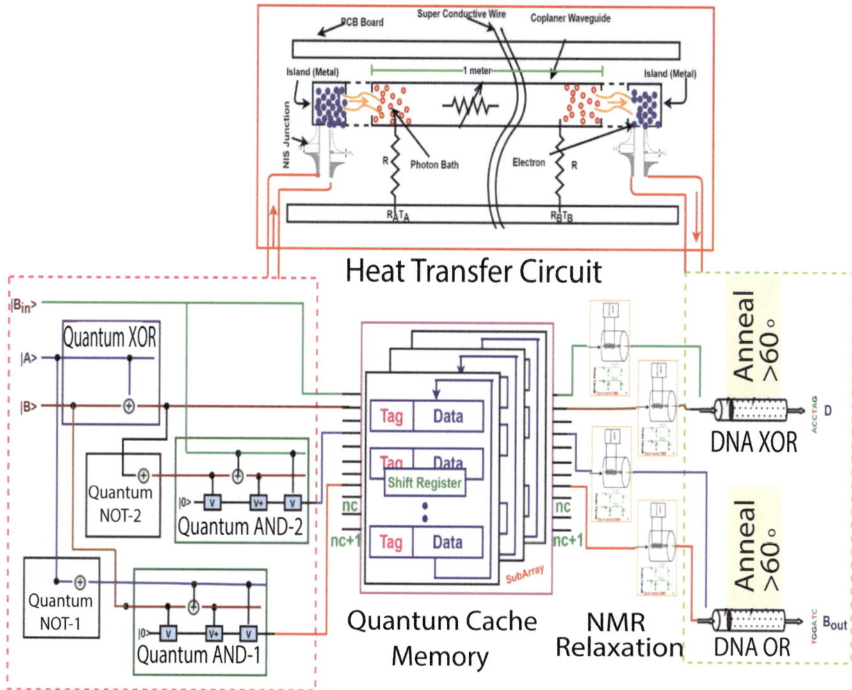

Fig. 15.9 Circuit of quantum-DNA full subtractor with cache memory

Table 15.4 Outputs of quantum-DNA full subtractor operation

|A0>	|A1>	|Q>	SUM	Carry
|0>	|0>	|0>	TGGATC	TGGATC
|0>	|0>	|1>	ACCTAG	ACCTAG
|0>	|1>	|0>	ACCTAG	ACCTAG
|0>	|1>	|1>	TGGATC	ACCTAG
|1>	|0>	|0>	ACCTAG	TGGATC
|1>	|0>	|1>	TGGATC	TGGATC
|1>	|1>	|0>	TGGATC	TGGATC
|1>	|1>	|1>	ACCTAG	ACCTAG

= Quantum AND (1, 1)
= 1
DI$_4$ = Quantum AND (NOT (A), B)
= Quantum AND (NOT (1), 1)
= Quantum AND (0, 1)
= 0

Output of Quantum cache memory will be converted into DNA sequence using NMR relaxation at room temperature. For qubit |0> DNA sequence is TGGATC and for |1> DNA sequence is ACCTAG.

$\mathbf{B}_{out}$ = DNA OR (DO_2, DO_3)
= DNA OR (ACCTAG, TGGATC)
= ACCTAG
D = DNA XOR (DO_0, DO_1)
= DNA XOR (ACCTAG, ACCTAG)
= TGGATC

15.3.5 Data Management in Quantum-DNA Three-Qubit Parity Bit Checker

A circuit that checks the parity in the receiver is called a Parity Checker. A combined circuits or device of parity generators and parity checkers are commonly used in digital systems to detect the single-bit errors in the transmitted data. To create a three-qubit even parity qubit checker, three XOR operations are required. Figure 15.10 describes the digital and quantum circuit of a three-qubit even parity qubit checker. A three-qubit even parity qubit checker receives four inputs and produces one output containing "E".

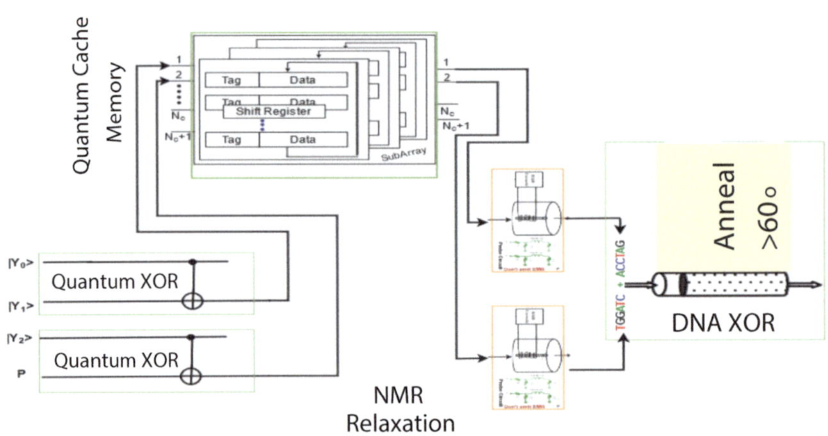

Fig. 15.10 Circuit of quantum-DNA three-qubit even parity qubit checker with cache memory

15.3.5.1 Design Procedure

To design a quantum-DNA three-qubit even parity qubit checker, quantum operations and DNA operations are used to operate the input qubit for their corresponding outputs. The quantum operations will be used for receiving the input qubits and the DNA operations will be used to produce the final output against the corresponding set of inputs. Each time, the quantum-DNA three-qubit even parity qubit checker will receive four qubits as input. After operating in a certain number of quantum operations, the qubit will be stored in cache memory. To store 2 qubits, a 4-to-2 quantum cache memory is needed.

After getting the qubits from cache memory, it will be turned into the corresponding DNA sequence using NMR relaxation room temperature probe. By using a room temperature probe and corresponding components of NMR relaxation, the excited qubit turns into a ground state and produces a DNA sequence. Then the DNA sequence is processed through DNA operations and outputs are received. Here, Fig. 15.10 describes quantum-DNA three-qubit even parity qubit checker using quantum operations and DNA operations.

From Fig. 15.10, it is found that the quantum-DNA three-qubit even parity qubit checker consists of two quantum operations, one DNA operations and a 4-to-2 cache memory. Here, two quantum XOR operations are used in quantum operation and stored qubits in cache memory and one DNA XOR operation is needed to provide the output.

15.3.5.2 Working Procedure

The working procedure of the quantum-DNA three-qubit even parity qubit checker is given below for each pattern of input qubits. Here, DNA sequence ACCTAG = TRUE for qubit | 1 > and DNA sequence TGGATC = FALSE for qubit | 0 > are used.

To store qubits in cache memory, 2 locations select line (A and B), one CLK input line, one |R/W>| input line and 2-input qubit lines in cache memory (DI_1, DI_2, DI_3, and DI_4). The output from the cache memory will be DO_0 and DO_1 are needed.

The working procedure of the quantum-DNA 3-bit even parity bit checker within cache memory is given below for one pattern of input qubits. Here, the inputs of cache memory from the quantum gates are found as follows:

1. For inputs A, B, C and P = 1, 1, 0, 1 $\mathbf{DI}_0$ = Quantum XOR (A, B)
   = Quantum XOR (1, 1)
   = 0
   $\mathbf{DI}_1$ = Quantum XOR (C, P)
   = Quantum XOR (0, 1)
   = 1

The output of quantum cache memory will be converted into DNA sequence using NMR relaxation at room temperature. For qubit |0> DNA sequence is TGGATC and for |1> DNA sequence is ACCTAG.

Output = DNA XOR (DO_0, DO_1)
= DNA XOR (TGGATC, ACCTAG)
= ACCTAG

15.4 Data Management in DNA-Quantum Circuits

In a DNA-quantum computing system, input will be received in DNA sequences and after performing a certain number of DNA operations, these DNA sequences will be turned into qubits by NMR process. Before entering into NMR process they will stay for a short time in DNA cache memory.

15.4.1 DNA Cache Memory to Control DNA to Quantum Data Flow

An intermediary system where the output data from the DNA system will be stored and retrieved for further processes. Figure 15.11 shows a cache memory cell to store one DNA sequence.

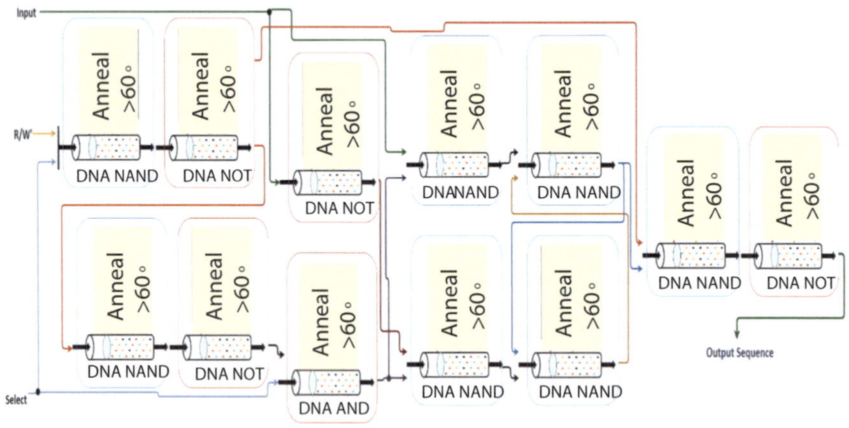

Fig. 15.11 The circuit diagram of one-molecular sequence DNA cache memory

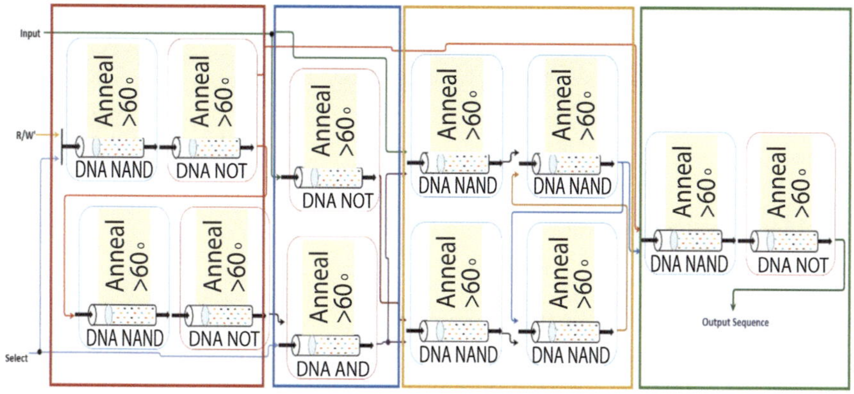

Fig. 15.12 The circuit diagram of one-molecular sequence DNA cache memory is divided into four blocks

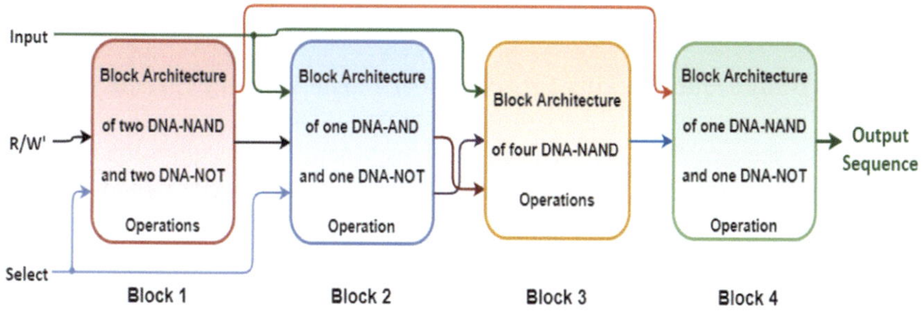

Fig. 15.13 The block architecture of one-molecular DNA cache memory

15.4.1.1 Design and Working Procedures of Cache Memory RAM Cell

RAM cell has three inputs and one output. First R/W' and select line go through the AND gate. Inverted R/W' and select line go through another AND gate. This AND gate output and an input bit will go to D flip-flop. The output of the D flip-flop AND with the first AND gate to produce output.

Using this one-bit DNA cache memory cell, DNA cache memory circuits can be constructed to store more DNA information. To do that easily, let's first divide the one-molecular DNA cache memory into four blocks. In the designed Fig. 15.12, where the one-molecular DNA cache memory divides into four blocks. The block diagram of one-molecular DNA cache memory is shown in Fig. 15.13.

Now, to construct a 4-to-2 DNA cache memory 8 one-molecular sequence DNA cache memory cells are required. Therefore, it can be expressed that the block architecture of one-bit sequence DNA cache memory is in a shorter form than Fig. 15.14, which will make the architecture of 4-to-2 DNA cache memory more straightforward. Based on Fig. 15.14, a one-molecular sequence DNA cache memory cell is depicted below.

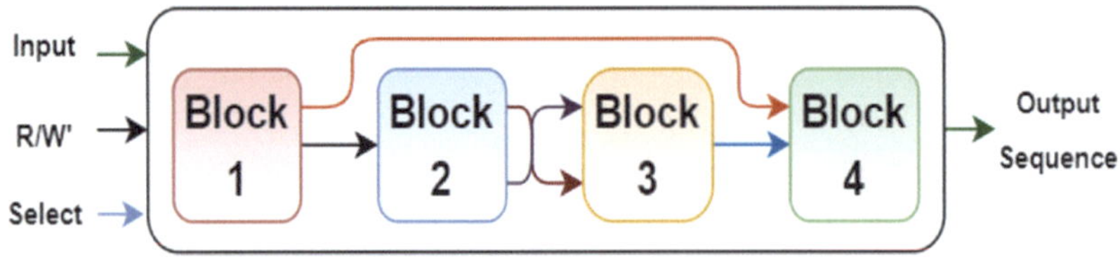

Fig. 15.14 DNA one-molecular cache memory cell

Here, Block 1 refers to the Block Architecture of two DNA NAND and two DNA NOT operations, Block 2 refers to the Block Architecture of one DNA AND and one DNA NOT operation, Block 3 refers to the Block Architecture of four DNA NAND operations, and Block 4 refers to the Block Architecture of one DNA NAND and one DNA NOT operation, and all of the block components are connected in a way shown in Figs. 15.13 and 15.14.

15.4.1.2 Design and Working Procedures of Cache Memory Circuit

Figure 15.15 shows a 4-to-2 DNA cache memory where 2-molecular sequence data can be stored and retrieved when necessary. From the 4-to-2 cache memory circuit diagram it is seen that, a 2-to-4 DNA decoder, eight DNA RAM cells, and two DNA OR operations are needed to be designed. Each DNA RAM cell has three inputs and one output. Three inputs are the select, R/W', and one input molecule. A 2-to-4 decoder's outputs are used as select input for RAM cells. A DNA OR operation is used to OR all the RAM cells output those are connected to the DI_0 line and perform DO_0 as cache memory output. Another OR operation is used to OR all RAM cells output those are connected to the DI_1 line and perform DO_1 as cache memory output.

The circuit diagram of a DNA 2-to-4 decoder operation is shown in Fig. 15.16. From Fig. 15.16, it is found that to construct a DNA 2-to-4 decoder, six DNA NOT and four DNA NAND operations are needed. The outputs of the decoder O_0, O_1, O_2, and O_3 from equations are given as follows.

$$O_0 = \overline{A_1}\,\overline{A_0}$$

$$O_1 = \overline{A_1}\ A_0$$

$$O_2 = A_1\,\overline{A_0}$$

$$O_3 = A_1\ A_0$$

Therefore, the circuit diagram for the DNA 4-to-2 cache memory has been constructed. Similarly, the circuit diagram for higher-order DNA cache memory can be constructed using the higher-order DNA decoders and more DNA one-molecular

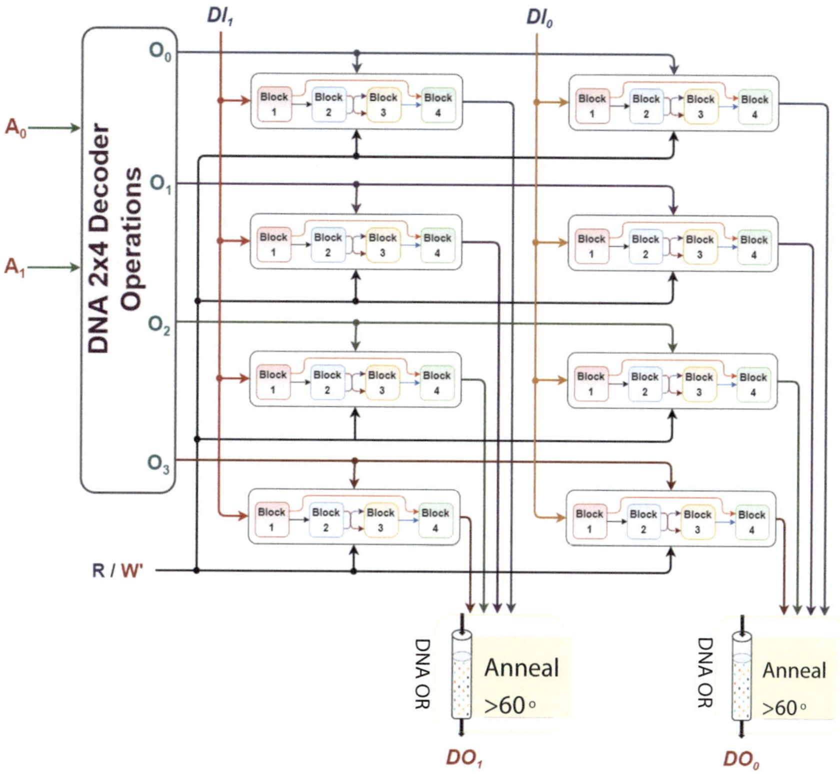

Fig. 15.15 The circuit diagram of DNA 4-to-2 cache memory

sequence cache memory cells. The circuits will be larger with the increase of their storage capacity.

It is not possible to use this massive circuit of DNA cache memory in the design, rather a block diagram will be used instead of the actual circuit. The block diagram of the DNA cache memory is shown in Fig. 15.17.

15.4.2 DNA-Quantum Full-Adder Operation

The data conversion circuit has been constructed to convert the DNA sequences to the quantum qubits, and store them to the DNA cache memory. A heat transfer circuit doesn't need to develop because the heat transfer circuit does not belong to the DNA system. So, now it is required to construct the full-adder circuit in DNA-quantum cross-platform. The architecture of the DNA-quantum full adder is shown in Fig. 15.18.

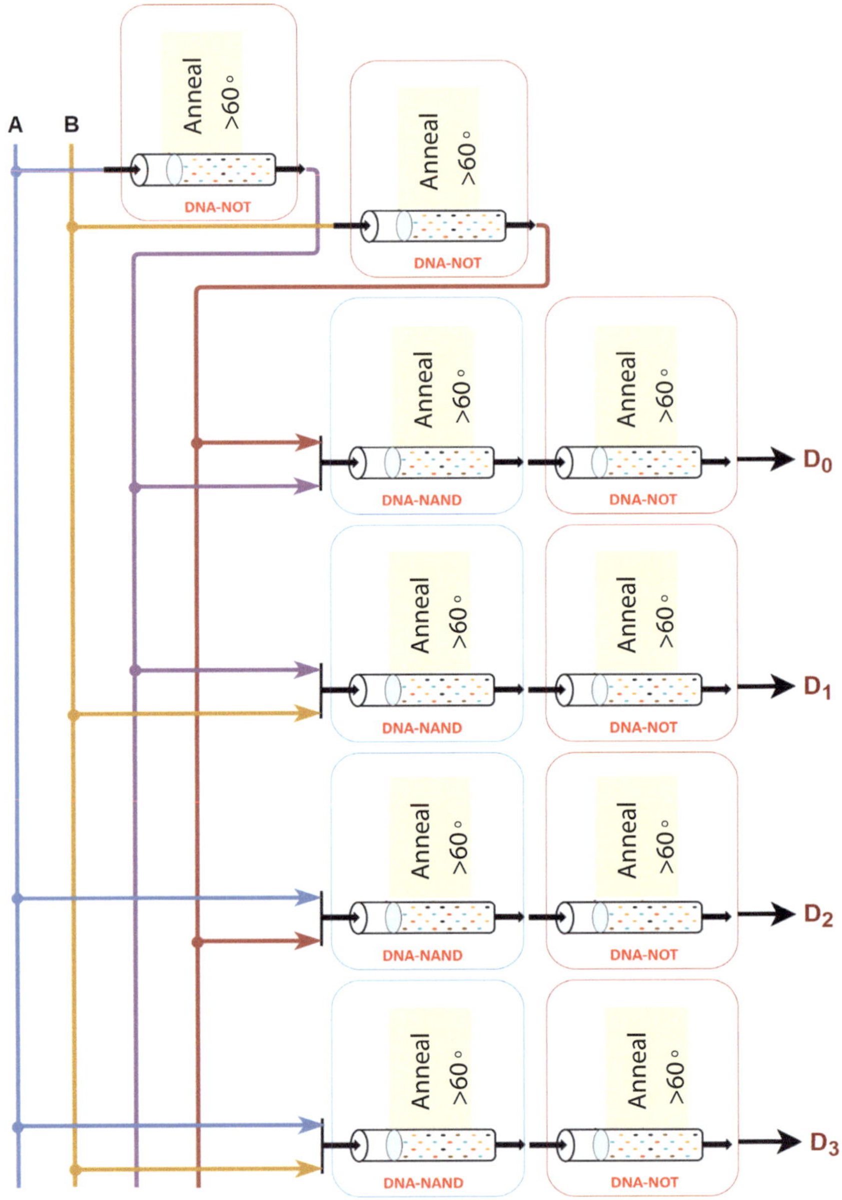

Fig. 15.16 DNA 2-to-4 decoder operation

So, the above figure is the absolute circuit diagram of the DNA-quantum full-adder. This operational diagram will be able to perform the full-adder operations in the DNA-quantum cross-platform.

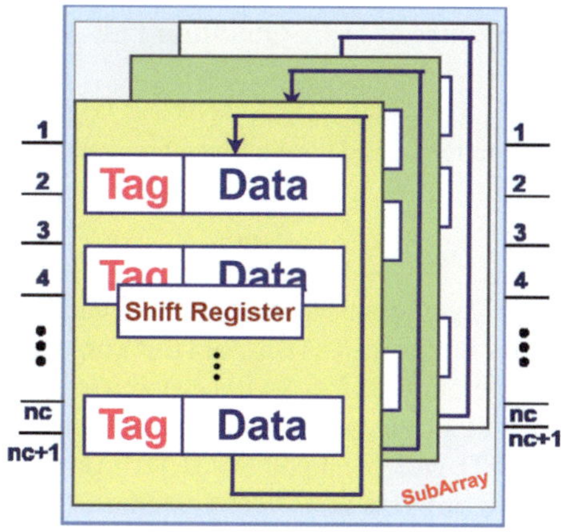

Fig. 15.17 The block diagram of DNA cache memory

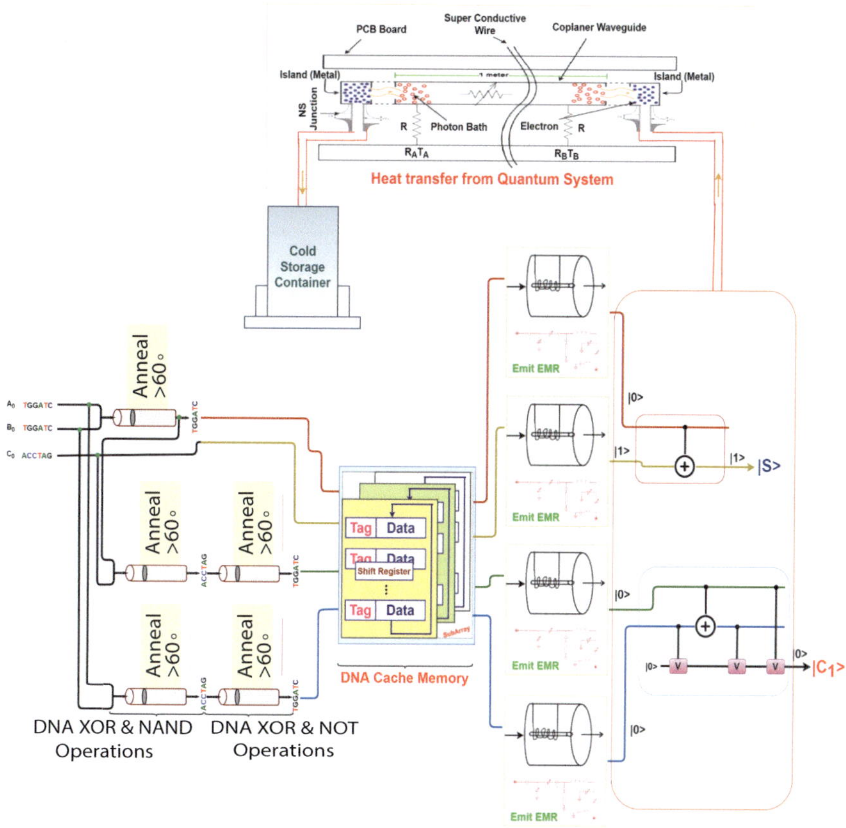

Fig. 15.18 The final circuit diagram of the DNA-quantum full adder

15.4.2.1 Working Procedure of DNA-Quantum Full-Adder Operation

This operation requires 3-input qubits. Thus, there will be 8 patterns of inputs. Consider some of those input patterns, and have a look at how they behave in the designed circuit in Fig. 15.18.

1. Consider the inputs A_1, B_1, C_0 = TGGATC,
   (a) The DNA XOR will get inputs A_1 = TGGATC, and B_1 = TGGATC. Therefore it will produce output TGGATC. And this output will be the input of the first DNA AND operation and also will be stored in the DNA cache memory. The value of C_0 will also be stored in the cache memory.
   (b) The first DNA AND will get input TGGATC from both C_0 and from Step a. Thus, it will also produce TGGATC as output which will be stored in the DNA cache memory.
   (c) The 2nd DNA AND will get input TGGATC from both A_1 and B_1. Thus, it will also produce TGGATC as output which will be stored in the DNA cache memory.
   (d) The NMR will get the DNA sequences from the DNA cache memory and will convert them to the equivalent quantum bits. As all the DNA system's outputs are TGGATC, they all will be converted into |0> and will pass to the Quantum systems.
   (e) Therefore, the Quantum XOR will receive input |0> from both the output of Step a and the value of C_0. Thus, it will produce the sum output qubit as |0>.
   (f) And the Quantum OR will get also |0> from both the output of Step b and Step c. Therefore it will also generate |0> as the carry output qubit.

2. Consider the inputs A_1 = ACCTAG, B_1 =, and C_0 = TGGATC,
   (a) The DNA XOR will get inputs A_1 = ACCTAG, and B_1 = ACCTAG. Therefore it will produce output TGGATC. And this output will be the input of the first DNA AND operation and also will be stored in the DNA cache memory. The value of C_0 will also be stored in the cache memory.
   (b) The first DNA AND will get input TGGATC from both C_0 and from Step a. Thus, it will also produce TGGATC as output which will be stored in the DNA cache memory.
   (c) The 2nd DNA AND will get input ACCTAG from both A_1 and B_1. Thus, it will also produce ACCTAG as output which will be stored in the DNA cache memory.
   (d) The NMR will get the DNA sequences from the DNA cache memory and will convert them to the equivalent quantum-bits. As the outputs from Steps a and b are TGGATC, they all will be converted into |0> and will pass to the Quantum Systems. The value of C_0 will also be converted as qubit |0>. And the output of Step c will be converted as qubit |1>.
   (e) Therefore, the Quantum XOR will receive input |0> from both the output of Step a and the value of C_0. Thus, it will produce the sum output qubit as |0>.

(f) And the Quantum OR will get also $|0>$ from the output of Step b and $|1>$ from the output of Step (c). Therefore it will generate $|1>$ as the carry output qubit.

3. Consider the inputs as A_1 = ACCTAG, B_1= ACCTAG, and C_0 = ACCTAG,

   (a) The DNA XOR will get inputs A_1 = ACCTAG, and B_1 = ACCTAG. Therefore it will produce output TGGATC. And this output will be the input of the first DNA AND operation and also will be stored in the DNA cache memory. The value of C_0 will also be stored in the cache memory.
   (b) The first DNA AND will get input ACCTAG from C_0 and TGGATC from Step a. Thus, it will also produce TGGATC as output which will be stored in the DNA cache memory.
   (c) The 2nd DNA AND will get input ACCTAG from both A1 and B_1. Thus, it will also produce ACCTAG as output which will be stored in the DNA cache memory.
   (d) The NMR will get the DNA sequences from the DNA cache memory and will convert them to the equivalent quantum bits. As the outputs from Steps a and b are TGGATC, they all will be converted into $|0>$ and will pass to the Quantum systems. The value of C_0 will be converted as qubit $|1>$. And the output of Step c will be converted as qubit $|1>$.
   (e) Therefore, the Quantum XOR will receive input $|0>$ from the output of Step a and $|1>$ from the value of C_0. Thus, it will produce the sum output qubit as $|1>$.
   (f) And the Quantum OR will get also $|0>$ from the output of Step b and $|1>$ from the output of Step c. Therefore it will generate $|1>$ as the carry output qubit.

Therefore, the designed architecture for the DNA-Quantum full-adder runs quite efficiently.

15.5 Applications

This section discusses several real-world applications of DNA and quantum operations, including how they are used and how they outperform traditional computing systems. Classical computing systems may fail in certain scenarios; however, quantum and DNA computing systems can demonstrate their capacity.

Logistics Optimization: A wide range of companies will be able to improve their logistics and scheduling procedures related to supply-chain management to improved data analysis and sophisticated modeling. The operating models must calculate and recalculate ideal routes for traffic management, fleet operations, air traffic control, shipping, and allocation on a constant basis, which may have a significant impact on applications. Normally, traditional computing is employed to complete these jobs; however, some of them may become too complex for an ideal computer solution, whereas a quantum technique may be able to complete them. Quantum annealing and universal quantum computers are two common quantum techniques that can be utilized to solve such challenges. Quantum annealing is a cutting-edge optimization technology that promises to outperform regular computers. Universal quantum computers, on the other hand, are capable of addressing any form of the computational issue and are not yet commercially available.

15.6 Summary

Quantum-DNA computing and DNA-quantum computing are new ways of computings in modern science and technology which are introduced here in this book for the first time. In this way of computing the system needs a cache memory. In quantum-DNA computing, quantum cache memory is needed to store qubits which are so fast and cannot enter directly to the conversion circuits. In the DNA-quantum cache memory, DNA cache memory is needed to perform smooth computations. This chapter has presented the details of quantum cache memories and DNA cache memories and showed the working processes during computations.

Concluding Remarks

In this book, a novel concept called computation using quantum biology (QB) is presented. Quantum biology is a burgeoning interdisciplinary field that explores how quantum mechanical principles influence biological processes that are not fully explained by classical physics. It focuses on how quantum phenomena, like quantum coherence and tunneling, can enhance or regulate biological functions. Here, a quantum biocomputer construction effort has been made. In terms of security, parallel computing power, and computation speed, quantum computers excel. Quantum computing, an intriguing idea in recent years, can solve many insoluble classical computer issues. Because of its parallel processing, enormous storage capacity, and capacity for nano-level computation, DNA (Deoxyribose Nucleic Acid) computing stands out from conventional computer systems. DNA computing or biocomputing or biological computing requires less power than conventional computer systems. Logic gates in DNA computing offer special qualities including stability and reusability.

The DNA molecule's properties help to induce quantum phenomena such as superposition, tunneling, coherence, and entanglement. The novel computer concepts introduced in this book, such as quantum-DNA computing or quantum biocomputing or quantum biological computing which explores the potential for biological systems to perform quantum computations, or the use of biological materials and processes to build quantum computers. It's a multidisciplinary field at the intersection of quantum physics, biology, and computer science and DNA-quantum computing or bioquantum computing or biological quantum computing which would instead involve quantum states (like spin states, electronic states) in biological entities. If we consider the brain as a quantum computer argument: the brain (if quantum) would be an analog, highly parallel processor, possibly leveraging quantum principles that we don't utilize yet. These two computing systems combine the advantages of quantum physics with molecular biology.

NMR is necessary when building a qubit out of a DNA sequence, and it can be done at two different temperatures: 0 Kelvin and room temperature. In this book, quantum-DNA and DNA-quantum Nanoprocessors are built for these two temperatures. To obtain the appropriate DNA sequence, the qubit is relaxed using RNR. While DNA computing uses DNA sequences as inputs, quantum computing uses

H. Md. Hasan Babu, *Quantum Biocomputing in Quantum Biology Volume II*,
https://doi.org/10.1007/978-981-97-5349-9

qubits to carry out computations. In quantum-DNA computing, a buffer is employed to match the speed of quantum cache memory. The qubits needed to transfer it to DNA circuits are present in cache memory. The storage of data that can be resolved by quantum-DNA computing is a restriction of quantum computing. DNA sequences are capable of securely storing a lot of data. For this reason, quantum-DNA computing is useful. Additionally, the processing of data by DNA processes generates heat, and all operations carried out by quantum systems do the same. The terms "DNA-quantum computing" and "quantum-DNA computing" facilitate the transfer of generated heat from quantum computing to DNA computing.

In Quantum biocomputing, which refers to quantum computing, DNA computing, DNA-quantum computing, and quantum-DNA computing, all arithmetic operations, combinational circuits, and speed of the operations, applications, and produced temperature are explained in this book.

References

1. G.A. Barbosa, Quantum half-adder. Phys. Rev. A **73**(5), 052321 (2006)
2. L. Diósi, *A Short Course in Quantum Information Theory: An Approach From Theoretical Physics*, vol. 827 (Springer, Berlin, 2011)
3. N. Isailovic, Y. Patel, M. Whitney, J. Kubiatowicz, Interconnection networks for scalable quantum computers, in *33rd International Symposium on Computer Architecture (ISCA'06)* (IEEE, 2006), pp. 366–377
4. L.B. Levitin, T. Toffoli, Z. Walton, Operation time of quantum gates (2022). arXiv:quant-ph/0210076
5. S.T. Marella, H.S.K. Parisa, Introduction to quantum computing, in *Quantum Computing and Communications* (IntechOpen, 2020)
6. T.S. Metodi, D.D. Thaker, A.W. Cross, F.T. Chong, I.L. Chuang, A quantum logic array microarchitecture: scalable quantum data movement and computation, in *38th Annual IEEE/ACM International Symposium on Microarchitecture (MICRO'05)* (IEEE, 2005), p. 12
7. M. Mohammadi, M. Eshghi, Behavioral description of quantum v and v+ gates to design quantum logic circuits, in *2008 5th International Multi-Conference on Systems, Signals and Devices* (IEEE, 2008), pp. 1–5
8. M. Morrison, Design of a reversible ALU based on novel reversible logic structures (University of South Florida, 2012)
9. A. Muthukrishnan, Classical and quantum logic gates: an introduction to quantum computing quantum information seminar (1999)
10. D.D. Thaker, T.S. Metodi, A.W. Cross, I.L. Chuang, F.T. Chong, Quantum memory hierarchies: efficient designs to match available parallelism in quantum computing, in *33rd International Symposium on Computer Architecture (ISCA'06)* (IEEE, 2006), pp. 378–390
11. H. Thapliyal, N. Ranganathan, Design of reversible sequential circuits optimizing quantum cost, delay, and garbage outputs. ACM J. Emerg. Technol. Comput. Syst. (JETC) **6**(4), 1–31 (2010)
12. S. Arunachalam, V. Gheorghiu, T. Jochym-O'Connor, M. Mosca, P.V. Srinivasan, On the robustness of bucket brigade quantum ram. New J. Phys. **17**(12), 123010 (2015)
13. M. Blencowe, Quantum ram. Nature **468**(7320), 44–45 (2010)
14. E. Blum, M. Castillo-Martin, M. Rosenberg, Survey on the security of the quantum rom (2019)
15. E.S. Boyden, Quantum computation: theory and implementation. Ph.D. thesis, Massachusetts Institute of Technology, Department of Physics; and (S. B. and S. M ..., 1999)

H. Md. Hasan Babu, *Quantum Biocomputing in Quantum Biology Volume II*,
https://doi.org/10.1007/978-981-97-5349-9

16. D. Deutsch, R. Jozsa, Rapid solution of problems by quantum computation. Proc. R. Soc. Lond. Ser. A: Math. Phys. Sci. **439**(1907), 553–558 (1992)
17. V. Giovannetti, S. Lloyd, L. Maccone, Architectures for a quantum random access memory. Phys. Rev. A **78**(5), 052310 (2008)
18. J.R. Goodman, Using cache memory to reduce processor-memory traffic, in *Proceedings of the 10th Annual International Symposium on Computer Architecture* (1983), pp. 124–131
19. M.M. Mano, *Digital Logic and Computer Design* (Pearson Education India, 2017)
20. P.W. Shor, Algorithms for quantum computation: discrete logarithms and factoring, in *Proceedings 35th Annual Symposium on Foundations of Computer Science* (IEEE, 1994), pp. 124–134
21. B.C. Travaglione, M.A. Nielsen, H.M. Wiseman, A. Ambainis, Rom-based computation: quantum versus classical (2001). arXiv:quant-ph/0109016
22. L.M. Adleman, Molecular computation of solutions to combinatorial problems. Science **266**(5187), 1021–1024 (1994)
23. K.J. Breslauer, R. Frank, H. Blöcker, L.A. Marky, Predicting DNA duplex stability from the base sequence. Proc. Natl. Acad. Sci. **83**(11), 3746–3750 (1986)
24. J. Watada, *DNA computing and its application, in Computational Intelligence: A Compendium* (Springer, Berlin, 2008), pp.1065–1089
25. X. Zheng, J. Yang, C. Zhou, C. Zhang, Q. Zhang, X. Wei, Allosteric Dnazyme-based DNA logic circuit: operations and dynamic analysis. Nucleic Acids Res. **47**(3), 1097–1109 (2019)
26. R.E. March, Quadrupole ion traps. Mass Spectrom. Rev. **28**(6), 961–989 (2009)
27. A. Steane, Quantum computing. Rep. Prog. Phys. **61**(2), 117 (1998)
28. J. Antony, D.M. Medvedev, A.A. Stuchebrukhov, Theoretical study of electron transfer between the photolyase catalytic cofactor FADH-and DNA thymine dimer. J. Am. Chem. Soc. **122**(6), 1057–1065 (2000)
29. S.C. Benjamin, N.F. Johnson, Entangled electronic states in multiple-quantum-dot systems. Phys. Rev. B **51**(20), 14733 (1995)
30. D. Bouwmeester, A. Zeilinger, *The physics of quantum information: basic concepts, in The Physics of Quantum Information* (Springer, Berlin, 2000), pp.1–14
31. E. Braun, Y. Eichen, U. Sivan, G. Ben-Yoseph, DNA-templated assembly and electrode attachment of a conducting silver wire. Nature **391**(6669), 775–778 (1998)
32. J. Chen, E. Antipov, B. Lemieux, W. Cedeño, D.H. Wood, In vitro selection for a max 1s DNA genetic algorithm. DNA Based Comput. V 23–37 (1999)
33. J.L. Coffer, S.R. Bigham, X. Li, R.F. Pinizzotto, Y.G. Rho, R.M. Pirtle, I.L. Pirtle, Dictation of the shape of mesoscale semiconductor nanoparticle assemblies by plasmid DNA. Appl. Phys. Lett. **69**(25), 3851–3853 (1996)
34. D. Deutsch, Quantum theory, the church-turing principle and the universal quantum computer. Proc. R. Soc. Lond. A. Math. Phys. Sci. **400**(1818), 97–117 (1985)
35. N.A. Gershenfeld, I.L. Chuang, Bulk spin-resonance quantum computation. Science **275**(5298), 350–356 (1997)
36. A. Kamaraj, P. Marichamy, Design and implementation of arithmetic and logic unit (ALU) using novel reversible gates in quantum cellular automata, in *2017 4th International Conference on Advanced Computing and Communication Systems (ICACCS)* (IEEE, 2017), pp. 1–8
37. Y. Kanamori, S.-M. Yoo, Quantum computing: principles and applications. J. Int. Technol. Inf. Manag. **29**(2), 43–71 (2020)
38. J.D. McCalpin, IEEE computer society technical committee on computer architecture (TCCA) newsletter (1995)
39. V. Scarani, M. Ziman, P. Štelmachovič, N. Gisin, V. Bužek, Thermalizing quantum machines: dissipation and entanglement. Phys. Rev. Lett. **88**(9), 097905 (2002)
40. F. Schmidt et al., Realization of the Cirac-Zoller controlled-not quantum gate. Nature **422**, 408–11 (2003)
41. M.K. Thomsen, R. Glück, H.B. Axelsen, Reversible arithmetic logic unit for quantum arithmetic. J. Phys. A: Math. Theor. **43**(38), 382002 (2010)

42. M. Arndt, T. Juffmann, V. Vedral, Quantum physics meets biology. HFSP J **3**(6), 386–400 (2009)
43. M. Asano, I. Basieva, A. Khrennikov, M. Ohya, Y. Tanaka, I. Yamato, Quantum information biology: from information interpretation of quantum mechanics to applications in molecular biology and cognitive psychology. Found. Phys. **45**(10), 1362–1378 (2015)
44. G. Balasubramanian, I.Y. Chan, R. Kolesov, M. Al-Hmoud, J. Tisler, C. Shin, C. Kim, A. Wojcik, P.R. Hemmer, A. Krueger et al., Nanoscale magnetic sensing with an individual electronic spin in diamond. Nat Nanotechnol **455**, 648–52 (2008)
45. C. Chatgilialoglu, L.A. Eriksson, M.G. Krokidis, A. Masi, S. Wang, R. Zhang, Oxygen dependent purine lesions in double-stranded oligodeoxynucleotides: kinetic and computational studies highlight the mechanism for 5, 8-cyclopurine formation. J. Am. Chem. Soc. **142**(12), 5825–5833 (2020)
46. H. Chen, S. Krishnamachari, J. Guo, L. Yao, P. Murugan, C.J. Weight, R.J. Turesky, Quantitation of lipid peroxidation product DNA adducts in human prostate by tandem mass spectrometry: a method that mitigates artifacts. Chem. Res. Toxicol. **32**(9), 1850–1862
47. H.B. Gray, J.R. Winkler, Electron tunneling through proteins. Q. Rev. Biophys. **36**(3), 341–372 (2003)
48. M.C. Jecklin, D. Touboul, C. Bovet, A. Wortmann, R. Zenobi, Which electrospray-based ionization method best reflects protein-ligand interactions found in solution? a comparison of ESI, nanoESI, and ESSI for the determination of dissociation constants with mass spectrometry. J. Am. Soc. Mass Spectrom. **19**(3), 332–343 (2008)
49. Y.N. Lambert, C. Li, G. Chen, F. Nori, Quantum biology. Nat. Phys. **9**, 10–18 (2013)
50. A. Marais, B. Adams, A.K. Ringsmuth, M. Ferretti, J.M. Gruber, R. Hendrikx, M. Schuld, S.L. Smith, I. Sinayskiy, T.P.J. Krüger et al., The future of quantum biology. J. R. Soc. Interface **15**(148), 20180640 (2018)
51. J. McFadden, J. Al-Khalili, The origins of quantum biology. Proc. R. Soc. A **474**(2220), 20180674 (2018)
52. A.A. Stuchebrukhov, Long-distance electron tunneling in proteins: a new challenge for time-resolved spectroscopy. Laser Phys. **20**(1), 125–138 (2010)
53. Y. Wang, Q. Zhang, Y. Wang, Tandem mass spectrometry for the determination of the sites of DNA interstrand cross-link. J. Am. Soc. Mass Spectrom. **15**(11), 1565–1571 (2004)
54. T.-C. Yena, Y.-C. Cheng, Electronic coherence effects in photosynthetic light harvesting. Procedia Chem. **3**(1), 211–221 (2011)
55. E.E. Bessette, S.D. Spivack, A.K. Goodenough, T. Wang, S. Pinto, F.F. Kadlubar, R.J. Turesky (2010) Identification of carcinogen DNA adducts in human saliva by linear quadrupole ion trap/multistage tandem mass spectrometry. Chem. Res. Toxicol. **23**(7), 1234–1244 (2010)
56. S.A. El-Seoud, R. Mohamed, S. Ghoneimy, DNA computing: challenges and application. Int. J. Interact. Mob. Technol. **11**(2) (2017)
57. V. Gabelica, T. Tabarin, R. Antoine, F. Rosu, I. Compagnon, M. Broyer, E. De Pauw, P. Dugourd, Electron photodetachment dissociation of DNA polyanions in a quadrupole ion trap mass spectrometer. Anal. Chem. **78**(18), 6564–6572 (2006)
58. P.P. Gariaev, P.J. Marcer, K.A. Leonova-Gariaeva, U. Kaempf, V.D. Artjukh, DNA as basis for quantum biocomputer. DNA Decipher J. **1**(1), 025–046 (2011)
59. H. Häffner, C.F. Roos, R. Blatt, Quantum computing with trapped ions. Phys. Rep. **469**(4), 155–203 (2008)
60. A.D. Córcoles, A. Kandala, A. Javadi-Abhari, D.T. Mcclure, A.W. Cross, K. Temme, P.D. Nation, M. Steffen, J.M. Gambetta, Challenges and opportunities of near-term quantum computing systems. Proc. IEEE **10** (2019)
61. R.E. March, An introduction to quadrupole ion trap mass spectrometry. J. Mass Spectrom. **32**(4), 351–369 (1997)
62. V. Nebendahl, H. Häffner, C.F. Roos, Optimal control of entangling operations for trapped-ion quantum computing. Phys. Rev. A **79**(1), 012312 (2009)
63. A.A. Tulub, V.E. Stefanov, Triplet-singlet spin communication between DNA nucleotides serves the basis for quantum computing. Chem. Phys. Lett. **436**(1–3), 258–262 (2007)

64. C. Xin, Shor's quantum algorithm: large number factoring and period finding
65. D.R. Alcoba, R.C. Bochicchio, L. Lain, A. Torre, On the measure of electron correlation and entanglement in quantum chemistry based on the cumulant of the second-order reduced density matrix. J. Chem. Phys. **133**(14), 144104 (2010)
66. L. Diósi, *Qubit thermodynamics, in A Short Course in Quantum Information Theory* (Springer, Berlin, 2011), pp.123–133
67. S.M. Freier, R. Kierzek, J.A. Jaeger, N. Sugimoto, M.H. Caruthers, T. Neilson, D.H. Turner, Improved free-energy parameters for predictions of RNA duplex stability. Proc. Natl. Acad. Sci. **83**(24), 9373–9377 (1986)
68. L.A. Marky, K.J. Breslauer, Calculating thermodynamic data for transitions of any molecularity from equilibrium melting curves. Biopolym.: Orig. Res. Biomol. **26**(9), 1601–1620 (1987)
69. D.H. Mathews, J. Sabina, M. Zuker, D.H. Turner, Expanded sequence dependence of thermodynamic parameters improves prediction of RNA secondary structure. J. Mol. Biol. **288**(5), 911–940 (1999)
70. J. SantaLucia Jr., A unified view of polymer, dumbbell, and oligonucleotide DNA nearest-neighbor thermodynamics. Proc. Natl. Acad. Sci. **95**(4), 1460–1465 (1998)
71. T.S. Metodi, F.T. Chong, Quantum computing for computer architects. Synth. Lect. Comput. Arch. **1**(1), 1–154 (2006)
72. C. Monroe, R. Raussendorf, A. Ruthven, K.R. Brown, P. Maunz, L.-M. Duan, J. Kim, Large-scale modular quantum-computer architecture with atomic memory and photonic interconnects. Phys. Rev. A **89**(2), 022317 (2014)
73. M. Whitney, N. Isailovic, Y. Patel, J. Kubiatowicz, Automated generation of layout and control for quantum circuits, in *Proceedings of the 4th International Conference on Computing Frontiers* (2007), pp. 83–94
74. J. Anders, D.K.L. Oi, E. Kashefi, D.E. Browne, E. Andersson, Ancilla-driven universal quantum computation. Phys. Rev. A **82**(2), 020301 (2010)
75. S. Anferova, V. Anferov, M. Adams, P. Blümler, N. Routley, K. Hailu, K. Kupferschläger, M.J.D. Mallett, G. Schroeder, S. Sharma, et al., Construction of a NMR-mouse with short dead time. Concepts Magn. Reson.: Educ. J. **15**(1), 15–25 (2002)
76. D. Auguin, V. Catherinot, T.E. Malliavin, J.L. Pons, M.A. Delsuc, Superposition of chemical shifts in NMR spectra can be overcome to determine automatically the structure of a protein. Spectroscopy **17**(2–3), 559–568 (2003)
77. F. Hobo, M. Takahashi, H. Maeda, S33 NMR cryogenic probe for taurine detection. Rev. Sci. Instrum. **80**(3), 036106 (2009)
78. A.M. Iliyasu, P.Q. Le, F. Dong, K. Hirota, A framework for representing and producing movies on quantum computers. Int. J. Quantum Inf. **9**(06), 1459–1497 (2011)
79. J.A. Jones, Quantum computing and nuclear magnetic resonance. Phys. Chem. Commun. **4**(11), 49–56 (2001)
80. V.D. Kodibagkar, M.S. Conradi, Remote tuning of NMR probe circuits. J. Magn. Reson. **144**(1), 53–57 (2000)
81. T.D. Ladd, F. Jelezko, R. Laflamme, Y. Nakamura, C. Monroe, J.L. O'Brien, Quantum computers. Nature **464**(7285), 45–53 (2010)
82. A. Marin, T.E. Malliavin, P. Nicolas, M.-A. Delsuc, From NMR chemical shifts to amino acid types: investigation of the predictive power carried by nuclei. J. Biomol. NMR **30**(1), 47–60 (2004)
83. R. Marx, A.F. Fahmy, J.M. Myers, W. Bermel, S.J. Glaser, Approaching five-bit NMR quantum computing. Phys. Rev. A **62**(1), 012310 (2000)
84. Y. Takahashi, S. Tani, Power of uninitialized qubits in shallow quantum circuits. Theoret. Comput. Sci. **851**, 129–153 (2021)
85. C.M. Tesch, R. de Vivie-Riedle, Quantum computation with vibrationally excited molecules. Phys. Rev. Lett. **89**(15), 157901 (2002)
86. Q. Wei, S. Kais, B. Friedrich, D. Herschbach, Entanglement of polar molecules in pendular states. J. Chem. Phys. **134**(12), 124107 (2011)

87. G. Werth, *Principles of ion traps, in Trapped Charged Particles and Fundamental Interactions* (Springer, Berlin, 2008), pp.1–37
88. S.S. Zalesskiy, E. Danieli, B. Blumich, V.P. Ananikov, Miniaturization of NMR systems: desktop spectrometers, microcoil spectroscopy, and "NMR on a chipâŁž for chemistry, biochemistry, and industry. Chem. Rev. **114**(11), 5641–5694 (2014)
89. G. Chatterjee, N. Dalchau, R.A. Muscat, A. Phillips, G. Seelig, A spatially localized architecture for fast and modular DNA computing. Nat. Nanotechnol. **12**(9), 920–927 (2017)
90. J. Elbaz, O. Lioubashevski, F. Wang, F. Remacle, R.D. Levine, I. Willner, DNA computing circuits using libraries of Dnazyme subunits. Nat. Nanotechnol. **5**(6), 417–422 (2010)
91. M. Hirvensalo, *Quantum Computing* (Springer Science & Business Media, 2003)
92. V. Mavroeidis, K. Vishi, M.D. Zych, A. Jøsang, The impact of quantum computing on present cryptography (2018). arXiv:1804.00200
93. National Academies of Sciences Engineering, Medicine, et al., Quantum computing: progress and prospects (2019)
94. J. Tao, R. Zhang, Y. Zhu, *DNA Computing Based Genetic Algorithm* (Springer, Berlin, 2020)
95. Hafiz Md. Hasan Babu, "Quantum Computing: A Pathway to Quantum Logic Design", IOP (Institute of Physics) Publishing, 2020, Bristol, UK
96. Hafiz Md. Hasan Babu, "Multiple-Valued Computing in Quantum Molecular Biology", Volume I, CRC Press, 2023, USA
97. Hafiz Md. Hasan Babu, "Multiple-Valued Computing in Quantum Molecular Biology", Volume II, CRC Press, 2023, USA
98. Hafiz Md. Hasan Babu, "Reversible and DNA Computing," Wiley Publishers, 2021, UK
99. Hafiz Md. Hasan Babu, "VLSI Circuits and Embedded Systems,âŁž CRC Press (A Publication of Taylor & Francis Group), July 2022, USA
100. Md. Jahangir Alam, Guoqing Hu, Hafiz Md. Hasan Babu and Huazhong Xu, "Control Engineering Theory and Applications," CRC Press (A Publication of Taylor & Francis Group), September 2022, USA
101. Hafiz Md. Hasan Babu, "DNA Logic Design: Computing with DNA", World Scientific Publishing Company, May 2024, Singapore

Index

H. Md. Hasan Babu, *Quantum Biocomputing in Quantum Biology Volume II*,
https://doi.org/10.1007/978-981-97-5349-9